**MODERN METHODS IN
TOPOLOGICAL VECTOR SPACES**

McGRAW-HILL
INTERNATIONAL
BOOK COMPANY

New York
St. Louis
San Francisco
Auckland
Bogotá
Guatemala
Hamburg
Johannesburg
Lisbon
London
Madrid
Mexico
Montreal
New Delhi
Panama
Paris
San Juan
São Paulo
Singapore
Sydney
Tokyo
Toronto

ALBERT WILANSKY

Department of Mathematics
Lehigh University

Modern Methods in Topological Vector Spaces

This book was set in Times Series 327

British Library Cataloging in Publication Data

Wilansky, Albert
  Modern methods in topographical vector spaces.
  1. Linear topographical spaces
  I. Title
  515'.73   QA322   78-40110

ISBN 0-07-070180-6

MODERN METHODS IN TOPOLOGICAL VECTOR SPACES

2 3 4 5 MPMP 8 0 7 9

Printed and bound in the United States of America

To Rosie, with love.
If Winter comes, can Spring be far behind?

CONTENTS

PREFACE

This book, designed for a one-year course at the beginning graduate level, displays those properties of topological vector spaces which are used by researchers in classical analysis, differential and integral equations, distributions, summability, and classical Banach and Fréchet spaces. In addition, optional examples and problems (with hints and references) will set the reader's foot on numerous paths such as non-locally convex (e.g., ultrabarrelled) spaces, Köthe–Toeplitz spaces, Banach algebra, sequentially barrelled spaces, and norming subspaces.

The prerequisites are laid out in Chapter 1, which is a rapid sketch of vector spaces and point set topology. The central theme of the book is duality, which is taken up in Chapter 8. In an ideal world the course would begin with this chapter, the material of the preceding seven being known to all educated persons. The climax is reached in Chapter 12, which presents completeness theorems in this setting: a function space is complete when membership in it is secured by continuity on a certain family of sets. (See the beginning of Chapter 12.) The remaining three chapters treat special topics such as inductive limits, distributions, weak compactness, and barrelled spaces, by means of the tools developed in Chapter 12. In particular, the separable quotient problem for Banach spaces (Section 15-3), as special and as classical as it appears, requires much of this material for its fullest understanding.

The style is that of a beginning text in which concepts are explained and motivated, and every theorem is delineated by examples which show that its hypotheses are minimal and which illustrate how the theorem is used, how it fits into the theory, and how it forms a step in some general program. For example, the equivalence program is a body of results of the form $P \equiv (Q \Rightarrow R)$ where P is a property of a space and Q, R are properties of sets in its dual. [See, for example, Theorem 9-3-4(*b*) and (*c*) and the beginning of Section 9-4.] Moreover, the book is more completely cross-referenced than most others that I have seen.

Both nets and filters are introduced and used whenever appropriate.

A set of 33 tables is given at the end of the book, allowing quick reference to theorems and counterexamples. There are also 1500 problems which are arranged in four sequences at the end of each section; the few problems whose numbers are below 100 are considered part of the text. (See Section 1-1.)

REMARKS TO THE EXPERIENCED READER

A possibly unfamiliar concept is that of property of a dual pair. A dual pair (X, Y) is said to have a property P if X has a compatible topology with property P. (See Remark 8-6-8.) Thus (c_0, l) is a complete dual pair while (φ, l) is not.

The open mapping and closed graph theorems (in the primitive case) are proved without use of quotients; completeness of the dual of a bornological space is given a simple direct proof (Corollary 8-6-6); and the more sophisticated deduction from Grothendieck's theorem is given later (Example 12-2-20.) Five unique features of the book are:

1. Boundedness is proved to be a duality invariant in an easy way (Theorem 8-4-1), long before the appearance of the Banach–Mackey theorem (10-4-8). This method is due to H. Nakano.
2. Relatively strong topologies and an easy version of the Mackey–Arens theorem (8-2-14) are given before the standard identification of the Mackey topology (Theorem 9-2-3).
3. The convex compactness property and sequential completeness are emphasized and shown to have the same consequences in many cases (e.g., Theorems 10-4-8 and 10-4-11). Since bounded completeness implies both of these properties, this represents an improvement of the usual treatment.
4. F linked topologies (Definition 6-1-9) are featured, with the consequent upper heredity of all forms of completeness. This ties in with the aforementioned properties of dual pairs, e.g., Mazur (Definition 8-6-3) is downward hereditary so a dual pair (X, Y) is Mazur if and only if $\sigma(X, Y)$ is (Problem 8-6-101).
5. Emphasis is placed on converse theorems (Section 12-6). This is the "Bourbaki program" of finding the natural setting for classical Banach space theorems. This includes the first textbook appearance of a recently discovered simple proof of Mahowald's characterization of barrelled spaces (Theorem 12-6-3).

ACKNOWLEDGMENTS

I have made extensive use of earlier texts [8, 14, 26, 37, 38, 58, 82, 88, 116, 138], the book of H. H. Schaefer, and lecture notes of D. A. Edwards.

Many results in the book are attributed—others are not because they have passed into the lore of the subject. Some are unattributed by oversight; an example is Theorem 14-4-11 which I recently found out occurs in [24].

My book would have been impossible to write without constant consultation of *Mathematical Reviews*. Science owes a vast debt to the dedicated staff of this most excellent and complete publication.

A great stroke of fortune brought D. J. H. Garling, G. Bennett, and N. J. Kalton to Lehigh University for a year or two under the laughable misapprehension that they had something to learn from me. Any virtue this book may happen to possess is largely due to their generosity and patience during their visit and in subsequent correspondence.

A special word of thanks goes to J. Diestel for many fruitful conversations in bars and cocktail lounges throughout the United States. I had many occasions to consult A. K. Snyder and also received much help from E. G. Ostling, D. B. Anderson, W. G. Powell, C. L. Madden, and W. H. Ruckle.

The excellent Mathematics Department of Lehigh University provides an ambience in which scholarship and mutual assistance prevails.

My daughter Carole Wilansky helped with many organizational chores. Judy Arroyo typed the whole book with unbelievable speed and accuracy—I greatly appreciate her dedication to the project. My thanks also to the staff of McGraw-Hill for their helpfulness at every step.

Albert Wilansky,
January 1978

ONE

INTRODUCTION

1-1 EXPLANATORY

We use the notations of elementary set theory such as $A \subset B$ (A is a subset of B). The only possibly unfamiliar one is the useful symbol $\pitchfork$; $A \pitchfork B$ (A does not meet B) means that A, B are disjoint. When $A \subset X$, $X \backslash A$ or (when X is understood) $\tilde{A}$ is the complement of A (in X).

We use R^n, $\mathscr{C}^n$ for the spaces of n-tuples of real, respectively complex, numbers; $R = R^1$, $\mathscr{C} = \mathscr{C}^1$. Every vector space X has for its space of scalars the space $\mathscr{K}$ and, in this book, $\mathscr{K}$ is always R or $\mathscr{C}$; all statements made will be correct for either interpretation except when we specifically mention real vector space or complex vector space. (To avoid duplication, proofs are written using complex scalars.)

Problems

Those numbered from 1 to 99 are basic for further developments and form part of the text. Problems numbered > 200 are more difficult, and those numbered > 300 are really notes with references to the literature.

Proof Brackets

When part of a discussion is enclosed thus $[\![\ldots]\!]$ it means that the immediately preceding statement is being proved. For example, suppose that the text reads: "Since $x \neq 0$ $[\![$if $x = 0$, $\cos x = 1$, contradicting the hypotheses$]\!]$ we may cancel x." The reader should first absorb "Since $x \neq 0$, we may cancel x." He may then, if required, consult the proof in brackets.

Notation

δ^k is the sequence x where $x_k = 1$, $x_n = 0$ for $n \neq k$; that is, δ^k_n is the Kronecker delta. For $x \in \mathcal{K}$, sgn x is defined to be $|x|/x$ if $x \neq 0$; and sgn $0 = 1$.

1-2 TABLE OF SPACES

Several spaces will be used to illustrate the developments of the text. They are all vector spaces (Sec. 1-5) and each, with a few exceptions, has a distinguished real function defined on it, called *paranorm* (Sec. 2-1) or *norm* (Sec. 2-2), and is denoted by $\|x\|$, its value at x. Whenever such a sentence occurs as "show that c has a certain property," the reader may consult this table. Unless otherwise stated, the space is supposed to be endowed with its paranorm or norm.

1. Definition If f is a scalar-valued function on a set X, $\|f\|_\infty = \sup\{|f(x)| : x \in X\}$. This is called the sup *norm*.

In particular if x is a sequence, $\|x\|_\infty = \sup\{|x_n| : n = 1, 2, \ldots\}$.

2. Definition If f is a scalar-valued function on $[0, 1]$, $\|f\|_p = (\int_0^1 |f(t)|^p \, dt)^{1/p}$ if $p \geq 1$, $\|f\|_p = \int_0^1 |f(t)|^p \, dt$ if $0 < p < 1$. For a sequence x, $\|x\|_p = (\sum |x_n|^p)^{1/p}$ if $p \geq 1$, $\|x\|_p = \sum |x_n|^p$ if $0 < p < 1$.

Table 1-2-1 Table of spaces, where all functions and sequences are scalar valued

bfa (N)	See Example 2-3-14.
bfa (H, A)	See Sec. 14-4.
$B(X, Y)$	See Definition 2-3-2.
c	Convergent sequences, with $\|x\|_\infty$.
c_0	Null sequences (i.e., converging to 0), with $\|x\|_\infty$.
$\mathscr{C}$	The complex numbers.
$C(H)$	Continuous functions on H, with $\|f\|_\infty$ if H is a compact Hausdorff space. For a general Hausdorff space, see Prob. 4-1-105.
$C^*(H)$	Bounded continuous functions on H, with $\|f\|_\infty$.
$C_0(H)$	Continuous functions vanishing at infinity (that is, $\{x \in H : |f(x)| \geq \varepsilon\}$ is compact for each $\varepsilon > 0$), with $\|f\|_\infty$.
cs	Convergent series, with $\|x\| = \|s\|_\infty$ where $s_n = \sum_{k=1}^n x_k$.
Disc algebra	Members of $C(D)$ which are analytic in U, with $\|f\|_\infty$ where $U = \{z \in \mathscr{C} : |z| < 1\}$, $D = \bar{U}$.
$\mathscr{K}$	The scalar field; either $\mathscr{C}$ or R.
l^p	The set of sequences x, with $\|x\|_p < \infty$ (Definition 1-2-2).
l^∞	The bounded sequences, with $\|x\|_\infty$.
L^p	(Equivalence classes of) measurable functions on $[0, 1]$, with $\|f\|_p < \infty$ (Definition 1-2-2).
$\mathscr{M}$	(Equivalence classes of) all measurable functions on $[0, 1]$, with $$\|f\| = \int_0^1 |f(t)|/[1 + |f(t)|] \, dt$$
$M(H)$	See Sec. 14-6.
m_0	See Example 2-3-15.

Table 1-2-1 *Continued*

N The positive integers.
ω All sequences, with $\|x\| = \sum (1/2^n)|x_n|/(1 + |x_n|)$.
φ All finite sequences, that is, x, such that $x_n = 0$ eventually.
R The real numbers.

1-3 SOME COMPUTATIONS

A few useful results from classical analysis are presented in this section.
Suppose that $f''(x) \geq 0$ for $x > 0$. Then, for $0 < a < x < b$,

$$\frac{f(x) - f(a)}{x - a} = \frac{1}{x - a} \int_a^x f' \leq f'(x) \leq \frac{1}{b - x} \int_x^b f' = \frac{f(b) - f(x)}{b - x}$$

Hence

$$f(x) \leq \frac{b - x}{b - a} f(a) + \frac{x - a}{b - a} f(b)$$

Apply this to the function $f = -\log$. With $\theta = (b - x)/(b - a)$ we have

$$a^\theta b^{1-\theta} \leq \theta a + (1 - \theta)b \qquad (1\text{-}3\text{-}1)$$

By symmetry, (1-3-1) holds for all positive a, b and $0 \leq \theta \leq 1$.
   Now with $\{a_n\}$, $\{b_n\}$ nonnegative real sequences, $A = \sum a_n$, $B = \sum b_n$, we have
$\sum (a_n/A)^\theta (b_n/B)^{1-\theta} \leq (\theta/A)\sum a_n + [(1 - \theta)/B]\sum b_n = 1$, and so

$$\sum a_n^\theta b_n^{1-\theta} \leq (\sum a_n)^\theta (\sum b_n)^{1-\theta} \qquad (1\text{-}3\text{-}2)$$

Let u_n, v_n be complex sequences, $p > 1$, $1/p + 1/q = 1$, and, in (1-3-2), set $\theta = 1/p$, $a_n = |u_n|^p$, $b_n = |v_n|^q$. We obtain *Hölder's inequality*:

$$\sum |u_n v_n| \leq (\sum |u_n|^p)^{1/p}(\sum |v_n|^q)^{1/q} \qquad (1\text{-}3\text{-}3)$$

that is, $\|uv\|_1 \leq \|u\|^p \|v\|^q$, $1/p + 1/q = 1$.
   Applying (1-3-3) to partial sums we see that convergence of the series on the right implies convergence of the left-hand series. The same remark applies to the following arguments.
   For $p > 1$, $1/p + 1/q = 1$, applying (1-3-3) gives $(a_n \geq 0, b_n \geq 0)$

$$\sum (a_n + b_n)^p = \sum a_n (a_n + b_n)^{p-1} + \sum b_n (a_n + b_n)^{p-1}$$

$$\leq (\sum a_n^p)^{1/p}[\sum (a_n + b_n)^{(p-1)q}]^{1/q} + (\sum b_n^p)^{1/p}[\sum (a_n + b_n)^{(p-1)q}]^{1/q}$$

Dividing the first and last terms by $[\sum (a_n + b_n)^p]^{1/q}$ and using $(p - 1)q = p$, we obtain

$$[\sum (a_n + b_n)^p]^{1/p} \leq (\sum a_n^p)^{1/p} + (\sum b_n^p)^{1/p}$$

and so

$$(\sum |u_n + v_n|^p)^{1/p} \leq (\sum |u_n|^p)^{1/p} + (\sum |v_n|^p)^{1/p} \qquad p \geq 1 \qquad (1\text{-}3\text{-}4)$$

Now let a, b be complex numbers, set $u_1 = |a|^{1/p}$, $v_2 = |b|^{1/p}$, all other u_i and $v_j = 0$. Then $(|a| + |b|)^{1/p} \le |a|^{1/p} + |b|^{1/p}$. Thus, for $0 < p < 1$, $|a + b|^p \le |a|^p + |b|^p$ and so

$$\sum |u_n + v_n|^p \le \sum |u_n|^p + \sum |v_n|^p \qquad 0 < p < 1 \qquad (1\text{-}3\text{-}5)$$

We shall refer to both (1-3-4) and (1-3-5) as *Minkowski's inequality*. Each shows that $\|u + v\|_p \le \|u\|_p + \|v\|_p$.

Next is given an important theorem proved by I. Schur in 1920. Let $A = (a_{nk})$ be a matrix of complex numbers. For $x \in \omega$, let $(Ax)_n = \sum_k a_{nk} x_k$, $Ax = \{(Ax)_n\}$, if these series converge.

1. Definition For a matrix A, $\|A\| = \sup_n \sum_k |a_{nk}|$.

This is called norm A (it may be ∞) and the reason for the notation is explained in Prob. 3-3-103. It will be seen in Remark 15-2-3 that the assumption $\|A\| < \infty$ is redundant in Theorem 1-3-2.

2. Theorem Suppose that $\|A\| < \infty$ and $Ax \in c_0$ for every sequence x of zeros and ones. Then $\sum_k |a_{nk}| \to 0$ as $n \to \infty$.

PROOF If the result is false we may assume that $\sum_k |a_{nk}| \to 1$. [Choose a sequence $\{i(n)\}$ of integers such that $\sum_k |a_{i(n)k}| \to t > 0$. Let $b_{nk} = a_{i(n)k}/t$. Then $\|B\| < \infty$, $Bx \in c_0$ for each sequence x of zeros and ones and $\sum |b_{nk}| \to 1$. It is sufficient to prove that B cannot exist.] It follows that

$$\lim_{n \to \infty} a_{nk} = 0 \qquad \text{for each } k$$

and

$$(1\text{-}3\text{-}6)$$

$$\lim_{n \to \infty} \sum_{k=m}^{\infty} |a_{nk}| = 1 \qquad \text{for each } m$$

The second part follows from the first, which is proved by setting $x = \delta^k$ in the hypothesis. Now choose $r(1)$ so that $\sum_k |a_{r(1)k}| > \frac{1}{2}$, then $m(1)$ so that $\sum_{k=1}^{m(1)} |a_{r(1)k}| > \frac{1}{2}$, and then $\sum_{k=m(1)+1}^{\infty} |a_{r(1)k}| < \frac{1}{4}$. [Choose separate m's to satisfy each inequality and let $m(1)$ be the larger.] Now by (1-3-6) we may choose $r(2) > r(1)$ so that $\sum_{k=1}^{m(1)} |a_{r(2)k}| < \frac{1}{8}$ and $\sum_{k=m(1)+1}^{\infty} |a_{r(2)k}| > \frac{1}{2}$. Next choose $m(2) > m(1)$ so that $\sum_{k=m(1)+1}^{m(2)} |a_{r(2)k}| > \frac{1}{2}$ and $\sum_{k=m(2)+1}^{\infty} |a_{r(2)k}| < \frac{1}{8}$. Continuing, we obtain $\sum_{k=1}^{m(i-1)} |a_{r(i)k}| < \frac{1}{8}$, $\sum_{k=m(i-1)+1}^{\infty} |a_{r(i)k}| > \frac{1}{2}$; and $\sum_{k=m(i-1)+1}^{m(i)} |a_{r(i)k}| > \frac{1}{2}$, $\sum_{k=m(i)+1}^{\infty} |a_{r(i)k}| < \frac{1}{8}$. Now define $x_k = \operatorname{sgn} a_{r(1)k}$ for $k = 1, 2, \ldots, m(1)$, $x_k = \operatorname{sgn} a_{r(2)k}$ for $k = m(1) + 1, m(1) + 2, \ldots, m(2)$, and so on. Then $x_k = \pm 1$ for each k and so $Ax \in c_0$. [Let $y = (x + 1)/2$. Then y is a sequence of zeros and ones and $x = 2y - 1$.] But

$$|(Ax)_{r(i)}| = \left| \sum_{k=1}^{m(i-1)} a_{r(i)k} x_k + \sum_{k=m(i-1)+1}^{m(i)} |a_{r(i)k}| + \sum_{k=m(i)+1}^{\infty} a_{r(i)k} x_k \right|$$

$$\geq \tfrac{1}{2} - \sum_{k=1}^{m(i-1)} - \sum_{k=m(i+1)}^{\infty} |a_{r(i)k}|$$

$$> \tfrac{1}{2} - \tfrac{1}{8} - \tfrac{1}{8} = \tfrac{1}{4}$$

Thus $Ax \notin c_0$.

PROBLEMS

1 If $x \in l^p$, $y \in l^q$, $1/p + 1/q = 1$, show that $\{x_n y_n\} \in l$.

---

101 Show that equality holds in (1-3-3) if and only if there exist constants A, B such that $A|u_n|^p = B|v_n|^q$.

102 Show that equality holds in (1-3-4) if and only if there exist constants A, B such that $Au_n = Bv_n$.

103 Show that (1-3-4) is false for $0 < p < 1$.

104 State and prove Hölder's and Minkowski's inequalities for integrals.

---

201 Show that, for $p > 0$, $t(p) = (\sum |u_n|^p)^{1/p}$ is a decreasing function of p. 〚See [153], p. 7.〛

202 Show that (Prob. 1-3-201) $\lim \{t(p): p \to +\infty\} = \inf \{t(p): p > 0\} = \max |u_n|$. 〚See [153], p. 7.〛

1-4 NETS

We begin with the concept of a *partially ordered* set, abbreviated *poset*. This is a set X together with a relation $\geq$, with $x \geq y$ true or false for each $x, y \in X$. We assume that the relation is *reflexive*, that is, $x \geq x$ for all x, and *transitive*, that is, $x \geq y \geq z$ imply $x \geq z$. Some authors also require that it be *antisymmetric*, that is, $x \geq y \geq x$ imply $y = x$, but this rules out certain posets which rise naturally in convergence discussions (Prob. 1-4-1). Introducing such a relation into a set is called *ordering the set*.

Extremely important examples are the set P of all subsets of a set X with

(*a*) *order by inclusion*: $A \geq B$ means $A \subset B$;
(*b*) *order by containment*: $A \geq B$ means $A \supset B$.

A *directed set* is a poset with the additional property that for each x, y there exists z with $z \geq x, z \geq y$. For example, R with its usual order is a directed set.

A *chain* is a poset which is antisymmetric and satisfies $x \geq y$ or $y \geq x$ for each pair of members x, y; that is, any two members are *comparable*. For example, R is a chain.

Any subset of a poset is a poset with the same ordering and might possibly be a chain. For example, give R^2 the order $(a, b) \geq (x, y)$ means $a \geq x$ and $b \geq y$ in the ordinary sense. Then the X axis is a chain. Indeed, it is a *maximal chain* in that it is properly contained in no other chain, although there are other chains such as $\{(x, x): x \in R\}$.

We shall now state an axiom of set theory. This axiom will be an unstated hypothesis in all theorems where the phrase "let C be a maximal chain" occurs in the proof. The first such is Theorem 1-5-5.

1. Axiom: Maximal axiom Every nonempty poset includes a maximal chain.

Some references for a discussion of the place of this axiom in mathematics, and alternate forms of the axiom, may be found in [156], Sec. 7-3.

A *net* is a function defined on some directed set. For example, a sequence is a net defined on the positive integers. Just as there are sequences of points, numbers, functions, so there are nets of points, numbers, functions. For example, a net of real numbers is a function $x : D \to R$ where D is some directed set. Such a net is written $(x_\delta : D)$, and in this case x_δ is a real number for each $\delta \in D$.

Now suppose that $(x_\delta : D)$ is a net in some set X, that is, $x : D \to X$ is a map. Let $S \subset X$. We say that $x \in S$ *eventually* if there exists $\delta_0 \in D$ such that $x_\delta \in S$ for all $\delta \geq \delta_0$.

2. Example Let $D = \{\delta \in \mathscr{C} : |\delta| \leq 1\}$. Order D by $\delta \geq \delta'$ if $|\delta| \leq |\delta'|$. Let $u = (u_\delta : D)$ be the net of complex numbers given by $u_\delta = e^\delta$. Let $S = \{z \in \mathscr{C} : |z - 1| < 10^{-2}\}$. Then $u \in S$ eventually. This is just a special case of the familiar fact that $e^\delta \to 1$ as $\delta \to 0$.

We also say that a net has certain properties eventually; for example, "$|x_\delta - 2| < 1$ eventually" means "$x_\delta \in \{x : |x - 2| < 1\}$ eventually."

PROBLEMS

1 Although any two members of D, Example 1-4-2, are comparable, D is not a chain. [It is not antisymmetric.]

2 Show that the set of all subsets of a set X ordered by inclusion is a directed set. Show the same for containment. (For this reason we use the phrase "directed by inclusion" to mean "ordered by inclusion.")

3 Show that the directed set in Prob. 1-4-2 is not a chain if X has more than one point.

4 Let D be a directed set and S a nonempty finite subset. Show that there exists x with $x \geq s$ for all $s \in S$.

5 Let X be a directed set and $U_1, U_2, \ldots, U_n$ subsets of X. Suppose that x is a net with, for each i, $x \in U_i$ eventually. Show that $x \in \cap U_i$ eventually [Prob. 1-4-4].

101 Give an example of a poset which is not a directed set.

102 The *discrete order* on a set X is defined by $x \geq y$ if and only if $x = y$. The *indiscrete order* has $x \geq y$ for all x, y. Which of these is directed? antisymmetric?

103 The set of discs in the plane, ordered by containment, is a directed set but not a lattice. (A *lattice* is an antisymmetric poset such that each pair has a least upper bound and a greatest lower bound.)

104 Show that Prob. 1-4-4 becomes false if S is allowed to be infinite.

105 Describe the ordering of names in a telephone book. This is called the *lexicographic* order. Show how to order R^n lexicographically.

106 Let $D = (0, 1)$ with the usual order. Let $f_\delta(t) = (\cos t)^{1-\delta}$ for $\delta \in D$, $t \in I$ where I is some closed interval in R. Show that $\| f_\delta \|_\infty < 10^{-2}$ eventually if $I = [\frac{1}{4}, 1]$, but not if $I = [0, 1]$.

---

201 The maximal axiom for countable posets is equivalent to induction.

1-5 VECTOR SPACE

As mentioned in Sec. 1-1, our vector spaces have the scalar field $\mathscr{K}$, which is R or $\mathscr{C}$. Throughout this section, X denotes a fixed vector space. For $A \subset X$, the *span of* A is the set of all (finite) linear combinations of A; it is a vector subspace of X. For a vector subspace S and point x, $S + [x]$ denotes the span of $S \cup \{x\}$. If the span of A is equal to X we say that A *spans* X.

A subset $A \subset X$ is called *convex* if $sA + tA \subset A$ for $0 \le t \le 1$, $s + t = 1$; *balanced* if $tA \subset A$ for $|t| \le 1$; and *absorbing* if for every $x \in X$ there exists $\varepsilon > 0$ such that $tx \in A$ for $|t| < \varepsilon$. A vector subspace S of X is called *maximal* if $S \ne X$ and $X = S + [x]$ for some x.

A function $f: X \to \mathscr{K}$ is called a *functional* and $X^\#$ denotes the vector space of all *linear* functionals on X, that is, those satisfying $f(sx + ty) = sf(x) + tf(y)$ for $s, t \in \mathscr{K}$, $x, y \in X$.

There is a natural correspondence between linear functionals and maximal subspaces as follows. *For each nonzero $f \in X^\#$, $f^\perp = \{x: f(x) = 0\}$ is a maximal subspace. For each maximal subspace S there exist many $f \in X^\#$ such that $f^\perp = S$ but only one whose value at any specified $a \notin S$ is a specified nonzero scalar u.* [Fix $a \notin f^\perp$. Then $x - [f(x)/f(a)]a \in f^\perp$. Conversely, fix $a \notin S$. Every x is $s + ta$, $s \in S$, and we may set $f(x) = tu$.]

1. Theorem Let $f, f_1, f_2, \ldots, f_n \in X^\#$ and $f^\perp \supset \cap \{f_i^\perp : i = 1, 2, \ldots, n\}$. Then $f = \sum t_i f_i$.

PROOF For $n = 1$, write $f_1 = g$. We may assume $g \ne 0$. Say $g(a) = 1$. Then for each x, $x - g(x)a \in g^\perp \subset f^\perp$ so $0 = f(x) - g(x)f(a)$. Thus $f = [f(a)]g$. Proceeding by induction, let $f^\perp \supset \cap \{f_i^\perp : i = 1, 2, \ldots, n + 1\}$. Let $g = f | f_{n+1}^\perp$ (the *restriction* of f to the smaller subspace), $g_i = f_i | f_{n+1}^\perp$ for $i = 1, 2, \ldots, n$. Then $g^\perp \supset \cap \{g_i^\perp : i = 1, 2, \ldots, n\}$. By the induction hypothesis $g = \sum t_i g_i$ and so $f(x) = \sum_{i=1}^n t_i f_i(x)$ for $x \in f_{n+1}^\perp$. By the case $n = 1$, this implies that $f - \sum_{i=1}^n t_i f_i = t f_{n+1}$.

Now let X be a complex vector space and X_R the same space but using only real scalars; thus X_R is a real vector space. Let Rz denote the real part of the complex number z.

2. Theorem Let X be a complex vector space, $f \in X^\#$. Then $Rf \in (X_R)^\#$. Moreover, for each $g \in (X_R)^\#$ there exists unique $f \in X^\#$ such that $g = Rf$.

PROOF The first part is trivial. Next, to prove uniqueness, let $g \in (X_R)^{\#}$, $f \in X^{\#}$ with $g = Rf$. Write $f = g + ih$, $h \in (X_R)^{\#}$. Then $g(ix) + ih(ix) = f(ix) = if(x) = ig(x) - h(x)$. Equating real parts yields $h(x) = -g(ix)$ and so h, hence f, is uniquely determined if it exists. Finally, given g, define $h \in (X_R)^{\#}$ by $h(x) = -g(ix)$ (the only formula that could work!). Let $f = g + ih$, and we shall prove that f is linear. It is clear that $f(x + y) = f(x) + f(y)$ and $f(tx) = tf(x)$ for real t; but also $f(ix) = g(ix) + ih(ix) = g(ix) + ig(iix) = g(ix) + ig(x) = i[g(x) - ig(ix)] = if(x)$.

3. Definition A Hamel basis for X is a linearly independent set which spans X.

An *n-dimensional space* $(n < \infty)$ is one which has a Hamel basis with n members.

4. Theorem Let X be an *n*-dimensional vector space, $n < \infty$. Then $X^{\#}$ is also *n*-dimensional. Further, for each $F \in X^{\#\#}$ there exists $x \in X$ such that $F(u) = u(x)$ for all $u \in X^{\#}$.

PROOF Since X is isomorphic with $\mathcal{K}^n$ we may as well take $X = \mathcal{K}^n$. Let $P_i \in X^{\#}$ be defined by $P_i(x) = x_i$ for $i = 1, 2, \ldots, n$. For each $f \in X^{\#}$, $x \in X$, we have $f(x) = f(\sum x_k \delta^k) = \sum f(\delta^k) P_k(x)$. Thus $f = \sum f(\delta^k) P_k$ and so $P = (P_1, P_2, \ldots, P_n)$ spans $X^{\#}$. It is also linearly independent since if $\sum t_k P_k = 0$, for any i, $0 = \sum t_k P_k(\delta^i) = t_i$.

Next, given $F \in X^{\#\#}$, let $x = [F(P_1), F(P_2), \ldots, F(P_n)] \in X$. Then for each $u \in X^{\#}$, $u = \sum t_k P_k$ and so $F(u) = \sum t_k F(P_k) = \sum t_k x_k = \sum t_k P_k(x) = u(x)$.

5. Theorem Every vector space X has a Hamel basis.

PROOF Let P be the family of linearly independent subsets of X; order P by containment and let C be a maximal chain in P. (See Remark 1-5-6.) Let H be the union of the sets in C. This is the required basis. First, H is linearly independent. [This is the same as saying that every finite subset is linearly independent. But such a subset is contained in some $S \in C$ since C is a chain; hence it is linearly independent.] Also, H spans X. [H is maximal among linearly independent sets since, if not, a larger one could be adjoined to C contradicting its maximality. So for any $x \notin H$, $H \cup \{x\}$ is linearly dependent, that is, $tx + \sum t_k h_k = 0$ for some t, $t_1, t_2, \ldots, t_n$ not all 0, $h_i \in H$. Since H is linearly independent, $t \neq 0$ and so we can solve for x.]

6. Remark If $X = \{0\}$, the set P in Theorem 1-5-5 is empty (voiding use of the maximal axiom) unless we make some special convention. The one usually chosen is to say that ϕ, the empty set, is linearly independent, and span $\phi = \{0\}$. Thus X has a Hamel basis with no members, and is 0-dimensional. Hopefully, all the results given in this book are true in such special cases, but we shall not take the space to spell them out.

PROBLEMS

In this list X is a vector space.

1 Let x, A, B be a point in and two subsets of X. Show that $x + A$ meets B if and only if $x \in B - A$.

2 Every absorbing and every balanced set in X contains 0.

3 If A is convex, $sA + tA = (s + t)A$ whenever $s > 0, t > 0$.

4 If S is a maximal subspace of X and $x \notin X$, then $S + [x] = X$.

5 If A is balanced, $tA = |t| A$ for all scalar t.

6 A set which is both balanced and convex is called *absolutely convex*. Show that A is absolutely convex if and only if $sA + tA \subset A$ whenever $|s| + |t| \leq 1$.

7 Let A be absolutely convex and suppose that for every $x \in X$, $tx \in A$ for some $t \neq 0$. Show that A is absorbing.

8 Show that $\{\delta^n\}$ is a Hamel basis for φ.

9 Find all absorbing vector subspaces of X.

10 Every linearly independent set in X is a subset of some Hamel basis for X. [Let the members of P include the set.]

11 Let A be a linearly independent set in X and $f : A \to \mathscr{K}$. Show that there exists $F \in X^*$ with $F \mid A = f$. [By Prob. 1-5-10 you may assume that A spans X.]

12 Show that $c_0, c, l^p, l^\infty, \varphi, \omega$ are vector spaces.

13 The *balanced hull* B of a set S is defined to be the intersection of all the balanced sets which include S. Show that B is balanced and if B' is a balanced set which includes S then $B' \supset B$.

14 Let X be infinite-dimensional and $f_1, f_2, \ldots, f_n \in X^*$. Show that $\cap \{f_i^\perp : i = 1, 2, \ldots, n\} \neq \{0\}$. [Define $F : X \to \mathscr{K}^n$ by $F(x) = [f_1(x), f_2(x), \ldots, f_n(x)]$. If the conclusion were false, F would be one to one.]

15 Show that the subspace of Prob. 1-5-14 has finite codimension in X. (We say that a subspace A has *finite codimension* in X if there is a finite-dimensional subspace B such that $A + B = X$.)

16 The convex hull of a balanced set is balanced, hence absolutely convex. [If $x = \sum t_i x^i$, then $tx = \sum t_i (tx^i)$.]

17 Show that c_0 has a Hamel basis with $\mathfrak{c}$ (cardinality of R) members. [Consider $x(t) = \{t^n\}$ for $0 < t < 1$; also $c_0 \subset \omega = \mathscr{K}^N$.] Do the same for l^2.

18 Let $f \in X^*$ and suppose that f is bounded on a vector subspace S. Show that $f = 0$ on S.

19 Show that the sum of two absolutely convex sets is absolutely convex.

20 Every proper subspace of a vector space X is the intersection of all the maximal subspaces which include it. [For $x \notin S$, let a Hamel basis for S be extended to a Hamel basis for X which contains x. Now make $f = 0$ on S, $f(x) = 1$.]

---

101 (Schatz's apple). Let $A \subset R^2$ be $\{(x, y):(x + 1)^2 + y^2 \leq 1, y \geq 0\} \cup \{(x, y):(x - 1)^2 + y^2 \leq 1, y \geq 0\} \cup \{(x, y):x^2 + y^2 \leq 4, y \leq 0\} \cup \{(0, y):0 \leq y \leq 1\}$. Show that A is absorbing and that there exists no absorbing set B such that $B + B \subset A$.

102 For $n = 1, 2, \ldots,$ let $A_n = \{x \in \varphi : x_i = 0 \text{ for } i > n, |x_i| \leq 1/n \text{ for all } i\}$. Let $A = \cup A_n$. Show that A is absorbing and that there exists no absorbing set B such that $B + B \subset A$. [Let $t\delta^1 \in B$, $m > 1/t$, $s\delta^m \in B$, $z = t\delta^1 + s\delta^m \notin A$.]

103 Let g be any real function on X satisfying $g(0) = 0$. Suppose $f : X \to \mathscr{K}$ satisfies $\left| \sum_{i=1}^n t_i f(x_i) \right| \leq g(\sum t_i x_i)$ for all $t_i \in \mathscr{K}$, $x_i \in X$, $i = 1, 2, \ldots, n$. Show that f must be linear.

104 Let X be a vector space of real functions on $[0, 1]$ which contains at least one noncontinuous function. Suppose that every $f \in X$ satisfying $f(0) = 0$ is continuous. Show that every continuous $f \in X$ satisfies $f(0) = 0$.

105 Show that every maximal subspace of X_R includes a unique maximal subspace of X (see Theorem 1-5-2).

106 Let Y be the set of all subsets of X with the natural addition and multiplication by scalars. Is Y a vector space?

107 Find an absorbing set A such that $A - A$ does not include any convex absorbing set.

108 Find a nonconvex set A satisfying $A + A = 2A$.

201 Any two Hamel bases for a vector space can be put into one-to-one correspondence. ⟦See [153], Theorem 2-3-2.⟧

1-6 TOPOLOGY

A *pseudometric* is a real function d on a set $X \times X$ such that $d(x, y) = d(y, x) \geq 0$, $d(x, x) = 0$, $d(x, z) \leq d(x, y) + d(y, z)$. If, also, $d(x, y) = 0$ implies $x = y$, d is called a *metric*. We assume known the definition of topological space and standard ideas as in [156]. In particular, a topology on X is a collection of subsets of X, the open sets. A pseudometric space is a topological space in the usual way and the topology is Hausdorff (T_2) if and only if the pseudometric is a metric. A *neighborhood* of x is any set which includes an open set containing x. Thus *neighborhoods need not be open.*

A set S in a topological space X is said to be of *first category in X* if it is the union of a sequence of closed sets each of which has empty interior; if not, it is said to be of *second category in X*. For example, Z, the integers, is of first category in R; however, Z is of second category in itself since it has no subsets with empty interior (except ϕ).

The Baire category theorem, which follows, is an important tool in functional analysis.

1. Theorem A complete pseudometric space X is of second category in itself.

PROOF Let $\{A_n\}$ be a sequence of closed sets with empty interior. Let $G_n = X \backslash A_n$ for each n, a dense open set. Then G_1 includes a disc $D(x_1, r_1) = \{x : d(x, x_1) \leq r_1\}$, $r_1 > 0$. Since G_2 is dense, its intersection with $D(x_1, r_1)$ includes a disc $D(x_2, r_2)$ and we may assume $0 < r_2 < r_1/2$. Continuing, we obtain $\{r_n\}$ with $0 < r_{n+1} < r_n/2$, $D(x_{n+1}, r_{n+1}) \subset D(x_n, r_n) \subset G_n$. For $m > n$, $x_m \in D(x_m, r_m) \subset D(x_n, r_n)$ so that $d(x_m, x_n) < r_n$; hence $\{x_n\}$ is a Cauchy sequence, so it is convergent, say $x_n \to x$. Since $x_m \in D(x_n, r_n)$ for $m > n$ and the latter set is closed, $x \in D(x_n, r_n)$ for each n. Thus $x \in G_n$ for each n. We have proved that $\cap G_n \neq \phi$; it follows that $\cup A_n \neq X$.

We record a few properties which a topological space X may have: X is called *regular* if each neighborhood of any point x includes a closed neighborhood of x; T_1 if each singleton is a closed set; T_3 if X is regular and T_1; *completely regular* if for each closed set $F \subset X$ and $x \notin F$ there is a continuous real map f on X with $f = 0$ on F, $f(x) = 1$; $T_{3\frac{1}{2}}$ if X is completely regular and T_1.

2. Definition The set of neighborhoods of a point x in a topological space will be written $\mathcal{N}_x$.

3. Remark $\mathcal{N}_x$ is directed by inclusion and we shall always regard $\mathcal{N}_x$ as a directed set.

4. Definition Let x be a net in a topological space X, $a \in X$. We say $x \to a$ if for each $U \in \mathcal{N}_a$, $x \in U$ eventually.

5. Theorem Let X be a topological space, $a \in X$. For each $U \in \mathcal{N}_a$ choose $x_U \in U$. Then the net $x = (x_U : \mathcal{N}_a) \to a$.

PROOF See Remark 1-6-3. Let $U \in \mathcal{N}_a$. Then for $V \in \mathcal{N}_a$, $V \geq U$ means $V \subset U$ which implies $x_V \in V \subset U$. Thus $x \in U$ eventually.

6. Theorem Let X, Y be topological spaces and $f : X \to Y$. Then f is continuous at $a \in X$ if and only if $x \to a$ implies $f(x) \to f(a)$ for each net x in X.

PROOF $\Rightarrow$: With $x \to a$, let U be a neighborhood of $f(a)$ in Y. Then $x \in f^{-1}[U]$ eventually since this is a neighborhood of a. Thus $f(x) \in U$ eventually. Hence $f(x) \to f(a)$. $\Leftarrow$: Suppose f is not continuous at a. Then there exists a neighborhood U of $f(a)$ such that $f^{-1}[U]$ is not a neighborhood of a. For each $V \in \mathcal{N}_a$, $V \not\subset f^{-1}[U]$ so we may choose $x_V \in V \backslash f^{-1}[U]$. By Theorem 1-6-5, $x_V \to a$, but $f(x_V) \not\to f(a)$ since $f(x_V) \notin U$.

7. Corollary Let T, T' be topologies for a set X such that for any net x in X, $x \to a$ in (X, T) implies $x \to a$ in (X, T'). Then $T \supset T'$.

PROOF By Theorem 1-6-6, $i : (X, T) \to (X, T')$ is continuous.

In particular, a topology is characterized by its convergent nets and their limits.

We now describe three important methods of combining topologies and topological spaces. These three processes (and taking quotients) will be continually recurring ideas, to be tested and applied at every opportunity.

Problem 1-6-7 explains the choice of the name given to the topology in Theorem 1-6-8.

8. Theorem Let Φ be a collection of topologies for a set X. Then there exists a unique topology, called $\vee \Phi$ (sup Φ) such that for any net x in X, $x \to a$ in $(X, \vee \Phi)$ if and only if $x \to a$ in (X, T) for each $T \in \Phi$. For any topological space Z, a function $f : Z \to (X, \vee \Phi)$ is continuous if and only if $f : Z \to (X, T)$ is continuous for each $T \in \Phi$.

PROOF Namely $\vee\Phi$ is the set of all unions of finite intersections of members of $\cup\Phi$. Clearly $\vee\Phi \supset T$ for each $T \in \Phi$, so $x \to a$ in $(X, \vee\Phi)$ implies $x \to a$ in (X, T) by Theorem 1-6-8. Conversely, if $x \to a$ in (X, T) for each T, let U be a neighborhood of a in $(X, \vee\Phi)$. There exist $U_1, U_2, \ldots, U_n$, each U_i a neighborhood of a in some (X, T) with $a \in \cap U_i \subset U$. For each i, $x \in U_i$ eventually, so, eventually, $x \in \cap U_i \subset U$ [Prob. 1-4-5]. Thus $x \to a$ in $(X, \vee\Phi)$. Uniqueness is by Corollary 1-6-7. To prove the continuity criterion: $\Rightarrow$: This is trivial since $\vee\Phi \supset T$. $\Leftarrow$: Let x be a net in Z with $x \to a$. Then $f(x) \to f(a)$ in (X, T) for each T. Thus $f(x) \to f(a)$ in $(X, \vee\Phi)$.

9. Remark *A local base of neighborhoods of $x \in X$ is any set $\mathscr{B} \subset \mathscr{N}_x$ such that each $U \in \mathscr{N}_x$ includes some member of $\mathscr{B}$.* From the proof of Theorem 1-6-8 we see that $\vee\Phi$ has as a local base of neighborhoods of x the set of all finite intersections of neighborhoods of x in the various (X, T).

10. Theorem Let X be a set and F a family of functions $f: X \to Y_f$ where, for each f, Y_f is a topological space. Then there exists a unique topology for X called wF (the weak topology by F) such that for any net x in X, $x \to a$ in (X, wF) if and only if $f(x) \to f(a)$ in Y_f for each $f \in F$. For any topological space Z, a function $g: Z \to (X, wF)$ is continuous if and only if $f \circ g$ is continuous for each $f \in F$.

PROOF We first assume that F has one member $f: X \to Y$. Let $wf = \{f^{-1}(G): G$ an open set in $Y\}$. Clearly $f: (X, wf) \to Y$ is continuous, so Theorem 1-6-6 yields half the result as concerns net convergence. Conversely, suppose $f(x) \to f(a)$ and let U be an open neighborhood of a in (X, wf). Then $U = f^{-1}[V]$ where V is a neighborhood of $f(a)$ in Y. Now $f(x) \in V$ eventually, so $x \in U$ eventually. To prove the continuity criterion: $\Rightarrow$: This is trivial since g and f are continuous. $\Leftarrow$: Let x be a net in Z with $x \to a$. Then $f[g(x)] \to f[g(a)]$ and so $g(x) \to g(a)$ in (X, wf).

In the general case let $wF = \vee\{wf: f \in F\}$. The result follows from the one-function case and Theorem 1-6-8. Uniqueness is by Corollary 1-6-7.

We remark that the use of the word "weak" stems from the result of Prob. 1-6-7 and the fact that "weaker" is sometimes used instead of "smaller" for topologies.

Now suppose that a family $(X_\alpha: \alpha \in \mathscr{A})$ of topological spaces is given. The *product* πX_α is the set of all functions $x: \mathscr{A} \to \cup X_\alpha$ such that $x_\alpha \in X_\alpha$ for each $\alpha \in \mathscr{A}$. [We wrote x_α for $x(\alpha)$.] For two spaces X, Y we write the product as $X \times Y$; similarly for any finite number. By $X^{\mathscr{A}}$ we mean $\pi\{X_\alpha: \alpha \in \mathscr{A}\}$ with $X_\alpha = X$ for each $\alpha \in \mathscr{A}$. Let $\pi = \pi X_\alpha$ and for each $\alpha \in \mathscr{A}$ define $P_\alpha: \pi \to X_\alpha$ by $P_\alpha(x) = x_\alpha$. This is called the *projection* on the αth factor.

11. Theorem Let $\{X_\alpha: \alpha \in \mathscr{A}\}$ be a family of topological spaces. There exists a unique topology for πX_α (the product topology) such that for any net

$(x^\delta:D)$ in πX_α, $x^\delta \to a$ if and only if $x^\delta_\alpha \to a_\alpha$ for each $\alpha \in \mathscr{A}$. For any topological space Z and function $g:Z \to \pi X_\alpha$, g is continuous if and only if $P_\alpha \circ g$ is continuous for each $\alpha \in \mathscr{A}$.

PROOF Note that $(x^\delta:D)$ is a map from D to πX_α; hence for each $\delta \in D$, x^δ is a map from $\mathscr{A}$ to $\cup X_\alpha$ such that $x^\delta_\alpha \in X_\alpha$ for each $\alpha \in \mathscr{A}$. The product topology is simply $w\{P_\alpha : \alpha \in \mathscr{A}\}$ and the result follows from Theorem 1-6-10.

12. Theorem Let X, Y be topological spaces, $Z = X \times Y$, T the product topology for Z. Then for each $z = (x, y) \in Z$, $\{U \times V : U, V$ neighborhoods of x, y in X, Y respectively$\}$ is a local base of neighborhoods of z in (Z, T).

PROOF By $U \times V$ we mean $\{(u, v) : u \in U, v \in V\}$. By Remark 1-6-9 and the proof of Theorem 1-6-10, a local base of neighborhoods of z is given by $\{P_1^{-1}[U] \cap P_2^{-1}[V] : U, V$ neighborhoods of x, y in X, Y, respectively$\}$. Here $P_1 : Z \to X$, $P_2 : Z \to Y$ are the projections. But $P_1^{-1}[U] \cap P_2^{-1}[V] = U \times V$.

13. Theorem Let X, Y, Z be topological spaces and $T: X \times Y \to Z$ a function. Then T is continuous at $(x, y) \in X \times Y$ (with the product topology) if and only if for each directed set D and nets $(x^\delta:D)$, $(y^\delta:D)$ in X, Y with $x^\delta \to x$, $y^\delta \to y$ it follows that $T(x^\delta, y^\delta) \to T(x, y)$.

PROOF The point of this result is that it is sufficient to consider two nets defined on the same directed set. $\Rightarrow$: If $x^\delta \to x$, $y^\delta \to y$ it follows that $(x^\delta, y^\delta) \to (x, y)$ [Theorem 1-6-11] so $T(x^\delta, y^\delta) \to T(x, y)$ [Theorem 1-6-6]. $\Leftarrow$: Let $(z^\delta:D)$ be a net in $X \times Y$ with $z^\delta \to (x, y)$. Then $z^\delta = (x^\delta, y^\delta)$ and $x^\delta \to x$, $y^\delta \to y$ [Theorem 1-6-11]. By hypothesis $Tz^\delta \to T(x, y)$ and so, by Theorem 1-6-6, T is continuous at (x, y).

PROBLEMS

In this list, X, Y, Z are topological spaces.

1 Let x, y be nets in $\mathscr{K}$ with $x \to a$, $y \to b$. Show that $x + y \to a + b$, $xy \to ab$.

2 Let f, g be continuous scalar functions on X, that is, functions $X \to \mathscr{K}$. Show that $f + g$ and tf are continuous for any fixed $t \in \mathscr{K}$ [Prob. 1-6-1]. Hence show that $C(H)$, $C^*(H)$, and $C_0(H)$ are vector spaces (Sec. 1-2).

3 Let $u: X \times Y \to Z$ be continuous and fix $y \in Y$. Define $f: X \to Z$ by $f(x) = u(x, y)$. Show that f is continuous. (*Joint continuity* implies *separate continuity*.)

4 Show that every completely regular space is regular.

5 Show that every T_3 space is a Hausdorff space.

6 Show that every local base of neighborhoods of x is directed by inclusion.

7 Show that $\vee \Phi$ (Theorem 1-6-8) is the smallest topology larger than every $T \in \Phi$, and that wF (Theorem 1-6-10) is the smallest topology such that every $f \in F$ is continuous. [Apply Theorem 1-6-6 to the identity map.]

8 Let $\{x_n^\delta : \delta \in D\}$ be a net in $\mathcal{K}$ for each $n = 1, 2, \ldots$. Suppose that $|x_n^\delta| \le M_n$ for all δ, n, that $\sum M_n < \infty$, and that $x_n^\delta \to x_n$ for each n. Show that $\sum x_n^\delta \to \sum x_n$. $\big[\big|\sum x_n^\delta - \sum x_n\big| \le \big|\sum_{n=1}^m (x_n^\delta - x_n)\big| + \sum_{n=m}^\infty (|x_n^\delta| + |x_n|)$. Apply Prob. 1-6-1.$\big]$

9 Show that $\omega = \mathcal{K}^N$ (as sets).

10 Suppose that in Theorem 1-6-6, X has a countable base of neighborhoods of each of its points. Show that net may be replaced by sequence.

11 Let X, Y be complete metric spaces. Show that $X \times Y$ is a complete metric space with $d[(x_1, y_1), (x_2, y_2)] = d_1(x_1, x_2) + d_2(y_1, y_2)$.

12 Let X be a topological space, Y a set, $f: X \to Y$ onto. Let T, T' be topologies for Y such that $f: X \to (Y, T)$ is continuous; $f: X \to (Y, T')$ is open. Show that $T' \supset T$.

13 Show that each $P_\beta : \pi X_\alpha \to X_\beta$ is an open map, that is, $\rho_\beta[G]$ is open if G is.

14 Let X be compact, Y Hausdorff, and $f: X \to Y$ continuous. Show that f preserves closed sets. Deduce that f is a homeomorphism (into) if it is also one to one.

---

101 Let $S \subset X$, $x \in X$. Show that $x \in \bar{S}$ if and only if some net in S converges to x $[$Theorem 1-6-5$]$.

102 Extend the result of Prob. 1-6-8 to an arbitrary uniformly convergent series.

103 Show that the open interval $(0, 1)$ is of second category in itself.

104 Give an example to show that separate continuity in every variable does not imply joint continuity (see Prob. 1-6-3).

105 Let G be a nonempty open set in $\pi\{X_\alpha : \alpha \in A\}$. Show that
 (a) there exists a finite subset F of A such that, if $x \in G$ and $y_\alpha = x_\alpha$ for $\alpha \in F$, then $y \in G$; hence
 (b) $P_\alpha[G] = X_\alpha$ for all but finitely many α.

106 Let S be a collection of sets. A *cobase* for S is a subset S' of S such that for each $A \in S$ there exists $B \in S'$ with $B \supset A$. A topological space is called *hemicompact* if it has a countable cobase for its compact sets. Show that every open interval in R is hemicompact.

107 Show that the space of rational real numbers is not hemicompact. (Although it is σ compact, i.e., the union of countably many compact subsets.)

108 A G_δ is a countable intersection of open sets. Show that a dense G_δ must be residual. (A set is called *residual* if its complement is of first category.)

109 The set of rationals is not a G_δ in R $[$Prob. 1-6-108$]$.

110 Is the set of irrational numbers σ compact?

111 Let S be a dense subspace of a pseudometric space X such that every Cauchy sequence in S is convergent in X. Show that X is complete.

112 Let X be a set, $\{T_\alpha : \alpha \in A\}$ a collection of topologies for X, and $X_\alpha = (X, T_\alpha)$ for each α. The *diagonal* D in πX_α is $\{x : x_\alpha$ is a constant member of X independent of $\alpha\}$. Show that D, with the (restriction of the) product topology is homeomorphic with $(X, \vee T_\alpha)$. Thus the sup topology could have been defined from the product topology. $[$Use nets to apply Theorems 1-6-8 and 1-6-11.$]$

113 Show that every infinite compact Hausdorff space H has an infinite subspace on which the relative topology is discrete. $[$If H is a convergent sequence this is trivial. Otherwise every point has a neighborhood whose complement is infinite and an induction may be used.$]$

---

201 Deduce the result of Prob. 1-6-8 from Lebesgue's dominated convergence theorem.

202 Let G be an absorbing set in a product of vector spaces. Show that, in Prob. 1-6-105, (b) is true but (a) need not be.

---

301 Let X be a metric space and S a subspace which has a complete metric giving the same topology as that induced by the metric of X. Then S is a G_δ in X. $[$See $[80]$, Prob. 6 K(c); $[93]$, Sec. 31-III.$]$

METRIC IDEAS

2-1 PARANORMS

The concept of paranorm is a generalization of that of absolute value. The *paranorm of* x, written $\| x \|$ (and, by the vagaries of custom, usually pronounced "norm x"), is a real number defined for all x in a vector space X and satisfying, for all $x, y \in X$,

(a) $\| 0 \| = 0$.

(b) $\| x \| \geq 0$.

(c) $\| -x \| = \| x \|$.

(d) $\| x + y \| \leq \| x \| + \| y \|$ (triangle inequality).

(e) *If* $\{t_n\}$ *is a sequence of scalars with* $t_n \to t$ *and* $\{x_n\} \subset X$ *with* $\| x_n - x \| \to 0$, *then* $\| t_n x_n - tx \| \to 0$ (continuity of multiplication).

The paranorm is called total if, in addition, we have

(f) $\| x \| = 0$ *implies* $x = 0$.

The most famous paranorm is the ordinary absolute value function. Others are given in Probs. 2-1-1 and 2-1-4 and in Sec. 2-2.

1. Example Let p be a paranorm and define $\| x \| = p(x)/[1 + p(x)]$. This defines a paranorm; for example, $\| x + y \| \leq [p(x) + p(y)]/[1 + p(x) + p(y)]$ $[\text{since } t/(1 + t) = 1 - 1/(1 + t) \text{ is an increasing function of } t \geq 0] \leq p(x)/$ $[1 + p(x)] + p(y)/[1 + p(y)] = \| x \| + \| y \|$.

2. Theorem Let $\{p_n\}$ be a sequence of paranorms on a vector space X. Then there exists a paranorm $\|\cdot\|$ such that for any net x in X, $\|x\| \to 0$ if and only if $p_n(x) \to 0$ for each n.

PROOF There are several possibilities. (See Probs. 2-1-8 and 2-1-107.) The one we give here is called the *Fréchet combination*; namely, set

$$\|x\| = \sum_{n=1}^{\infty} \frac{1}{2^n} \frac{p_n(x)}{1 + p_n(x)}$$

Parts (a), (b), (c) of the definition of paranorm are obviously true. Part (d) follows from Example 2-1-1. Part (e) will follow when we prove the equivalence stated in the theorem. Suppose, first, that $\|x\| \to 0$, say $x = \{x_\delta : \delta \in D\}$. For each δ, n we have $\|x_\delta\| \geq 2^{-n} p_n(x_\delta)/[1 + p_n(x_\delta)]$; hence $p_n(x_\delta) \leq 2^n \|x_\delta\| / (1 - 2^n \|x_\delta\|) \to 0$ for each n. Conversely, if $p_n(x) \to 0$ for each n, $\|x\| \to 0$ since the series is uniformly convergent by comparison with $\sum 2^{-n}$ [Prob. 1-6-8].

3. Definition A pseudometric vector space is a pair (X, d) where X is a vector space and d is a pseudometric such that the operations are (jointly) continuous.

This means that if $x_n \to x$, $y_n \to y$ in X and $t_n \to t$ in $\mathcal{K}$, then $t_n x_n + y_n \to tx + y$. (See Theorem 1-6-13.) It is immediate that if X is a vector space and a paranorm p is selected, X becomes such a space with the pseudometric $d(x, y) = p(x - y)$. We shall prove (Theorem 4-5-5) that the converse is true; namely, any pseudometric vector space can be given an equivalent pseudometric which is derived from a paranorm. Temporarily we ignore this and consider only pseudometric vector spaces of the following type. (To see how much generality is lost, consult Remark 4-5-6.)

4. Definition A paranormed space is a pair (X, p) where X is a vector space and p is a paranorm. It is called complete if (X, d) is complete, where $d(x, y) = p(x - y)$.

Note that if (X, d) is a pseudometric vector space such that $d(x, y) = d(x - y, 0)$, then $p(x) = d(x, 0)$ defines a paranorm.

5. Remark The function $p:(x, p) \to R$ is continuous. For $|p(x) - p(y)| \leq p(x - y) = d(x, y)$.

6. Definition We say of paranorms p, q that p is stronger than q and q is weaker than p if $p(x_n) \to 0$ implies $q(x_n) \to 0$. If each is stronger than the other we call them equivalent. If p is stronger than and not equivalent with q it is said to be strictly stronger. If $p(x) \geq q(x)$ for all x we say that p is larger than q and q is smaller than p.

Note that each paranorm is smaller than itself. (See also Prob. 2-1-3.)

PROBLEMS

In this list $\|\cdot\|$ denotes a paranorm on a vector space X.

1 Define p, q, r, s on R^2 by $p(x, y) = |x| + |y|$, $q(x, y) = |x|$, $r(x, y) = |y|$, $s(x, y) = |x| + |y|/(1 + |y|)$. Show that these are paranorms. Which ones are total?

2 Show that the paranorms in Example 2-1-1 are equivalent and that in Prob. 2-1-1; p, s are equivalent and strictly stronger than q, r; and that q, r are not comparable.

3 If there exists a number M such that $q(x) \le Mp(x)$ for all x, then p is stronger than q. The converse is false [Prob. 2-1-1].

4 Check that the formulas given in Sec. 1-2 define paranorms on ω, l^p, $0 < p < 1$ [Theorem 2-1-2 and inequality (1-3-5)].

5 Show that $\|nx\| \le n\|x\|$, $\|x/n\| \ge \|x\|/n$ when n is a positive integer.

6 Suppose that $\|tx\| \le |t| \|x\|$ for all scalar t, $x \in X$. Show that equality must hold. (Compare Prob. 2-1-5.)

7 Show that $\{x : \|x\| = 0\}$ is a vector subspace of X.

8 Suppose that in Theorem 2-1-2 only finitely many $p_n \ne 0$. Show that $\sum p_n$ satisfies the conditions of the theorem.

9 A continuous linear map between paranormed spaces must be uniformly continuous. [$\|x - y\| < \delta$ implies $\|T(x - y)\| < \varepsilon$ by continuity at 0.] Hence show that such a map preserves Cauchy sequences. Deduce that a paranormed space which is linearly homeomorphic with a complete paranormed space is complete.

10 The results of Prob. 2-1-9 are false for pseudometric spaces in general. [Let $T : R \to (0, 1)$ be a homeomorphism onto. Define $d(x, y) = |Tx - Ty|$ for $x, y \in R$. Then (R, d) is linearly homeomorphic with R.]

11 Let S be a dense subspace of X, Y a complete paranormed space, and $T : S \to Y$ continuous, linear. Show that T can be extended to a continuous linear map: $X \to Y$ [Prob. 2-1-9].

12 Show that a countable product of pseudometric spaces is pseudometrizable [Fréchet combination].

---

101 Show that $\{x : \|x\| < \varepsilon\}$ is absorbing for each $\varepsilon > 0$.

102 Define $p(x) = 1$ if $x \ne 0$, $p(0) = 0$, $q(x) = 0$ for all x. Are p, q paranorms?

103 Show that $\|y\| = 0$ implies $\|x + y\| = \|x\|$ for all x.

104 Let $\{t_n\} \subseteq \mathscr{K}$ be bounded and $\|x_n\| \to 0$. Show that $\|t_n x_n\| \to 0$.

105 Let $p(x + iy) = |x|$. Is p a paranorm on $\mathscr{C}$? (Compare q in Prob. 2-1-1.)

106 If p is stronger than q, show that $p(x) = 0$ implies $q(x) = 0$.

107 Show that the following are paranorms suitable for the proof of Theorem 2-1-2: $p(x) = \inf \{\sum_{k=1}^n p_k(x) + 1/n : n = 1, 2, \ldots\}$; $q = \sum(p_k \wedge 2^{-k})$. [*Note*: $(f \wedge g)(x) = \min \{f(x), g(x)\}$.]

108 Give an example to show that (in the definition of paranorm) (a), (b), (c), (d), and "$\|x_n\| \to 0$ implies $\|t_n x_n\| \to 0$ for every convergent $\{t_n\}$" do not imply (e).

109 In the definition of paranorm, (e) may be replaced by (i) and (ii) or by (iii) and (iv), where

    (i) $\|x_n - x\| \to 0$, $t_n \to 0$ imply $\|t_n x_n\| \to 0$,

    (ii) $\|x_n\| \to 0$ implies $\|tx_n\| \to 0$,

    (iii) $\|x_n\| \to 0$, $t_n \to t$ imply $\|t_n x_n\| \to 0$,

    (iv) $t_n \to 0$ implies $\|t_n x\| \to 0$.

110 A paranorm on $\mathscr{C}$ is either total or identically zero.

111 Let $p(x, y) = |x^2 - y^2|^{1/2}$, $q(x, y) = |x^2 - y^2|^{3/2}$ for $(x, y) \in R^2$. Show that p, q are not paranorms.

112 For $x \in \varphi$, $x \ne 0$, let $p(x) = |x_n|$ where x_n is the last nonzero term of x; $p(0) = 0$. Show that p is not a paranorm.

113 Let H be a Hamel basis for a vector space X. Define $p(x) = \sum |r_i|^{1/2}$ if $x = \sum r_i h^i$. Show that p is a paranorm. [Note that $p(tx) = |t|^{1/2} p(x)$.]

114 If a vector subspace S of a complete paranormed space is a G_δ it must be closed. [We may assume S is dense. For any $x \notin S$, $x + S \subset \tilde{S}$; hence by Prob. 1-6-108, S and $\tilde{S}$ are of first category. This contradicts the Baire category theorem.]

115 In Prob. 2-1-11, T may be one to one and its extension not.
    [(a) $x \to \{x_{n+1} - x_n/n\}$ mapping φ into c_0;
      (b) let S be Y with a larger noncomplete metric (Example 5-2-8), $X = $ completion of S.]

116 In Prob. 2-1-11, the range of the extension of T may be a dense proper subset of Y.

---

201 Show that the set of paranorms on X is a conditionally complete lattice. [For $p \wedge q$ use $h(x) = \inf \{p(y) + q(z) : y + z = x\}$.]

202 Let H in Prob. 2-1-113 be uncountable. Show that $\{x : p(x) < 1\}$ cannot include a convex absorbing set.

203 Suppose (Theorem 2-1-2) that $q(x) = \sum p_i(x)$ is convergent for each $x \in X$. Is q a paranorm and, if so, is it equivalent with the Fréchet combination?

204 Let p be a paranorm defined on ω. Let $A_n = \{x : p(x) < 1\} + \mathcal{K}^n$. Show that there exists an integer n such that A_n is dense; if p is continuous, such that $A_n = \omega$.

205 Must $\sup \{\|x\| / \|2x\|\} < \infty$?

---

301 Give an example of a paranorm for R which is not monotone on $[0, \infty)$. [See [153], Prob. 4-1-22.]

302 Give an example of a pointwise bounded sequence $\{p_n\}$ of paranorms such that $q(x) = \sup p_n(x)$ does not define a paranorm. [See [104], p. 335.]

2-2 SEMINORMS

The *seminorm* of x, written $\|x\|$, is a real number defined for all x in a vector space X and satisfying, for all $x, y \in X$, $t \in \mathcal{K}$ (the scalars),

    (a) $\|x\| \geq 0$
    (b) $\|x + y\| \leq \|x\| + \|y\|$
    (c) $\|tx\| = |t| \|x\|$

If the seminorm is *total* ($\|x\| = 0$ implies $x = 0$) it is called a *norm. Every seminorm is a paranorm.* [For example, $\|(t + \varepsilon)(x + y) - tx\| \leq |t| \|y\| + |\varepsilon| \|x\| + |\varepsilon| \|y\| \to 0$ if $\varepsilon \to 0$, $\|y\| \to 0$, proving part (e).]

When a particular seminorm is specified, X becomes a pseudometric space (as in Sec. 2-1) and is called a *seminormed space*, or *normed space* if appropriate.

1. Theorem Let A be an absolutely convex absorbing set in a vector space X. Then there exists a unique seminorm $\|\cdot\|$ such that $\{x : \|x\| < 1\} \subset A \subset \{x : \|x\| \leq 1\}$. The seminorm is a norm if and only if A includes no one-dimensional vector subspace.

PROOF Uniqueness is immediate from Prob. 2-2-4. Now define the seminorm, known as the *gauge* (or *Minkowski gauge*) of A, by the formula

$$\|x\| = \inf \{t \geq 0 : x \in tA\}$$

To prove part (c) of the definition of seminorm suppose $\|x\| < s$. Then $x \in sA$ and so $tx \in stA = s|t|A$. Thus $\|tx\| \le |t|s$. This proves that $\|tx\| \le |t| \cdot \|x\|$ and we may apply Prob. 2-1-6. For part (b), let $s > \|x\|$, $t > \|y\|$. Then $x \in sA$, $y \in tA$ and so $x + y \in sA + tA = (s + t)A$ since A is convex. Thus $\|x + y\| \le s + t$. If the seminorm is not a norm the last part follows by Prob. 2-1-7. Conversely, if, for some $x \ne 0$, $tx \in A$ for all scalar t, we have $|t| \|x\| = \|tx\| \le 1$ for all t; so $\|x\| = 0$.

The three fundamental tools for the application of functional analysis are the Hahn–Banach theorem, the uniform boundedness theorem, and the closed graph theorem. The first of these now makes its initial appearance. The Hahn–Banach theorem appears in many forms and with several proofs; the Index may be consulted for these. An independent proof of Theorem 2-2-2 is given in Remark 7-3-3.

2. Theorem Let X be a vector space and p a seminorm. Let f be a linear functional defined only on a vector subspace S of X and such that $|f(x)| \le p(x)$ for $x \in S$. Then f can be extended to $F \in X^\#$ with $|F(x)| \le p(x)$ for $x \in X$.

PROOF
Case (a). X is a real vector space.
Let $P = \{g : g \in T^\# \text{ for some vector subspace } T = T_g \text{ of } X, T \supset S, g|S = f, |g(x)| \le p(x) \text{ for } x \in T\}$.

Note that T depends on g; P is not empty since $f \in P$. Make P into a poset by agreeing that $g_2 \ge g_1$ means that g_2 is an extension of g_1. Let M be a maximal chain in P, $D = \cup\{T_g : g \in M\}$. D is a vector subspace of X. ⟦For $t \in R$, $x, y \in D$ we must have $x, y \in$ some T_g since M is a chain. Thus $tx + y \in T_g \subset D$⟧ Define $F : D \to R$ by $F(x) = g(x)$ whenever $x \in T_g$ with $g \in M$; F is well defined since M is a chain; F is linear by the same argument which showed D to be a vector subspace. Also for $x \in D$, $x \in T_g$ for some g; hence $|F(x)| = |g(x)| \le p(x)$. *So we have only to prove that $D = X$.* If this is false, let $y \in X\backslash D$. Let $u = \sup\{F(x) - p(y - x) : x \in D\}$, $v = \inf\{F(x) + p(y - x) : x \in D\}$. Then $u \le v$. ⟦For any $a, b \in D$, $F(b) - F(a) = F(b - a) \le p(b - a) \le p(b - y) + p(y - a)$. Thus $F(b) - p(y - b) \le F(a) + p(y - a)$.⟧ Let D_1 be $D + [y]$ and define $F_1 \in D_1^\#$ by $F_1(d + ty) = F(d) + tu$. A contradiction will be reached when we show that $F_1 \in P$ since $M \cup \{F_1\}$ is strictly larger than M. First, F_1 extends F and hence extends f. Finally, let $x \in D_1$, $x = d + ty$. If $t = 0$, $|F_1(x)| = |F(d)| \le p(d) = p(x)$. If $t \ne 0$,

$$F\left(-\frac{d}{t}\right) - p\left(y + \frac{d}{t}\right) \le u \le v \le F\left(-\frac{d}{t}\right) + p\left(y + \frac{d}{t}\right)$$

If $t > 0$, multiplication by t leads to

$$-F(d) - p(ty + d) \le tu \le -F(d) + p(ty + d) \tag{2-2-1}$$

while if $t < 0$, multiplication by $-t$, which is positive, leads to

$$F(d) - p(-ty - d) \leq -tu \leq F(d) + p(-ty - d)$$

which is the same as (2-2-1). From (2-2-1), $-p(x) = -p(d + ty) \leq F(d) + tu = F_1(x) \leq p(d + ty) = p(x)$; thus $|F_1(x)| \leq p(x)$.

Case (b). X is a complex vector space.
Let g be the real part of f and consider X as a real linear space. ⟦Here and in the succeeding steps we use Theorem 1-5-2.⟧ Since $|g(x)| \leq |f(x)| \leq p(x)$, we may, by case (a), extend g to a real linear G on X with $|G(x)| \leq p(x)$. Let $G = RF$, $F \in X$ *. On S, $RF = G = g = Rf$ so $F | S = f$; finally, $|F(x)| \leq p(x)$ by Prob. 2-2-5.

3. Example : Generalized limits Let l^∞ be the *real* bounded sequences. *We shall prove the existence of a linear functional L satisfying* $\lim \inf x_n \leq L(x) \leq \lim \sup x_n$ *for all* $x \in l^\infty$. Apply Theorem 2-2-2 with $S = c$, the convergent sequences, $f(x) = \lim x_n$. Let L be the extension satisfying $|L(x)| \leq \|x\|_\infty$ for all x. Then L satisfies $L \in (l^\infty)$ *, $L(1) = 1$, $L = 0$ on φ, $|L(x)| \leq \|x\|_\infty$; this is all we use. We first show that L is a *positive* map (that is, $x_n \geq 0$ for all n implies $L(x) \geq 0$). ⟦Let $t = \|x\|_\infty$, $y_n = x_n - t/2$. Then $\|y\| = t/2$ so $|L(y)| \leq t/2$. Thus $-t/2 \leq L(y) = L(x) - t/2$.⟧ Now for any $x \in l^\infty$ let $t > \lim \sup x_n$. Then $x_n < t$ for $n > m$, say. Let $y_n = x_n$ for $n \leq m$, $y_n = t$ for $n > m$, and $z = y - x$. Then $z_n \geq 0$ for all n; $0 \leq L(z) = L(y) - L(x) = t - L(x)$ so that $L(x) \leq t$. Thus $L(x) \leq \lim \sup x$, and so $L(x) = -L(-x) \geq -\lim \sup(-x) = \lim \inf x$. In Prob. 2-3-106 we shall show that L can be made to have an additional property.

PROBLEMS

In this list p, q are seminorms.

1 Show that p is stronger than q if and only if there exists M such that $q(x) \leq Mp(x)$ for all x. Compare Prob. 2-1-3. ⟦If $q(x_n) > np(x_n)$, set $y_n = x_n/q(x_n)$. Then $p(y_n) < 1/n$, $q(y_n) = 1$.⟧

2 Show that p is stronger than q if and only if q is bounded on $\{x : p(x) < 1\}$.

3 Show that a bounded seminorm must be identically zero.

4 Suppose a, b are positive real numbers such that $p(x) < a$ implies $q(x) \leq b$. Show that $aq(x) \leq bp(x)$ for all x. ⟦If $aq(x) > t > bp(x)$, let $y = (a/t)x$.⟧

5 Let $F \in X$ * and suppose $RF(x) \leq p(x)$ for all x. Show that $|F(x)| \leq p(x)$. ⟦With $F(x) = r \exp(i\theta)$, let $y = x \exp(-i\theta)$. Then $r = F(y) \leq p(y)$.⟧

6 In Table 1-2-1 check that the given formulas define norms on l^∞, L^∞, L^p, l^p for $p \geq 1$ ⟦Inequality (1-3-4)⟧.

7 Suppose a, b are positive numbers, $f \in X$ *, and $p(x) < a$ implies $f(x) \neq b$. Show that $|f(x)| \leq (b/a)p(x)$ for all x. ⟦If $a|f(x)| > bp(x)$, let $y = bx/f(x)$.⟧

8 Show that every vector space has a norm defined on it. ⟦$\|x\| = \sum |t|$ where $x = \sum th$, $h \in$ Hamel basis.⟧

---

101 Give an example which shows that both inclusions in Theorem 2-2-1 may be strict.

102 Let $q \neq 0$; then p is stronger than q if and only if $\inf \{p(a): q(a) = 1\} > 0$.

103 Suppose that A in Theorem 2-2-1 is assumed to be absorbing only, or absorbing and convex only. What properties does its gauge have in each case?

104 Let S be a vector subspace of X and $z \in X$. Suppose that p is bounded on $z + S$. Show that p is constant on $z + S$. (See Prob. 2-2-3.)

105 For each absolutely convex absorbing set A, there is another, B, such that $B + B \subset A$.

106 Let $S \subset X^*$ and assume that $h(x) = \sup \{|f(x)|: f \in S\} < \infty$ for each x. Show that h is a seminorm.

107 Let $D(a, r) = \{x: \|x - a\| \leq r\}$. Show that in a normed space, $D(a, r) = D(b, s)$ implies $a = b$, $r = s$; but in a totally paranormed space, this is not necessarily true.

108 Let q be a complete norm on a vector space S which is a dense proper subspace of a normed space (X, p). (So far, $S = l$, $X = c_0$ would do.) Now assume that $q \leq p$ on S. (An example of this is given in Prob. 3-1-105.) Show that if q is extended to X as a seminorm satisfying $q \leq p$, q cannot be a norm. [If $p(x^n - x) \to 0$, $q(x^n - s) \to 0$, then $q(s - x) = 0$.] Thus, *there is no Hahn–Banach extension theorem* (Theorem 2-2-2) *for norms*, even from a dense subspace.

109 Prove the Hahn–Banach theorem for seminorms. [Let $q \leq p$ on S as in Theorem 2-2-2. Let $A = \{s \in S: q(s) \leq 1\} \cup \{x \in X: p(x) \leq 1\}$. The gauge of the convex hull of A extends q to a seminorm q' with $q' \leq p$.]

---

201 Do Prob. 2-1-201 for the set of seminorms. Show also that $p \wedge g$ is the gauge of $A + B$, p, q being the gauges of A, B.

2-3 SEMINORMED SPACE

A seminormed space is a pair $(X, \|\cdot\|)$ as in Sec. 2-2. This is a topological space, so we may discuss such properties as continuity.

We define the *unit disc* to be $\{x: \|x\| \leq 1\}$, and for a map T between seminormed spaces, let $\|T\| = \sup \{\|Tx\|: \|x\| \leq 1\}$.

1. Theorem Let X, Y be seminormed spaces and $T: X \to Y$ a linear map. The following are equivalent:

(a) T is continuous at some $a \in X$.
(b) T is continuous.
(c) T is bounded on the unit disc, that is, $\|T\| < \infty$.
(d) There exists a number M such that $\|Tx\| \leq M\|x\|$ for all x.

Further, if these hold we have $\|Tx\| \leq \|T\| \|x\|$ for all x.

PROOF $(a) \Rightarrow (b)$. Let $x_n \to x$. Then $a + x_n - x \to a$ so $T(a) + T(x_n) - T(x) \to T(a)$. Thus $T(x_n) \to T(x)$. $(b) \Rightarrow (c)$. T is continuous at 0 so $\{x: \|Tx\| < 1\} \in \mathcal{N}_0$. Thus it includes some $U = \{x: \|x\| < \varepsilon\}$. Now apply Prob. 2-2-4 to obtain $\|Tx\| \leq (1/\varepsilon)\|x\|$; hence $\|T\| \leq 1/\varepsilon$. $(c) \Rightarrow (d)$. Since $\|x\| < 1$ implies $\|Tx\| \leq \|T\|$ the result follows from Prob. 2-2-4, with $a = 1$, $b = \|T\|$. This also yields the last statement of the theorem. $(d) \Rightarrow (a)$. Take $x = 0$.

Because of this result, people who work only in this context refer to "continuous" by the name "bounded" for linear maps.

2. Definition Let X, Y be vector spaces and topological spaces; $B(X, Y)$ is the set of continuous linear maps from X to Y. If $Y = \mathcal{K}$, $B(X, Y)$ is written X'; thus X' is the set of continuous linear functionals on X, called the dual space of X.

It is implicit in such a definition that X, Y have the same scalars; in the second case, Y is the scalar field of X. In general we could not expect $B(X, Y)$ to be a vector space, although it certainly is if Y is a paranormed space or, more generally, a topological vector space (Chap. 4).

The dual space is also called the conjugate space, and people who specialize in Banach and Hilbert space tend to write it X^*. The present usage is firmly established in the context of this book.

3. Theorem If X, Y are seminormed spaces, so also is $B(X, Y)$; moreover, it is a normed space if Y is.

PROOF The seminorm definition (Sec. 2-2) is easy to check. If Y is normed and $T \in B(X, Y)$ has $\| T \| = 0$, it follows from Theorem 2-3-1 that $T = 0$.

4. Example We shall evaluate c_0'. For $x \in c_0$, using the notations of Sec. 1-1, we have $x = \sum x_k \delta^k$ $[\| x - \sum_{k=1}^{m} x_k \delta^k \|_\infty = \| (0, 0, \ldots, 0, x_{m+1}, x_{m+2}, \ldots \|_\infty = \sup \{ |x_n| : n > m \} \to 0]$. So if $f \in c_0'$, $f(x) = \sum y_k x_k$ where $y_k = f(\delta^k)$. Now we shall show that $y \in l$. [Fix m. Let $x_n = \operatorname{sgn} y_n$ for $n \le m$, $x_n = 0$ for $n > m$. Then $\sum_{n=1}^{m} |y_n| = \sum_{n=1}^{m} y_n x_n = f(x) \le \| f \|$ since $\| x \| = 1$.] The proof just given establishes $\| f \| \ge \| y \|_1$. But $\| x \|_\infty \le 1$ implies $|f(x)| = |\sum y_n x_n| \le \sum |y_n|$, so we have $\| f \| = \| y \|_1$. We have just set up the map $c_0' \to l$ given by $f \to \{ f(\delta^k) \}$ and have seen that it is an isometry. [It is clearly linear.] But it is also onto l. [Let $y \in l$. Let $f(x) = \sum y_n x_n$; then $f \in c_0'$ since $|f(x)| \le \| x \|_\infty \sum |y_n|$ and $\{ f(\delta^n) \} = y$.] *The final result is $c_0' = l$, that is, these two normed spaces are equivalent.* (An *equivalence* is a one-to-one linear isometry onto. The condition "one to one" is not redundant in general since there may be $x \ne 0$ with $\| x \| = 0$.)

5. Example $l' = l^\infty$. As in Example 2-3-4, $x = \sum x_k \delta^k$ and $f(x) = \sum y_k x_k$ with $y \in l^\infty$ $[|y_k| = |f(\delta^k)| \le \| f \| \| \delta^k \|_1 = \| f \|]$. Also, $|f(x)| \le \| y \|_\infty \sum |x_k|$ so $\| f \| \le \| y \|_\infty$. Continue as in Example 2-3-4.

6. Example For $p > 1$, $(l^p)' = l^q$, $1/p + 1/q = 1$. As in Example 2-3-4, $f(x) = \sum y_k x_k$, $y_k = f(\delta^k)$. Now $y \in l^q$. [Assume $y \ne 0$. Fix m so that $y_i \ne 0$ for some $i < m$. Let $u = (\sum_{i=1}^{m} |y_i|^q)^{1/p}$, $x_k = |y_k|^q/(u y_k)$ for $k \le m$, $x_k = 0$ for $x > m$. Then $f(x) = (\sum_{k=1}^{m} |y_k|^q)^{1/q}$. But $f(x) \le \| f \|$ since $\| x \| = 1$.] The proof just given establishes $\| f \| \ge \| y \|_q$. Hölder's inequality (1-3-3) yields $|f(x)| \le \| y \|_q \| x \|_p$ and so $\| f \| \le \| y \|_q$. That the map $(l^p)' \to l^q$ is onto is shown as in Example 2-3-4. (See Prob. 1-3-1.)

We proceed with aspects of the Hahn–Banach theorem.

7. Theorem Let X be a seminormed space and $f \in S'$, where S is a vector subspace. Then f may be extended to $F \in X'$ without increase of norm; that is, $\| F \| = \| f \|$.

PROOF Let $p(x) = \| f \| \| x \|$. Applying Theorem 2-2-2 we obtain an extension F with $|F(x)| \leq \| f \| \| x \|$.

8. Remark Theorem 2-3-7 deals with extension of functionals. This possibility is a special property of R and $\mathscr{C}$ and does not hold for maps to other spaces. A map may have an extension at the cost of increasing its norm (Prob. 3-2-101) or may have no continuous extension (Example 14-4-9). Of course, extension from a dense subspace is a special case (Prob. 2-1-11). See Prob. 2-3-109 for yet another aspect.

9. Theorem Let X be a seminormed space, S a vector subspace, and $x \in X \backslash \bar{S}$. Then there exists $f \in X'$ with $f(x) = 1$, $f = 0$ on S, $\| f \| = 1/d(x, S)$.

PROOF Here $\bar{S}$ is the closure of S and $d(x, S) = \inf \{ \| x - s \| : s \in S \}$. By Theorem 2-3-7 it is sufficient to construct $f \in (S + [x])'$ with the required properties. This is accomplished merely by extending 0, that is, we define f to be 0 on S and $f(x) = 1$. Specifically, let $f(s + tx) = t$ for $s \in S$, $t \in \mathscr{K}$. Let $d = d(x, S)$. Now for $z = s + tx$, $t \neq 0$, we have $\| z \| = |t| \| x - (-s/t) \| \geq |t| d = |f(z)| d$. Thus, $\| f \| \leq 1/d$ and, in particular, f is continuous. Conversely, for $s \in S$, $1 = f(x - s) \leq \| f \| \| x - s \|$ and so $1 \leq \| f \| d$.

10. Remark A consequence of Theorem 2-3-9 is that a normed space which is at least one dimensional has a continuous linear functional $f \neq 0$. [Take $S = \{0\}$.] This is somewhat exciting in view of Probs. 2-3-119 and 2-3-120.

11. Remark If, given $f \in X'$, there exists y with $\| y \| = 1$, $f(y) = \| f \|$, then $|f|$ has its maximum on the unit disc D at y. Theorem 2-3-9 implies that *for each such y, there is an f such that $|f|$ has its maximum on D at y.* The dual result is false (Prob. 2-3-104).

12. Definition A set $S \subset X$ is called fundamental if the span of S is dense in X, or, equivalently, if the linear closure of S ($=$ closure of the span) is X.

Many writers use *total* for this concept.

13. Theorem Let X be a seminormed space, $S \subset X$. Suppose $f \in X'$, $f = 0$ on S implies $f = 0$. Then S is fundamental. If S is a vector subspace it is dense.

PROOF This is immediate from Theorem 2-3-9.

An amusing proof can be given of the fact that if g is a noncontinuous linear functional $g^{\perp}$ is dense. (A more instructive proof is given in Prob. 4-2-5.) Let

$f \in X'$, $f = 0$ on $g^\perp$. By Theorem 1-5-1, $f = kg$. So $k = 0$ and the result follows by Theorem 2-3-13.

The subject matter of the next two examples will figure prominently in Chaps. 14 and 15 but will not be used earlier.

14. Example bfa(N). This is the space of bounded finitely additive complex valued set functions μ defined on all subsets of N, the positive integers. The finitely additive assumption is that $\mu(\phi) = 0$ and $\mu(S \cup T) = \mu(S) + \mu(T)$ whenever $S \pitchfork T$. Define $\|\mu\| = \sup \{\sum_{i=1}^{n} |\mu(e_i)| : N = \cup e_i, e_i \pitchfork e_j$ for $i \neq j\}$ and bfa(N) is defined to be the set of finitely additive μ such that $\|\mu\| < \infty$. It is a normed space.

15. Example *The dual of l^∞ is* bfa(N). For each $f \in (l^\infty)'$ define $\mu(S) = f(\chi_S)$ where χ_S is the characteristic function of S. Then μ is finitely additive. $[\chi_{S \cup T} = \chi_S + \chi_T$ if $S \pitchfork T.]$ Further, μ is bounded, indeed $\|\mu\| \leq \|f\|$. [Let $N = \cup e_i$, $e_i \pitchfork e_j$ for $i \neq j$. Define $x \in l^\infty$ by $x_n = \operatorname{sgn} \mu(e_k)$ for $n \in e_k$. Then $\|x\| = 1$ and $\sum |\mu(e_i)| = \sum x_{n_i} \mu(e_i)$ (choose any $n_i \in e_i$) $= f(\sum x_{n_i} \chi_{e_i}) \leq \|f\|$ since $y = \sum x_{n_i} \chi_{e_i}$ has $|y_n| = 0$ or 1 for all n so that $\|y\| \leq 1.]$ The map $\varphi : f \to \mu$ is onto bfa(N). To see this, let m_0 be the span in l^∞ of the set of sequences of zeros and ones; thus m_0 is the set of simple sequences, i.e., sequences taking on only finitely many values. For each $\mu \in$ bfa(N), define $f \in m_0^{\#}$ by $f(x) = \sum x_i \mu(e_i)$ where each e_i is a set on which x takes on one of its values x_i. Indeed, $f \in m_0'$ and $\|f\| \leq \|\mu\|$. [If $\|x\|_\infty \leq 1$, $|f(x)| \leq \sum |x_i| |\mu(e_i)| \leq \|\mu\|.]$ Since m_0 is dense in l^∞ we may extend f to all of l^∞ [Prob. 2-1-11] and write $f \in (l^\infty)'$. Clearly, $f(\chi_S) = \mu(S)$, that is, $\varphi(f) = \mu$. In the course of proving $f \in m_0'$ just now we also saw that $\|f\| \leq \|\mu\|$. The reverse inequality was proved earlier and so the map from $(l^\infty)' \to$ bfa(N) is an equivalence onto. The function f corresponding to μ is written $f(x) = \int_N x \, d\mu$. This can be expressed in terms of the limit of a Riemann sum (Prob. 2-3-117).

PROBLEMS

1 Show that (a) and (b) of Theorem 2-3-1 are equivalent for a paranormed space and are implied by (d). The rest is false [Prob. 2-1-3].

2 Let X ($\neq \{0\}$) be a normed space, $x \in X$. Then there exists $f \in X'$ with $\|f\| = 1$, $|f(x)| = \|x\|$. [Assume $x \neq 0$; let $y = x/\|x\|$ and apply Remark 2-3-11.]

3 Show that $c' = l$. $[x = (\lim x)1 + \sum(x_k - \lim x)\delta^k$. The equivalence is $y \to f$ where $f(x) = y_1 \lim x + \sum y_{k+1} x_k$, that is, $f \to y$ where $y_{k+1} = f(\delta^k)$, $y_1 = f(1) - \sum f(\delta^k).]$

4 Let p, q be equivalent norms for a vector space X. Show that $(X, p)' = (X, q)'$ and that on this space p, q induce equivalent norms. $[|f(x)| \leq \|f\|_p M q(x)$ by Prob. 2-2-1. Thus $\|f\|_q \leq M \|f\|_p.]$ (The converse is Prob. 8-4-110.)

5 Show that $\omega' = \varphi$ in the sense that $f \in \omega'$ if and only if there exists $y \in \varphi$ with $f(x) = \sum y_k x_k$. [Show $x = \sum x_k \delta^k$ by means of Theorem 2-1-2.]

6 A topological space is called *separable* if it has a countable dense subset. Show that c_0 is separable. [Consider $\{x \in \varphi : x_n$ is rational for all $n\}$.] (The same is true of c, l^p, $0 < p < \infty$.)

7 Show that l^∞ is not separable. [Consider the set of sequences of zeros and ones.]

---

101 Let $X = R^2$ with $\|(x, y)\| = |x|$. Show that X' is equivalent with R.

102 With X as in Prob. 2-3-101, show that $B(X, R^2)$ is equivalent with R^2. (Here R^2 has the euclidean norm.)

103 Find $f \in (l^\infty)' \backslash l$, that is, there exists no $x \in l$ such that $f(y) = \sum x_n y_n$ for all $y \in l^\infty$ [Example 2-2-3].

104 The dual result to Remark 2-3-11 is false. Let $f \in c_0'$ be $f(x) = \sum x_n/n!$. Show that there exists no z with $\|z\|_\infty = 1$, $|f(z)| = \|f\|$.

105 Let $S = \{x \in l^\infty : x = \{y_n - y_{n-1}\}$ for some $y \in l^\infty\}$. (Take $y_0 = 0$.) Show that $d(1, S) = 1$.

106 Show that real l^∞ allows a *Banach limit*, that is, $L \in (l^\infty)'$ with $\liminf x_n \le L(x) \le \limsup x_n$; $L(x) = L(\{x_{n+1}\})$. Apply Theorem 2-3-9 to Prob. 2-3-105 and use the argument of Example 2-2-3.] Use L to show that if x is a bounded sequence in $\mathcal{K}$ with $x_{n+1} - x_n \to t$, t must be 0.

107 Let L be a Banach limit. Let $u(x) = L[\{(-1)^n x_n\}]$. Show that $u = 0$ on c.

108 A Banach limit cannot satisfy $L(xy) = L(x)L(y)$. [$x = \{(-1)^n\}$, $y = \{(-1)^{n+1}\}$, $x + y = 0$, $x^2 = 1$.]

109 Let X be a normed space, S a subspace, and $T : S \to l^\infty$ a continuous linear map. Then T has an extension $T_1 : X \to l^\infty$ with $\|T_1\| = \|T\|$. [Let $u_n \in S'$ be $u_n(x) = (Tx)_n$. Extend each u_n to X.]

110 Let $H = [0, 1]$, $X = C(H)$. Fix $g \in X$ and set $u(f) = \int_0^1 f(t)g(t) \, dt$ for $f \in X$. Show that $u \in X'$ and $\|u\| = \int |g|$.

111 Let $u(t) = \{t^n\} \subset c_0$, $A = \{u(t) : 0 < t < 1\}$. Show that every infinite subset of A is fundamental in c_0.

112 Show that each $x \in l^{1/2}$ is equal to $\sum x_n \delta^n$ [like Example 2-3-4].

113 Show that each $f \in (l^{1/2})'$ can be represented as $f(x) = \sum t_n x_n$, $t \in l^\infty$, and conversely each such functional is in $(l^{1/2})'$ [Prob. 2-3-112].

114 Let X be a normed space and f, $g \in X'$ with $f(x_0) = g(x_0) = 1$. Show that $\|f - g\| \le a\|g\|_{f^\perp}\|$ where $a = 1 + \|f\| \|x_0\|$.

115 Define $f \in (l^\infty)'$ by $f(x) = 3x_1 + x_2$. Find μ (Example 2-3-15).

116 With L as in Example 2-2-3, show that μ (Example 2-3-15) is not additive; that is, $\mu(\cup e_n) \ne \sum \mu(e_n)$, $e_i \pitchfork e_j$, is possible. (See Prob. 3-2-114.)

117 Prove all the assertions in the following discussion. A partition of N is $\pi = (e_1, e_2, \ldots, e_n)$, $e_i \pitchfork e_j$, $N = \cup e_i$, together with a distinguished point $h_k \in e_k$ if $e_k \ne \phi$. Let $\pi_1 \ge \pi_2$ mean each $e'_k \in \pi_2$ is included in some $e_i \in \pi_1$. Then D, the set of partitions, is a directed set. Fix $x \in l^\infty$. Let $\sum_\pi = \sum_{i=1}^n x(h_i)\chi_{e_i}$. The net $(\sum_\pi : D)$ converges to x. So if $f \in (l^\infty)'$, $f(\sum_\pi) \to f(x)$, that is, $\sum x(h_i)\mu(e_i) \to f(x)$.

118 Obtain the result of Example 2-3-15 for the set of bounded complex functions on an arbitrary set N.

119 Show that $\mathcal{M}' = \{0\}$. [It is sufficient to consider the subspace X of real piecewise continuous functions. Let $u \in X'$. Let $\|f\| \le 1/n$ imply $|u(f)| < 1$. Given $f \in X$, write $f = \sum_{k=1}^n f_k$ where $f_k = 0$ off $[(k-1)/n, k/n]$. Then $\|f_k\| \le 1/n$ so $|u(f)| \le \sum |u(f_k)| < n$. Since u is bounded, it is 0.]

120 Show that $(L^p)' = \{0\}$ for $0 < p < 1$. [The solution is similar to that of Prob. 2-3-119. See [116], supplement to chap. 2.]

121 Show that $\mathcal{M}$ and L^p, $0 < p < 1$, are not normed spaces, i.e., no norm is equivalent to the paranorm given in Sec. 1-2. [See Remark 2-3-10 and Probs. 2-3-119 and 2-3-120.]

122 Theorem 2-3-13 fails for metric linear space. [See Probs. 2-3-119 and 2-3-120.]

---

201 Show that $(L^p)' = L^q$, $1/p + 1/q = 1$, $p > 1$. [See [37], IV.8; [153], 7.4, Example 4; [57], Sec. 15.]

202 Deduce the result of Example 2-3-4 from Prob. 2-3-301 below.

203 Let $X = (l^{1/2}, \|\cdot\|_{1/2})$, $Y = (l^{1/2}, \|\cdot\|_1)$. Find a closed vector subspace of X which is not closed in Y.

204 Let S be a separable subspace of l^∞, $f \in S'$. Show that there exists a matrix A such that $f(x) = \lim Ax$ for all $x \in S$. [See [153], Example 6-4-9.]

---

301 (F. Riesz representation). Show that for a locally compact Hausdorff space H, $C_0(H)' = M(H)$, the set of regular Borel measures on H, $\|\mu\| = $ the variation of μ on H, $u \leftrightarrow \mu$ where $u(f) = \int_H f \, d\mu$. [See [37], Theorem IV.6.3, [57], Theorem 20.45, or [43].]

302 Let X, Y be real normed spaces and $f : X \to Y$ an isometry satisfying $f(0) = 0$. Then f is linear. [See [8], Chap. 11, Theorem 2; [52], Sec. 74, Probs. 8, 9; [7], [18], [40], [48], [114].] The result is false for complex spaces [reflection of $\mathscr{C}$].

303 Let H be a compact Hausdorff space. Then $C(H)$ has the property given for l^∞ in Prob. 2-3-109 if and only if H is extremally disconnected. [See [81].]

THREE

BANACH SPACE

The origins of functional analysis early in this century lay in the development of Banach spaces—first through specific examples, notably L^p, and later maturing into an abstract development. After 1920, generalizations were made to Fréchet spaces. For the sake of applications, even more generality was sought after 1935 with the introduction of topological vector spaces. Although it is sometimes too expensive a luxury to trace a subject through its historical development, there are strong reasons for motivating the general theory by a short study of parts of the classical theory of Banach spaces. One of these is that some of the more general results use the special results as a preliminary step.

3-1 BANACH SPACE

1. Definition A Banach space is a complete normed space.

This means that a vector space and a particular norm are chosen and the resulting metric (vector) space is complete.

Before introducing examples we mention an important type of extra structure which classical spaces usually have. The ideas will also simplify presentation of examples.

2. Definition A K space is a vector space of sequences which has a topology such that each P_n is continuous, where $P_n(x) = x_n$.

3. Example *The spaces $\mathscr{C}^n$, R^n, c_0, c, $l^p(p \geq 1)$, are K spaces.* For each of these it is trivial that $|P_n(x)| = |x_n| \leq \|x\|$ so that Theorem 2-3-1 applies. Also ω *is a K space* by Theorem 2-1-2. Non-K spaces are given in Probs. 3-1-104, 5-4-101, and 5-5-108.

4. Example *The first five spaces mentioned in* Example 3-1-3 *are Banach spaces.* For example, consider l^p, $1 \leq p < \infty$. Let $\{x^n\}$ be a Cauchy sequence. For each k, $\{x_k^n\}$ is a convergent sequence of complex numbers by Example 3-1-3 and Prob. 2-1-9. Let $x_k^n \to x_k$. For each m, $\sum_{k=1}^m |x_k^i - x_k^j|^p \leq \|x^i - x^j\|^p < \varepsilon^p$ for large i, j. Thus $\sum_{k=1}^m |x_k^i - x_k|^p \leq \varepsilon^p$ for large i, independent of m. This shows that $x \in l^p$, and then that $\|x^i - x\|_p \leq \varepsilon$ for large i. Thus $x^i \to x$ and our space is complete.

5. Definition A *K*-function space is a vector space of functions on a set S which has a topology such that each P_s is continuous, where $P_s(f) = f(s)$.

6. Example $C_0(H)$ *and the disc algebra are K-function spaces* since $|P_s(f)| = |f(s)| \leq \|f\|_\infty$. Moreover, *they are Banach spaces.* For the disc algebra this follows from the fact (Weierstrass double-series theorem) that it is a closed subspace of $C_0(H) = C(H)$ where $H = \{z : |z| \leq 1\}$. For $C_0(H)$, let $\{f_n\}$ be a Cauchy sequence. For each $t \in H$, $\{f_n(t)\}$ is convergent by Prob. 2-1-9. Let $f_n(t) \to f(t)$. Since $f_n \to f$ uniformly, f is continuous. Further, $\{x : |f(x)| \geq \varepsilon\} \subset \{x : |f_n(x)| \geq \varepsilon/2\}$ if $\|f_n - f\|_\infty < \varepsilon/2$, so the first set, as a closed subset of the second, is compact.

Note that L^p is not a K-function space since the functions P_s are not defined; knowing f almost everywhere does not tell us $f(s)$ for any particular s. (See also Prob. 3-1-104.)

7. Example *Let X be a seminormed space, Y a Banach space. Then $B(X, Y)$ is a Banach space.* To prove completeness, let $\{T_n\}$ be a Cauchy sequence. For each $x \in X$, $\{T_n(x)\}$ is convergent in Y. [Either directly, $\|T_m(x) - T_n(x)\| \leq \|T_m - T_n\| \|x\|$ by Theorem 2-3-1; or imitate Example 3-1-6 using $P_x(T) = T(x)$ for $x \in X$, $P_x \in B[B(X, Y), Y]$.] Let $T_n(x) \to T(x)$ for each x defining $T : X \to Y$; T is linear. [For example, $T(x + z) = \lim T_n(x + z) = \lim T_n(x) + \lim T_n(z) = T(x) + T(z)$; note that Y is a Hausdorff space so that "$= \lim$" makes sense.] Next suppose $\|T_m - T_n\| < \varepsilon$ for $m > n \geq N$. For any x with $\|x\| \leq 1$, $\|T_m(x) - T_n(x)\| < \varepsilon$ and so $\|T(x) - T_n(x)\| \leq \varepsilon$ for $n \geq N$, using Remark 2-1-5. Thus $\|T\| \leq \varepsilon + \|T_N\|$ so that $T \in B(X, Y)$, and $\|T - T_n\| \leq \varepsilon$ for $n \geq N$ so that $T_n \to T$. (Now see Theorem 2-3-3.)

8. Example In particular, *the dual X' of a seminormed space is a Banach space* $[X' = B(X, \mathcal{K})]$.

PROBLEMS

1 Show that l^∞ is a Banach space. [Hence bfa(N) is by Examples 3-1-8 and 2-3-15.]

2 Show that φ is dense in c_0; hence that $(\varphi, \|\cdot\|_\infty)$ is not a Banach space.

3 The *support* of $f: H \to \mathscr{K}$ is the closure of $\{x: f(x) \neq 0\}$. Show that the set of functions of compact support is dense in $C_0(\mathscr{K})$.

4 Let X be an infinite dimensional Banach space. Show that X has a proper subspace which is of second category in X. [Use a Hamel basis to write $X = \cup S_n$. Some S_n must be of second category.]

5 A second-category normed space need not be complete [Prob. 3-1-4].

---

101 Prove the result of Prob. 3-1-3 with $\mathscr{K}$ [in $C_0(\mathscr{K})$] replaced by an arbitrary $T_{3\frac{1}{2}}$ space. Show that Prob. 3-1-2 is a special case.

102 Let K be a compact set in c_0. Show that $x_n \to 0$ uniformly on K.

103 Let K be a compact set in l. Show that $\sum_{k=n}^{\infty} |x_k| \to 0$ uniformly on K.

104 Fix $p \geq 1$ and let $X = L^p$ without identification, i.e., each $f \in X$ is a function. Now $(\int |f|^p)^{1/p}$ defines a seminorm which is not a norm. Show that $P_s \in X^{\#} \backslash X'$, that is, P_s is not continuous, $0 \leq s \leq 1$. Thus X is not a K-function space.

105 Show that every maximal subspace M of a Banach space X can be given a smaller complete norm, $[M = f^{\perp}, f(a) = 1$. Let $g \in X'$, $g(a) = 1$. Define $u: g^{\perp} \to M$ by $u(x) = x - f(x)a$, $p(x) = \|u(x)\|$.]

106 (D. Amir). Let $S = [0, 1]$, $T = \{x \in \mathscr{C}: |z| = 1$ or $z = 0\}$. Define $u: C(S) \to C(T)$ by $u(f)(e^{2\pi i x}) = f(x) + (\frac{1}{2} - x)[f(1) - f(0)]$. Show that u is a linear homeomorphism onto, $\|u\| = 2$, $\|u^{-1}\| = \frac{3}{2}$.

---

201 Show that L^p is a Banach space for $p \geq 1$. [See [57], Theorems 13.11, 20.14.]

---

301 Let $LH =$ "linear homeomorphism onto" or "linearly homeomorphic" and let X, Y be LH Banach spaces. Then $d(X, Y)$, the *Banach–Mazur distance*, is defined to be $\log [\inf \{\|u\| \|u^{-1}\|: u \in LH\}]$. If S, T are compact T_2 spaces and $d[C(S), C(T)] < \log 2$, S, T must be homeomorphic and there are nonhomeomorphic S, T making $d = \log 2$. [See [20].] In Prob. 3-1-106, $d = \log 3$. In [15] it is shown that $d(c, c_0) = \log 3$.

302 If H is a noncountable compact metric space $C(H)$ LH $C([0, 1])$. [A. A. Milutin, see [101], p. 174.]

303 A vector space of infinite Hamel dimension a allows a complete norm if and only if $a^{\aleph_0} = a$. Further, assuming the generalized continuum hypothesis, this is true if and only if a is not the limit of an increasing sequence of cardinals. [See [92].]

3-2 THE SECOND DUAL

The *second dual* X'' is simply the dual of X'. Here we are assuming that X is a seminormed space; thus X' is a Banach space, Example 3-1-8, and so $X'' = (X')'$ is also a Banach space. For each $x \in X$, the functional $f \to f(x)$ which maps $X' \to \mathscr{K}$ is written $\hat{x}$; that is, $\hat{x}(f) = f(x)$ for each $f \in X'$. Clearly, for each x, $\hat{x}$ is linear; it is also continuous; indeed, $|\hat{x}(f)| = |f(x)| \leq \|f\| \|x\|$ and so $\|\hat{x}\| \leq \|x\|$. Thus $\hat{x} \in X''$. The map $x \to \hat{x}$ from X to X'' is called the *natural embedding*.

1. Example $c'_0 = l$, $c''_0 = l^{\infty}$ [Examples 2-3-4 and 2-3-5]. We are looking at l as a sequence space with typical member $y = \{y_n\}$ or as a function space with typical member $f: c_0 \to \mathscr{K}$. As we saw in the cited examples, y corresponds to f_y where $f_y(x) = \sum y_i x_i$. Now for $x \in c_0$, $\hat{x}(f_y) = f_y(x) = \sum y_i x_i = \sum x_i y_i$ so that $\hat{x}$ corresponds to the particular member $z \in l^{\infty}$ whose representation as a function on l is $y \to \sum x_i y_i$. This is x. Thus *the natural embedding of c_0 into l^∞ is the inclusion map.*

We now discuss the natural embedding; the full generality of seminormed (rather than normed) space is needed for later use.

2. Theorem Let X be a seminormed space. The natural embedding is a linear isometry of X into X'', that is, $\|\hat{x}\| = \|x\|$ for each x. In case X is a normed space the natural embedding is an equivalence (i.e., it is also one to one).

PROOF We saw in the discussion before Example 3-2-1 that $\|\hat{x}\| \leq \|x\|$. In particular, if $\|x\| = 0$, then $\hat{x} = 0$ since X'' is a normed space. Now let $\|x\| \neq 0$. We apply Theorem 2-3-9 with $S = \{0\}$, $y = x/\|x\|$. This gives $f \in X'$ with $\|f\| = 1 = f(y)$. Thus $|\hat{x}(f)| = |f(x)| = \|x\|$ and so $\|\hat{x}\| \geq \|x\|$.

3. Definition The set $\{\hat{x} : x \in X\}$ is written $\hat{X}$.

4. Example Let us evaluate $\hat{x}$ for $x \in c$. For $y \in l$ we have $f_y(x) = y_1 \lim x + \sum y_{k+1} x_k$ [Prob. 2-3-3]. So $\hat{x}(y) = \sum z_k y_k$ where $z_1 = \lim x$, $z_k = x_{k-1}$, that is, $\hat{x} = (\lim x, x_1, x_2, \ldots)$. Conversely, given $z \in l^\infty$ with $\lim z = z_1$ (in particular, z is convergent), let $x = (z_2, z_3, \ldots)$. Then $\hat{x} = z$. Thus $\hat{c} = \{z \in l^\infty : z \in c$ and $\lim z = z_1\}$, the correspondence being $x \to (\lim x, x_1, x_2, \ldots)$ from c to $\hat{c}$, or $z \to (z_2, z_3, \ldots)$ from $\hat{c}$ to c. This is not ordinary inclusion of c in l^∞.

5. Definition A Banach space X is said to be reflexive if $\hat{X} = X''$.

6. Example Example 3-2-1 shows that c_0 is not reflexive while the same sort of argument applied to l^p, $p > 1$, shows that $\hat{x} = x$, $(l^p)'' = l^p$ [Example 2-3-6] and so $l^p, p > 1$, is reflexive. (This is false for $p = \infty$; see Example 3-2-10.) Problem 2-3-103 shows that l is not reflexive, but we shall see a more elementary proof in Example 3-2-10. (Another sort of proof for c_0 derives from Probs. 3-2-2 and 2-3-104.)

7. Remark Use of "$=$" requires caution. For example, we write $c = \hat{c}$ in the sense of Theorem 3-2-2, but $\hat{c} \neq c$ in Example 3-2-4. In Definition 3-2-5, the equality is precise, i.e., it means that for each $F \in X''$ there exists $x \in X$ such that $F = \hat{x}$, that is, $F(u) = u(x)$ for all $u \in X'$. It is possible that $X = X''$ in the sense that these spaces are equivalent yet X is not reflexive. (See [65].) The equivalence in this case is not the natural embedding. In the cited article $\hat{X}$ has finite codimension in X''; spaces such that $\dim X''/X < \infty$ are called *quasireflexive*.

8. Remark We could have defined reflexivity for normed spaces. But a reflexive normed space is complete [Example 3-1-8] so it is a Banach space.

9. Theorem Let X be a Banach space. Then X is reflexive if and only if X' is.

PROOF $\Rightarrow$: Let $F \in X'''$. Define $f \in X'$ by $f(x) = F(\hat{x})$. For any $\hat{x} \in X''$, $F(\hat{x}) = f(x) = \hat{x}(f)$ so $F = \hat{f}$: $\Leftarrow$: Consider an arbitrary member of X''', which we may call $\hat{f}$ ($f \in X'$), which vanishes on $\hat{X}$. Then for any $x \in X$, $f(x) = \hat{x}(f) = \hat{f}(\hat{x}) = 0$. Thus $f = 0$. By Theorem 2-3-13, $\hat{X}$ is dense in X''. Since it is complete, it is closed so $\hat{X} = X''$.

10. Example l and l^{∞} are not reflexive by Theorem 3-2-9 and Example 3-2-6, with Examples 2-3-4 and 2-3-5.

11. Theorem A closed subspace S of a reflexive Banach space X is reflexive.

PROOF Let $F \in S''$. Define $G \in X''$ by $G(u) = F(u \mid S)$ for $u \in X'$. Now $G = \hat{x}$ and $x \in S$. [Suppose $u \in X'$ vanishes on S. Then $u(x) = G(u) = F(0) = 0$. Thus $x \in \bar{S} = S$ by Theorem 2-3-9.] Finally, $F = \hat{x}$. [Let $v \in S'$; extend v to $u \in X'$. Then $F(v) = G(u) = u(x) = v(x)$.]

12. Remark We now begin cautiously to write X for $\hat{X}$ when X is a normed space. By Theorem 3-2-2, they are equivalent. Thus we may write $F \in X'' \setminus X$ to mean $F \in X''$ and $F \notin \hat{X}$, that is, there exists no x with $F = \hat{x}$. A little caution is indicated by Prob. 3-2-106.

13. Remark A consequence of Theorem 3-2-2 (with Example 3-1-8) is that every normed space X has a (Banach) *completion*, called γX, that is, X is a dense subspace of a Banach space γX. [Take $\gamma X = \bar{X} \subset X''$.]

PROBLEMS

1 Let X, Y be vector spaces and A a set of maps from X to Y. Then A is called *total over* X if $f(x) = 0$ for all $f \in A$ implies $x = 0$. Let X be a normed space, $S \subset X$. Show that S is fundamental if and only if $\hat{S}$ is total over X' [Theorem 2-3-13].

2 Let X be reflexive, $f \in X'$. Then $|f|$ assumes a maximum on the unit disc (see Remark 2-3-11 and Prob. 2-3-104) [Prob. 2-3-2]. (The converse is true, see [68].)

3 Let D, D_2 be the unit discs in X, X''. Show that $\hat{D} = D_2$ if and only if X is reflexive.

4 If a Banach space is linearly homeomorphic with a reflexive Banach space, it is reflexive. [Say $u: X \to Y$; define $p(x) = \|ux\|$ and apply Prob. 2-3-4.]

5 A *projection* is a continuous linear map P from X onto $S \subset X$ satisfying $P^2 = P$. Show that for any normed space X there is a projection $P: X''' \to X'$ such that $\|P\| = 1$ [the map $F \to f$ given in Theorem 3-2-9].

101 Let $P: c \to c_0$ be a projection. Show that $\|P\| \geq 2$. [Let $P1 = v$, $x = 1 - 2\delta^n$, $\|Px\| \geq |v_n - 2| \to 2$.] Thus the map $i: c_0 \to c_0$ cannot be extended without an increase of norm.

102 Theorem 3-2-9 is false for normed spaces (see Remark 3-2-8).

103 For $H = [0, 1]$, $C(H)$ is not reflexive. [c_0 is a closed subspace via $x \to f$ where $f(1/n) = x_n$.] This is generalized by Prob. 10-2-120.

104 Suppose that for a Banach space X, X' has a closed proper subspace which is total over X. Show that X is not reflexive [Theorem 2-3-9].

105 Prove the converse of Prob. 3-2-104. [Consider $F^\perp$ where $F \in X''\backslash X$; Theorem 1-5-1.]

106 Define $\psi \in c''$ by $\psi(f) = f(1) - \sum f(\delta^k)$ (Prob. 2-3-3). Show that $\psi \in c''\backslash c$. [As a member of l^∞, $\psi = \delta^1$; thus $\psi \in c$ in one sense, but $\psi \notin \hat{c}$, which is what the statement in the problem means.]

107 Every finite-dimensional normed space X is reflexive. [With d for dimension, $d(X') \leq d(X)$ by Theorem 1-5-4, so $d(X'') \leq d(X') \leq d(X)$. But $X \subset X''$.]

108 Every linear functional on a finite-dimensional normed space is continuous. [$d(X^*) = d(X) = d(X')$ as in Prob. 3-2-107. But $X' \subset X^*$.]

109 Every finite-dimensional normed space is complete [Prob. 3-2-107]. Hence every finite-dimensional subspace of a normed space is closed.

110 Every n-dimensional normed space X is topologically isomorphic with $\mathscr{K}^n$. [$T: \mathscr{K}^n \to X$, $T(y) = \sum y_i b^i$ where $(b^1, b^2, \ldots, b^n)$ is a basis for X. Use Prob. 3-2-108.] Hence all norms on X are equivalent.

111 Let p, q be seminorms on a finite-dimensional vector space such that $p^\perp = q^\perp$. Show that p, q are equivalent [Prob. 3-2-110].

112 A Banach space cannot have a countably infinite Hamel basis [Prob. 3-2-109 and Theorem 1-6-1]. What does this say about φ?

113 Show that c_0 is not equivalent to a dual space. Compare Prob. 11-1-122. [Probs. 3-2-5 and 3-2-101.]

114 Let $u \in (l^\infty)' = l'$ correspond to $\mu \in \mathrm{bfa}(N)$, Example 2-3-15. Show that $u \in l$ if and only if μ is countably additive. [$\Leftarrow$: Let $x_n = \mu(\{n\})$.]

---

201 Let X be the real polynomials of degree ≤ 2, $\|f\| = \max\{|f(t)| : 0 \leq t \leq 1\}$ for $f \in X$. Let $u(f) = f'(0)$. Show that $\|u\| = 8$. [See [53].]

---

301 A Banach space is reflexive if and only if for every bounded set B there is a regular matrix A such that B contains an A-summable sequence. [See [145].]

302 Every uniformly convex Banach space is reflexive. [See [153], Theorem 7.4.1 or Prob. 13.4.23.]

3-3 UNIFORM BOUNDEDNESS

The second fundamental tool for the applications of functional analysis is the uniform boundedness principle. As with the Hahn–Banach theorem it takes many forms (see the Index). The main result is Theorem 3-3-6. It should be noted that the assumption of completeness cannot be dropped from the various hypotheses. This is shown in Probs. 3-3-4, 3-3-5, and 3-3-102.

We begin with a preliminary application of the Baire category Theorem 1-6-1.

1. Definition A barrel is an absolutely convex absorbing closed set.

2. Lemma Let X be a normed space of second category. Then every barrel A is a neighborhood of 0. In particular, this is true if X is a Banach space.

PROOF The second part follows from the first by the Baire category theorem. For each $x \in X$, $(1/n)x \in A$ for some n, so $X = \cup\{nA : n = 1, 2, \ldots\}$. Since X is of second category, some nA has nonempty interior, say $nA \supset \{x:$

$\| x - a \| < r\}$. Then $A \supset \{x: \| x - a/n \| < \varepsilon\}$ where $\varepsilon = r/n$, and so $A - a/n \supset \{x: \| x \| < \varepsilon\}$. Since A is absolutely convex, $A \supset \frac{1}{2}(A + A) = \frac{1}{2}(A - A) \supset \frac{1}{2}(A - a/n) \supset \{x: \| x \| < \varepsilon/2\}$. Thus A includes a neighborhood of 0.

See Prob. 3-3-4.

3. Definition A set S in a seminormed space is called bounded if $\{\| x \| : x \in S\}$ is bounded.

It is important that this is *not* the definition of bounded set in paranormed spaces in general.

4. Definition Let X, Y be seminormed spaces and $S \subset B(X, Y)$. We say that S is pointwise bounded if $\{ux: u \in S\}$ is a bounded set in Y for each $x \in X$; and that S is uniformly bounded, or norm bounded, if $\{\| u \| : u \in S\}$ is bounded.

Note that uniform boundedness is merely boundedness in the seminormed space $B(X, Y)$.

5. Lemma Let X, Y be seminormed spaces and $S \subset B(X, Y)$. For $r > 0$, let $A_r = \{x: \| ux \| \leq r \text{ for all } u \in S\}$. Then S is pointwise bounded if and only if A_1 is a barrel; S is uniformly bounded if and only if A_1 is a neighborhood of 0.

PROOF We shall prove only the half of this result that is needed below. Suppose that S is pointwise bounded. Given $x \in X$, there exists M such that $\| ux \| < M$ for all $u \in S$. Thus $(1/M)x \in A_1$ and so A_1 is absorbing, by Prob. 1-5-7. That A_1 is closed and absolutely convex follows from continuity and linearity. Next suppose that A_1 is a neighborhood of 0, say $\| x \| \leq \varepsilon$ implies $x \in A_1$. Then for $\| x \| \leq 1$, $\| u(\varepsilon x) \| \leq 1$ for all u, and so $\| u \| \leq 1/\varepsilon$ for all u.

6. Theorem: Uniform boundedness theorem Let X be a Banach space, Y a seminormed space, $S \subset B(X, Y)$. Then if S is pointwise bounded it is uniformly bounded.

PROOF By Lemmas 3-3-2 and 3-3-5.

See Prob. 3-3-5.
We sketch a few applications of this remarkable theorem. Direct proofs exist, but they involve (in some cases) clever ideas and complicated constructions. See, for example, Prob. 3-3-201.

7. Example *If $\{a_k\}$ is a sequence of complex numbers with the property that $\sum a_k x_k$ is convergent for all $x \in c_0$, then $\sum |a_k| < \infty$.* Let $f_n(x) = \sum_{k=1}^{n} a_k x_k$ for $x \in c_0$. Then $f_n \in c_0'$; indeed, $\| f_n \| = \sum_{k=1}^{n} |a_k|$ [Example 2-3-4]. The hypothesis implies that $\{f_n\}$ is pointwise bounded and Theorem 3-3-6 yields the result.

8. Example *Let A be a matrix (a_{nk}) of complex numbers, $n, k = 1, 2, \ldots$. Suppose that for each $x \in c_0$, $Ax = \{\sum_{k=1}^{\infty} a_{nk}x_k\} \in l^{\infty}$. Then $\|A\| < \infty$* (Definition 1-3-1). Let $f_n(x) = \sum_k a_{nk}x_k = (Ax)_n$ for $x \in c_0$. By Example 2-3-4, $\|f_n\| = \sum_k |a_{nk}| < \infty$ for each n by Example 3-3-7. Since $\{f_n\}$ is pointwise bounded, Theorem 3-3-6 yields the result.

9. Example Let $A: l^{\infty} \to c_0$, A a matrix as in Example 3-3-8. Then, in particular, $A[c_0] \subset l^{\infty}$ and so, by Example 3-3-8, $\|A\| < \infty$. It follows by Schur's Theorem 1-3-2 that $\sum_k |a_{nk}| \to 0$.

The next example may be omitted.

10. Example *There exists a continuous function with divergent Fourier series.* Consider the real Banach space $C(H)$ where $H = [-\pi, \pi]$. For $f \in C(H)$, let $u_n(f) = a_0/2 + \sum_{k=1}^{n} a_k$ where a_i are the Fourier coefficients of f. Thus $u_n(f)$ is the value at 0 of a segment of the Fourier series of f. If the series converges at 0 for all f we have $\lim u_n(f)$ exists for all $f \in C(H)$ and so, by Theorem 3-3-6, $\sup \|u_n\| < \infty$. Now $u_n(f) = \int f(t)g_n(t)\,dt$, where g_n is the Fourier kernel, and $\|u_n\| = \int |g_n|$ [Prob. 2-3-110]. A direct calculation shows that this is unbounded.

An important corollary of Theorem 3-3-6 is that a weak form of boundedness is sufficient to imply norm boundedness. It is for this reason that we do not define "weakly bounded." Here, paradoxically, completeness does not need to be assumed since it appears automatically in the dual space. We give the result for seminormed space and the reader is warned that this generality is needed.

11. Theorem Let E be a set in a seminormed space X such that each $f \in X'$ is bounded on E, that is, $f[E]$ is a bounded set in $\mathscr{K}$. Then E is bounded.

PROOF Let $S = \hat{E} \subset X'' = B(X', \mathscr{K})$. The hypothesis is that S is pointwise bounded, so by Theorem 3-3-6 it is uniformly bounded. [That X' is a Banach space was noted in Example 3-1-8.] By Theorem 3-2-2, E is bounded.

12. Remark Theorem 3-3-11 shows that if $S \subset X'$ and $F[S]$ is bounded for each $F \in X''$, then S is bounded. If X is a Banach space we have the stronger result that if $F(S)$ is bounded for each $F \in \hat{X}$, then S is bounded [Theorem 3-3-6, $Y = \mathscr{K}$]. In language to be developed in Chap. 8 these statements read: "weakly bounded" implies bounded (in any seminormed space), and "weak $*$ bounded" implies bounded (in the dual of a Banach space). As Prob. 3-3-5 shows, the second result fails for the dual of a normed space in general.

13. Theorem: Banach–Steinhaus closure theorem Let X be a Banach space, Y a normed space $\{u_n\} \subset B(X, Y)$ and suppose that $u(x) = \lim u_n(x)$ exists for each $x \in X$. Then $u \in B(X, Y)$.

PROOF Clearly u is linear. By Theorem 3-3-6, for some M, $\| u_n \| < M$ for all n. Now if $\| x \| \leq 1$, $\| ux \| = \lim \| u_n x \| \leq M$ so $\| u \| \leq M$.

Theorem 3-3-13 is interpreted as a sequential completeness theorem in Example 9-3-8.

Applying Theorem 3-3-13 to X' we see that noncontinuous linear functionals are difficult to construct. Any series or integral built up from continuous maps will be continuous and so will be useless for that purpose; any example, such as the one given in Example 3-3-14, would seem to require transfinite methods. This belief is strengthened by the proof [161] that in a certain set theory every linear functional on a Banach space must be continuous. For noncomplete spaces a construction may be easy (see Prob. 3-3-102).

14. Example *Every infinite dimensional paranormed space allows a noncontinuous linear functional.* (Example 8-2-13 shows a space for which this is false.) Let $\{x^n\}$ be a linearly independent sequence. By continuity of multiplication [part (*e*) of the definition of paranorm] we may find $t_n > 0$ with $\| t_n x^n \| < 1/n$. Let f be a linear functional with $f(t_n x^n) = 1$. $[\![\{t_n x^n\}$ is linearly independent; use Prob. 1-5-11.$]\!]$

PROBLEMS

1 Prove the converse of Theorem 3-3-11.

2 Let X be a seminormed space, $\{f_n\}$ a uniformly bounded sequence in X'. Let $u: X \to l^\infty$ be $ux = \{f_n(x)\}$. Show that u is continuous.

3 (*Convergence lemma.*) With X, f_n as in Prob. 3-3-2, show that $\{x: \lim f_n(x) \text{ exists}\}$ and $\{x: f_n(x) \to 0\}$ are closed vector subspaces of X. $[\![$They are $u^{-1}[c]$, $u^{-1}[c_0]$.$]\!]$

4 Let $X = (\varphi, \| \cdot \|_\infty)$. Let $A = \{x: |x_n| \leq 1/n \text{ for all } n\}$. Show that A is a barrel which is not a neighborhood of 0.

5 With X as in Prob. 3-3-4, show that $\{nP_n\}$ is pointwise but not uniformly bounded. (Here $P_n(x) = x_n$.)

---

101 Prove the other half of Lemma 3-3-5. If A_1 is a barrel, S is pointwise bounded; if S is uniformly bounded, A_1 is a neighborhood of 0.

102 Let $X = l$ with $\| x \| = \sup |x_n|$. (This is not the norm given in Sec. 1-2.) Let $f(x) = \sum x_n$. Show that f is not continuous. Deduce that in Theorem 3-3-13 it is not sufficient to assume that X is a normed space.

103 Let $u \in B(l^\infty, l^\infty)$ and suppose there exists a matrix A such that $(ux)_n = \sum_k a_{nk} x_k$ for all n, x. Show that $\| u \| = \| A \|$, Definition 1-3-1.

104 Suppose $f_n \in X'$, X a Banach space, $\| f_n \| = 1$. Show that possibly $f_n(x) \to 0$ for each $x \in X$; but for any $\varepsilon_n \to 0$ it must be false that $f_n(x) = 0(\varepsilon_n)$ for all x. (This shows a sense in which the Riemann–Lebesgue lemma [57], Theorem 16.35, is the best possible.) Apply Theorem 3-3-6 to $\{f_n/\varepsilon_n\}$. A classical proof is given in [137], Sec. 13.7.$]\!]$

105 Problem 3-3-104 is false without completeness of X. $[\![$See Probs. 3-3-4 and 3-3-5.$]\!]$

106 (Köthe–Toeplitz duals). For a sequence space X, let $X^\alpha = \{y: \sum |x_n y_n| < \infty \text{ for all } x \in X\}$, $X^\beta = \{y: \sum x_n y_n \text{ is convergent for all } x \in X\}$, $X^\gamma = \{y: \sum x_n y_n \text{ is bounded for all } x \in X\}$. Show, using the uniform boundedness theorem as much as possible, that $c_0^{\gamma} = c_0^{\alpha} = l$, $l^{\gamma} = l^{\beta} = l^{\infty}$, $l^{\alpha\alpha} = l$, $(l^p)^{\gamma} = l^q$, $p > 1$, $1/p + 1/q = 1$, $(l^{\infty})^{\gamma} \neq (l^{\infty})^{\alpha}$.

107 Show that $X^{\alpha\alpha} \supset X$; $X \subset Y$ implies $X^{\alpha} \supset Y^{\alpha}$; $X^{\alpha\alpha\alpha} = X^{\alpha}$. Do the same for X^{β} and X^{γ}.

108 A set S in a sequence space is called *solid* if $x \in S$, $|y_n| \leq |x_n|$ for all n implies $y \in S$. Show that X^{α} is always solid but X^{β} need not be.

109 (L. L. Silverman–O. Toeplitz). Let A be a matrix, $c_A = \{x : Ax \in c\}$. Call A *conservative* if $c_A \supset c$. Show that A is conservative if and only if $\|A\| < \infty$, $\lim_{n \to \infty} a_{nk} = a_k$ exists for each k, and $\lim \sum_k a_{nk} = t$ exists. [Example 3-3-8; $\delta^k \in c_A$; $1 \in c_A$. Conversely, let $f_n(x) = (Ax)_n$, $f_n \in c'$. By Prob. 3-3-3, $c_A \cap c$ is closed in c; by hypothesis it contains 1 and includes φ, so it is c.]

110 Let A be conservative (Prob. 3-3-109). Define $\lim_A \in c_A^{\#}$ by $\lim_A x = \lim (Ax)_n$. Denote $\lim_A |_c$ by $\lim_A$ also. Show that $\lim_A \in c'$ and for $x \in c$, $\lim_A x = \chi \lim x + \sum a_k x_k$ where $\chi = \lim_n \sum_k a_{nk} - \sum a_k$ [Prob. 2-3-3]. (A is called *coregular* if $\chi \neq 0$; otherwise *conull*.)

111 Show that (Prob. 3-3-110) $\chi(A) = \lim_m \lim_n \sum_{k=m}^{\infty} a_{nk}$. Deduce that $|\chi(A)| \leq \|A\|$. (This can also be deduced from Probs. 8-1-116 and 5-5-201.)

112 If $\|A\| < \infty$, show that $c_A \cap l^{\infty}$ is closed in l^{∞} (see Prob. 3-3-109). [See Prob. 3-3-3.]

---

201 Let $\{a_n\} \subset \mathscr{C}$ with $\sum |a_n| = \infty$. Show that $\sum |a_n / \sum_{k=1}^{n} a_k| = \infty$. Deduce the result of Example 3-3-7.

202 Let X be an infinite-dimensional paranormed space which has at least one x with $\|x\| \neq 0$. Show that there exists a noncontinuous linear bijection $T : X \to X$.

203 Show that $X^{\alpha} = X$ if and only if $X = l^2$ (Prob. 3-3-106).

FOUR

TOPOLOGICAL VECTOR SPACES

4-1 DEFINITIONS AND EXAMPLES

1. Definition A topological vector space X is a topological space and a vector space such that the vector operations are continuous. The abbreviation TVS will be used. If X is a vector space, any topology T which makes (X, T) a TVS will be called a vector topology.

The meaning of the assumption is that the maps $(t, x) \to tx$ and $(x, y) \to x + y$ which carry $\mathscr{K} \times X$ and $X \times X$ to X are continuous.

2. Example *A paranormed space is a* TVS. If $x_n \to x$, $y_n \to y$, then $d(x_n + y_n, x + y) = \| x_n - x + y_n - y \| \leq \| x_n - x \| + \| y_n - y \|$; continuity of tx is part (e) of the definition of paranorm. Sequences may be used here because $\mathscr{K} \times X$ is pseudometrizable $[\![$Probs. 1-6-10 and 1-6-11$]\!]$.

It turns out that the topology of a TVS can be localized at 0. This is the content of Theorem 4-1-4.

3. Definition The set of neighborhoods of 0 in a TVS (X, T) will be written $\mathscr{N}$, $\mathscr{N}(X)$, $\mathscr{N}(T)$, or $\mathscr{N}(X, T)$.

Thus the subscript 0 is dropped in Definition 1-6-2.

4. Theorem: Localization Let X be a TVS, $a \in X$, $G \subset X$. Then $G \in \mathcal{N}_a$ if and only if $G - a \in \mathcal{N}$. In other words, $a + U \in \mathcal{N}_a$ if and only if $U \in \mathcal{N}$.

PROOF For fixed a, the map $x \to x + a$ is continuous [Prob. 1-6-3]. Its inverse $x \to x - a$ is also continuous; hence the map is a homeomorphism of X onto itself and so preserves neighborhoods.

5. Remark Theorem 4-1-4 justifies the phrase: let $x + U$, $U \in \mathcal{N}$, be an arbitrary neighborhood of x. This style will be followed uniformly. Another consequence is that two vector topologies are equal if they have the same nets converging to 0.

Example 4-1-2 yields a plentiful supply of TVSs. Others may be constructed from them as follows.

6. Example *Let Φ be a collection of vector topologies for a vector space X. Then $\bigvee \Phi$ is a vector topology. Moreover, a net $x \to 0$ in this topology if and only if $x \to 0$ in T for each $T \in \Phi$.* The last part is by Theorem 1-6-8. Suppose that (t, x) is a net in $\mathcal{K} \times X$ with $t \to s$, $x \to a$ in $\bigvee \Phi$. Note that t, x are nets defined on the directed set on which the net (t, x) is defined. Thus $x \to a$ in (X, T) for each $T \in \Phi$ [Theorem 1-6-8]. Hence $tx \to sa$ in each (X, T) and so, again by Theorem 1-6-8, $tx \to sa$ in $\bigvee \Phi$ and so multiplication is continuous. A similar proof yields continuity of addition.

Example 4-1-6 may be applied to collections of paranorms.

7. Definition Let P be a collection of paranorms on a vector space X. Then σP denotes $\bigvee \{p : p \in P\}$ where p also refers to the topology induced by p.

The use of σ derives from the German word *schwach*, meaning weak.

8. Example *In Definition 4-1-7, σP is a vector topology. Moreover, $x \to 0$ in this topology if and only if $p(x) \to 0$ for each $p \in P$.* This follows from Examples 4-1-2 and 4-1-6.

A discussion of the inappropriateness of using wP (Theorem 1-6-10) is given in Probs. 4-1-103 and 4-1-104.

9. Example *Let X be a vector space and F a set of linear maps $f : X \to Y_f$ where each Y_f is a TVS. Then wF (Theorem 1-6-10) is a vector topology. Moreover, $x \to 0$ in this topology if and only if $f(x) \to 0$ for each $f \in F$.* It is sufficient to prove this for the case in which F has one member $f : X \to Y$ and apply Example 4-1-6. If (t, x) is a net in $\mathcal{K} \times X$ with $t \to s$, $x \to a$, then $f(x) \to f(a)$ in Y [Theorem 1-6-10] and so $f(tx) = tf(x) \to sf(a) = f(sa)$ in Y. Again by Theorem 1-6-10, $tx \to sa$.

10. Definition If F is a set of linear maps as in Example 4-1-9, $\sigma F = wF$.

The reason for this definition is that it is customary to use σ for vector topologies. For maps which are not linear (e.g., paranorms) it is necessary to distinguish between σ and w; for linear maps, they are the same by definition. See also Prob. 4-1-6.

11. Example *A product of TVSs is a TVS. Moreover, $x \to 0$ in the product topology if and only if $Px \to 0$ for each projection P.* This is a special case of Example 4-1-9 since the projections are linear.

Topologies of the form σF with $F \subset X^{\#}$ are called *weak topologies*. Their distinguishing feature is the largeness of their neighborhoods.

12. Theorem Let X be a vector space, $F \subset X^{\#}$. In the TVS $(X, \sigma F)$, every $U \in \mathcal{N}$ includes a vector subspace of X of finite codimension.

PROOF By Theorem 1-6-8 (and the fact that $\sigma = w$, Definition 4-1-10), U includes a finite intersection $\cap V_i$ where each $V_i \in \mathcal{N}(X, \sigma f_i)$, $f_i \in F$, $i = 1, 2, \ldots, n$. Now each $V_i \supset \{x : |f_i(x)| < \varepsilon_i\}$ for some $\varepsilon_i > 0$. A *fortiori*, $U \supset \cap \{f_i^{\perp} : i = 1, 2, \ldots, n\}$, a vector subspace of finite codimension $[\![\text{Prob. 1-5-15}]\!]$.

13. Example If $X = \mathcal{K}^A$ the result of Theorem 4-1-12 holds since the projections are functionals. In particular this is true for $\omega = \mathcal{K}^N$ $[\![\text{Prob. 4-1-3}]\!]$.

14. Remark If X is infinite dimensional, Theorem 4-1-12 shows that $(X, \sigma F)$ cannot be a normed space. It may be metrizable, however $[\![\text{Example 4-1-13}]\!]$.

15. Remark We select for emphasis the result given in the proof of Theorem 4-1-12 that if $U \in \mathcal{N}(X, \sigma F)$, then $U \supset \cap \{f_i^{\perp} : i = 1, 2, \ldots, n\}$. For a product $\mathcal{K}^A$ this says $U \supset \cap \{P_i^{\perp} : i = 1, 2, \ldots, n\}$.

PROBLEMS

1 Let x be a net in a TVS X, $a \in X$. Show that $x \to a$ if and only if $x - a \to 0$. Hence show that a linear map between TVSs which is continuous at 0 must be continuous everywhere.

2 The *discrete topology* for a set X is the collection of all subsets; the *indiscrete topology* is $\{\phi, X\}$. Show that the indiscrete topology on a vector space is a vector topology but that the discrete topology is not, with the exception $X = \{0\}$. Compare Prob. 2-1-102.

3 Show that $\omega = \mathcal{K}^N$, that is, the paranorm gives the product topology $[\![\text{Theorem 2-1-2}]\!]$.

4 A family F of functions on a set is called *separating* if for each $x \neq y$, F contains a function f with $f(x) \neq f(y)$. Suppose that F is a set of linear maps defined on a vector space. Show that F is separating if and only if it is total.

5 If P is a countable family of paranorms show that σP is pseudometrizable $[\![\text{Theorem 2-1-2}]\!]$.

6 Let X be a vector space, $F \subset X^*$. For each $f \in F$ let $p(x) = |f(x)|$ and let P be the set of such p. Show that $\sigma F = \sigma P$, Definitions 4-1-10 and 4-1-7 respectively.

7 Let X be a vector space, Y_n a paranormed space, and $f_n: X \to Y_n$ a linear map for each n. Show that $\sigma\{f_n\}$ is given by a paranorm. [Prove this for one map and use the Fréchet combination.]

8 Show that σP, Definition 4-1-7, is the smallest vector topology for X which makes all the members of P continuous.

---

101 Give an uncountable set P of seminorms on R such that $\sigma(P)$ is the ordinary topology. Thus the converse of Prob. 4-1-5 fails.

102 Show that every vector space has a largest vector topology. [Let Φ, Example 4-1-6, be the set of all vector topologies. Why is it nonempty?] Compare Prob. 4-1-2.

103 Let p be a paranorm on a vector space X. Show that $x \to a$ in wp if and only if $p(x) \to p(a)$. Hence show that wp need not be a vector topology, in particular $wp \neq \sigma p$ is possible. [Take $X = R$, $p(x) = |x|$.]

104 Show that wP is the smallest topology for X which makes all the members of P continuous, where P is a set of paranorms on a vector space X. Deduce that $\sigma P \supset wP$ and that whenever $wP \neq \sigma P$, wP is not a vector topology [Prob. 4-1-8].

105 Let X be a vector space of continuous complex functions on a topological space H. For $f \in X$, $K \subset H$ set $p_K(f) = \sup\{|f(h)| : h \in K\}$. Let $P = \{p_K : K$ a compact subset of $H\}$. Describe net convergence in $(X, \sigma P)$ [Example 4-1-8]. If H is hemicompact show that σP is metrizable [Theorem 2-1-2]. This topology is called the *topology of compact convergence*.

106 We may consider $X_\beta \subset \pi X_\alpha$ for each β by identifying $a \in X_\beta$ with $x \in \pi$, where $x_\beta = a$, $x_\alpha = 0$ for $\alpha \neq \beta$. Suppose that S is a closed subspace of πX_α and $S \supset X_\beta$ for each β. Show that $S = \pi X_\alpha$. Hence if $f \in \pi'$ and $f = 0$ on each X_β, then $f = 0$. [For each $x \in \pi$, apply Prob. 1-6-105 to show that $x \in \bar{S}$.]

107 Show that the assumption that S is closed may not be omitted in Prob. 4-1-106.

4-2 PROPERTIES

Some of the results of this section are valid in an arbitrary topological group and some of these, in turn, are valid in an arbitrary uniform space. Details may be found in [156].

Continuity of the vector operations is expressed in geometrical language in Theorems 4-2-2 to 4-2-4, Lemma 4-2-5, and Prob. 4-2-104.

1. Definition A collection $\mathscr{F}$ of subsets of a vector space is called additive if for each $U \in \mathscr{F}$ there exists $V \in \mathscr{F}$ with $V + V \subset U$.

In the proofs of Theorems 4-2-2 and 4-2-4 we make several unstated references to Theorem 1-6-12. A different sort of proof is given in Remark 4-2-14 (nets) and Prob. 4-3-105 (filters).

2. Theorem Let X be a vector space and a topological space. Define $u: X \times X \to X$ by $u(x, y) = x + y$. Then u is continuous at $0[= (0, 0)]$ if and only if $\mathscr{N}(X)$ is additive.

PROOF $\Leftarrow$: Given $U \in \mathcal{N}(X)$ let $V \in \mathcal{N}(X)$ with $V + V \subset U$. Then $V \times V \in \mathcal{N}(X \times X)$ and $u[V \times V] = V + V \subset U$. Thus u is continuous at 0. $\Rightarrow$: Given $U \in \mathcal{N}(X)$, there exists $W \in \mathcal{N}(X \times X)$ with $u[W] \subset U$. We may assume that $W = V_1 \times V_2$, $V_i \in \mathcal{N}(X)$. Let $V = V_1 \cap V_2$. Then $V + V \subset V_1 + V_2 = u[W] \subset U$.

Theorem 4-2-2 may be regarded as a generalization of the triangle inequality for metrics. It is used in TVS versions of classical arguments where $\varepsilon/2$ is used, such as "$A + B < \varepsilon/2 + \varepsilon/2 = \varepsilon$," which will appear in TVS situations as "$A + B \subset V + V \subset U$." See, for example, Remark 4-2-14.

3. Theorem Let X be a TVS and $U \in \mathcal{N}$. Then $tU \in \mathcal{N}$ for every $t \neq 0$.

PROOF Of course $t \in \mathcal{K}$, else tU is meaningless. The map $x \to tx$ is continuous, as is its inverse $x \to (1/t)x$, so this map is a homeomorphism of X onto itself and hence preserves open sets.

4. Theorem Every neighborhood U of 0 in a TVS X is absorbing.

PROOF Let $x \in X$ and define $u: \mathcal{K} \to X$ by $u(t) = tx$. Since u is continuous, there exists $\varepsilon > 0$ such that $|t| < \varepsilon$ implies $u(t) \in U$, that is, $tx \in U$.

5. Lemma Every neighborhood U of 0 in a TVS X includes a balanced neighborhood of 0.

PROOF The map $u: \mathcal{K} \times X$ given by $u(t, x) = tx$ is continuous. Thus there exists $W \in \mathcal{N}(\mathcal{K} \times X)$ with $u[W] \subset U$. We may assume that $W = V_1 \times V_2$ where V_1, V_2 are neighborhoods of 0 in $\mathcal{K}, X$. Then $V_1 \supset \{t : |t| \leq \varepsilon\}$ for some $\varepsilon > 0$. Let $V = \cup\{tV_2 : |t| \leq \varepsilon\}$. Then $V \in \mathcal{N}$ since $V \supset \varepsilon V_2$ [[Theorem 4-2-3]]. Also, V is balanced and $V \subset U$.

6. Theorem Let X be a TVS and $S \subset X$. Then $\bar{S} = \cap\{S + U : U \in \mathcal{N}\}$. In particular, $\bar{S} \subset S + U$ for every $U \in \mathcal{N}$.

PROOF Let $x \in \bar{S}, U \in \mathcal{N}$. We may, by Lemma 4-2-5, assume that U is balanced. Since $x + U$ is a neighborhood of x it meets S. So $x \in S - U = S + U$. Conversely, let $x \notin \bar{S}$. Then x has a neighborhood (which we may write $x + U$, U a balanced neighborhood of 0) which does not meet S; hence $x \notin S - U = S + U$. Thus $x \notin \cap\{S + U : U \in \mathcal{N}\}$.

7. Theorem Every neighborhood U of 0 in a TVS X includes a closed balanced neighborhood of 0.

PROOF Let $V \in \mathcal{N}$ with $V + V \subset U$ [[Theorem 4-2-2]]. By Lemma 4-2-5 let $W \in \mathcal{N}$ be balanced, $W \subset V$. Then $\bar{W} \subset U$ [[Theorem 4-2-6]] and it remains to show that $\overline{W}$ is balanced. Let $0 \leq |t| \leq 1$. By the argument of Theorem 4-2-3, $t\overline{W} \subset \overline{tW}$ and so $t\overline{W} \subset \overline{tW} \subset \overline{W}$ since W is balanced.

8. Theorem Every TVS is a regular topological space. The following conditions on a TVS X are equivalent:

(a) X is a T_3 space.
(b) $\{0\}$ is a closed set.
(c) For each $x \neq 0$, there is a neighborhood U of 0 with $x \notin U$.

PROOF Regularity follows from Theorem 4-2-7. If X is a T_3 space, all singletons are closed, and (b) implies (a) since a regular T_1 space is T_3. [All singletons are closed by the localization principle, Theorem 4-1-4.] Condition (b) is equivalent to $\overline{\{0\}} = \{0\}$, while (c) is equivalent to $\cap\{U : U \in \mathcal{N}\} = \{0\}$. These are the same by Theorem 4-2-6.

9. Definition A TVS is called separated if the three equivalent conditions of Theorem 4-2-8 hold. We also say that the topology is separated.

Thus a separated TVS is a Hausdorff space.
A useful consequence, Lemma 4-2-10, of the localization principle is that all convergence discussions may be carried out with nets defined on a single directed set, $\mathcal{N}$, the set of neighborhoods of 0 (directed by inclusion). An analogue of this result is quite generally true (Prob. 4-2-201).

10. Lemma Let X be a TVS, $S \subset X$ and $a \in \overline{S}$. Then there exists a net $(x_\delta : \mathcal{N})$ in S with $x_\delta \to a$.

PROOF For each $\delta \in \mathcal{N}$, $a + \delta$ is a neighborhood of a and so meets S. Let $x_\delta \in (a + \delta) \cap S$. Then $x_\delta \to a$ by Theorem 1-6-5.

11. Theorem Let X be a TVS, $A, B \subset X$. Then $\overline{A} + \overline{B} \subset \overline{A + B}$.

PROOF Let $x \in \overline{A} + \overline{B}$, $x = u + v$, $u \in \overline{A}$, $v \in \overline{B}$. We can find nets a, b in A, B respectively with $a \to u$, $b \to v$, both nets defined on the same directed set [Lemma 4-2-10]. Thus they may be added and so $a + b \to u + v = x$ [continuity of addition, see Theorem 1-6-13]. Since $a + b \in A + B$ we have $x \in \overline{A + B}$.

12. Theorem The closure of a vector subspace S of a TVS X is a vector subspace of X. The closure of a convex set is convex. The closure of a balanced set is balanced.

PROOF We first note that for any set S and scalar t, $t\overline{S} \subset \overline{tS}$ as in the proof of Theorem 4-2-7. Thus for scalars a, b, $a\overline{S} + b\overline{S} \subset \overline{aS} + \overline{bS}$ [Theorem

4-2-11$]\!] \subset \overline{S}$ if S is a vector subspace or if S is convex and $a + b = 1, 0 \leq a \leq 1$. The last part was given in the proof of Theorem 4-2-7.

On the other hand, it is easy to show that the convex hull of a closed set need not be closed, even in R^2.

13. Remark (very important)

(a) We shall often make arbitrary choices in a special form. For example, given $U \in \mathcal{N}$ we can choose $V \in \mathcal{N}$ with $V - V \subset U$. $[\![$By Theorem 4-2-2 we choose W with $W + W \subset U$. By Theorem 4-2-7 choose V balanced with $V \subset W$. Then $V - V = V + V \subset W + W \subset U.]\!]$

(b) Similarly, we may often say let $x - U$, $U \in \mathcal{N}$ be an arbitrary neighborhood of x.

(c) By the localization principle, Theorem 4-1-4, if T, T' are vector topologies for X we have $T \supset T'$ if and only if (for a net x in X) $x \to 0$ in T implies $x \to 0$ in T'. $[\![$We wrote $x \to 0$ in T to mean $x \to 0$ in $(X, T).]\!]$

The reader is put on notice that the contents of Remark 4-2-13 will be used without citation.

14. Remark (This may be omitted.) We give a different sort of proof of Theorem 4-2-2 which omits product considerations. Continuity of addition at 0 is equivalent to the condition $x \to 0$, $y \to 0$ implies $x + y \to 0$ where x, y are nets with the same directed set as domain. (See Theorem 1-6-13.) If $\mathcal{N}$ is additive and $x \to 0$, $y \to 0$, let $U \in \mathcal{N}$. Let $V \in \mathcal{N}$ with $V + V \subset U$. Let $\delta \geq \delta_1, \delta \geq \delta_2$ imply $x_\delta \in V$, $y_\delta \in V$ respectively. Let $\delta_0 \geq \delta_1, \delta_0 \geq \delta_2$. Then $\delta \geq \delta_0$ implies $x_\delta + y_\delta \in V + V \subset U$. Conversely, suppose that $\mathcal{N}$ is not additive. Let $U \in \mathcal{N}$ have the property $V + V \not\subset U$ for all $V \in \mathcal{N}$. Choose $x_V \in (V + V) \backslash U$ for each $V \in \mathcal{N}$. Then $x_V = y_V + z_V$ with $y_V, z_V \in V$. The net $(y_V, \mathcal{N})$ (Remark 1-6-3 and Theorem 1-6-5) converges to 0 as does $(z_V, \mathcal{N})$. But $y_V + z_V = x_V \notin U$, and so addition is not continuous at 0.

PROBLEMS

In this list X is a TVS.

1 In Example 4-1-9 (and Definition 4-1-10) show that σF is separated if and only if F is separating.

2 If at least one $T \in \Phi$ (Example 4-1-6) is separated, show that $\vee \Phi$ is separated. The converse is false (Prob. 4-2-3).

3 Give two nontotal paranorms p, q on R^2 such that $\sigma(p, q)$ is the euclidean topology.

4 Show that a product of separated spaces is separated $[\![$Prob. 4-2-1$]\!]$.

5 Show that a maximal subspace of a TVS is either closed or dense $[\![$Theorem 4-2-12$]\!]$.

6 Let $G \subset X$ be open. Show that $S + G$ is open for each $S \subset X$. $[\![$It is $\cup \{s + G\}.]\!]$

7 Let $\mathcal{F}$ be a local base for $\mathcal{N}(X), S \subset X$. Show that $\overline{S} = \cap \{S + U : U \in \mathcal{F}\}$. This extends Theorem 4-2-6.

8 Let A be an absolutely convex set with nonempty interior. Show that $A \in \mathcal{N}$. $[\![A = \frac{1}{2}(A - A) \supset \frac{1}{2}(A - x).]\!]$

9 Let A be balanced. Show that $A^i \cup \{0\}$ is balanced, where A^i is the interior of A. $[\![0 < |t| < 1 \text{ implies } tA^i = (tA)^i \subset A^i.]\!]$

10 Show that every neighborhood of 0 in a TVS includes a balanced open neighborhood of 0 $[\![\text{Prob. 4-2-9}]\!]$.

---

101 Give an example in R to show that equality need not hold in Theorem 4-2-11.

102 For $A, B \subset X$ show that $\overline{\overline{A} + \overline{B}} = \overline{A + B}$. Deduce that if $A + B$ is closed, $\overline{A} + \overline{B}$ is also closed.

103 For $S \subset X$, show that $\overline{S} = \cap \{S + U : U \in \mathcal{N}\}$.

104 Show that the map $(t, x) \to tx$ is continuous at $(0, 0) \in \mathcal{K} \times X$ if and only if for each $U \in \mathcal{N}$ there exists $V \in \mathcal{N}$ and $W \in \mathcal{N}(\mathcal{K})$ such that $WV \subset U$. $[\![\text{Part of this is in Lemma 4-2-5}.]\!]$

105 Carry out the program of Remark 4-2-14 for multiplication $[\![\text{Prob. 4-2-104}]\!]$.

106 Let $p(x) = |x|$ for $x \in R$. Deduce that wp is not a vector topology from Theorem 4-2-8, $(b) \Rightarrow (a)$.

107 Show that every TVS is connected.

108 Let $A \subset X$ with $A + A \subset 2\overline{A}$. Show that $\overline{A}$ is convex.

109 Let $A \subset \mathcal{K}, B \subset X$. How does $\overline{A}\,\overline{B}$ compare with $\overline{AB}$? In particular, show that if neither $\overline{A}$ nor $\overline{B}$ contains 0, $\overline{A}\,\overline{B} = \overline{AB}$.

110 Let X be a vector space, Y a separated TVS, and L the set of linear maps $X \to Y$. Show that L is a closed vector subspace of Y^X. $[\![\text{If } f \to g, g(a + b) = \lim f(a + b) = \lim f(a) + \lim f(b).]\!]$

111 Let X be a topological space, Y a separated TVS, and C the set of continuous maps $X \to Y$. Show that C is a vector subspace of Y^X and is not necessarily closed. $[\![\text{With } X = Y = R, C \text{ is dense}.]\!]$

112 Let $F \subset X^{\#}$. Show that $\sigma\{Rf : f \in F\} = \sigma F$ $[\![\text{Theorem 1-5-2}]\!]$.

113 Show that Prob. 4-2-5 is false for a subspace of codimension 2.

---

201 Let T be a topological space. Show that there exists a directed set D such that if $S \subset T$ and $t \in \overline{S}$, there is a net $(t_\delta : D)$ in S with $t_\delta \to t$. Compare Lemma 4-2-10.

202 Let C be a convex closed set in a normed space such that $C + D_1 \supset D_{1+\varepsilon}$ for some $\varepsilon > 0$, where $D_r = \{x : \|x\| \le r\}$. Show that C has a nonempty interior. $[\![\text{See [151]}.]\!]$

4-3 CONSTRUCTION

Theorem 4-3-5 shows the most general method of placing a vector topology on a vector space.

We have been using nets in convergence arguments, preferring them to filters because their language resembles that of sequences. Later on there will be some cases where filters will prove more convenient. For the present they will be introduced for purposes of abbreviation only.

1. Definition Let X be a set. A filter on X is a nonempty collection $\mathscr{F}$ of nonempty subsets of X such that $A \in \mathscr{F}$, $B \in \mathscr{F}$ implies $A \cap B \in \mathscr{F}$, and $B \supset A \in \mathscr{F}$ implies $B \in \mathscr{F}$.

It follows that $X \in \mathscr{F}$. If $\mathscr{F} = \{X\}$, $\mathscr{F}$ is called the *indiscrete filter*. For any fixed $x \in X$ let $\mathscr{F}$ be the collection of all subsets of X which contain x; $\mathscr{F}$ is called the *discrete filter at* x.

The most important type of filter is $\mathcal{N}_x$, the set of all neighborhoods of a point x in a topological space. This is called the *neighborhood filter at* x.

2. Definition Let X be a set. A filterbase on X is a nonempty collection $\mathcal{B}$ of nonempty subsets of X such that $\mathcal{B}$ is directed by inclusion. (See Prob. 1-4-2 for this terminology.)

Thus any filter is a filterbase. Conversely, given a filterbase $\mathcal{B}$ on X, let $\mathcal{F} = \{B \subset X : B \supset A$ for some $A \in \mathcal{B}\}$. Then $\mathcal{F}$ is a filter, and (very important) $\mathcal{F} \supset \mathcal{B}$; it is called *the filter generated by* $\mathcal{B}$. For example, if $\mathcal{B} = \{x\}$, $\mathcal{F}$ is the discrete filter at x.

3. Example Let $D_r = \{(x, y): x^2 + y^2 \leq r\} \subset R^2$ and $\mathcal{B} = \{D_r : r > 0\}$. This is a filterbase and it generates $\mathcal{N}$, the neighborhood filter at the origin for the euclidean topology of R^2.

4. Remark The notion "filterbase generating $\mathcal{N}_x$" is identical with "local base of neighborhoods of x."

We have seen in Lemma 4-2-5 and Theorem 4-2-4 that every TVS has a local base of neighborhoods of 0 consisting of balanced absorbing sets. Thus the next theorem gives the most general construction of a vector topology.

5. Theorem Let X be a vector space. Let $\mathcal{B}$ be an additive filterbase of balanced absorbing subsets of X. Then there is a unique vector topology for X for which $\mathcal{B}$ is a local base of neighborhoods of 0.

PROOF We shall call a set G open if (G is empty or) for each $x \in G$ there exists $U \in \mathcal{B}$ such that $x + U \subset G$. We first prove that the collection of open sets is a topology. The intersection of two open sets G_1, G_2 is open. $\llbracket$ Let $x \in G = G_1 \cap G_2$. Let $x + U_i \subset G_i$ for $i = 1, 2$. Let $U \in \mathcal{B}$ with $U \subset U_1 \cap U_2$. Then $x + U \subset G.\rrbracket$ The union of an arbitrary family (G_α) of open sets is open. $\llbracket$ Let $x \in G = \cup G_\alpha$. Then $x \in G_\alpha$ for some α. Thus for some $U \in \mathcal{B}$, $x + U \subset G_\alpha \subset G.\rrbracket$ Finally, X is open since $x + U \subset X$ for all $x \in X$, $U \in \mathcal{B}$ (and $\mathcal{B}$ is not empty).

Next we show that $\mathcal{B}$ is a local base of neighborhoods of 0. Certainly every neighborhood of 0 includes an open set G containing 0 hence includes $0 + U = U$ for some $U \in \mathcal{B}$. However, we must not forget to prove that $\mathcal{B} \subset \mathcal{N}$, that is, the members of $\mathcal{B}$ are neighborhoods of 0. $\llbracket$ Let $U \in \mathcal{B}$. Let $G = \{x : x + V \subset U$ for some $V \in \mathcal{B}\}$. We have $0 \in G$ since $0 + U \subset U$; $G \subset U$ since $x \in G$ implies $x \in x + V \subset U$ for some $V \in \mathcal{B}$ (using the fact that $0 \in V$ since V is absorbing); and so it remains to prove that G is open. Let $x \in G$. By definition $x + V \subset U$ for some $V \in \mathcal{B}$. Let $W \in \mathcal{B}$ with $W + W \subset V$. When we prove that $x + W \subset G$ we will know that G is open. So let $w \in W$. Then $x + w + W \subset x + W + W \subset x + V \subset U$ and so $x + w \in G.\rrbracket$ We now know that $\mathcal{B}$ generates the filter $\mathcal{N}$ of neighborhoods of 0. (See Remark 4-3-4.)

Our next task is to show that this topology is a vector topology. For continuity of addition let x, y be nets in X with $x \to a$, $y \to b$. Now $x - a \to 0$. $[\![$Let U be a neighborhood of 0; we may assume $U \in \mathscr{B}$. Then $a + U$ is a neighborhood of a so $x \in a + U$ eventually. Thus $x - a \in U$ eventually.$]\!]$ Similarly, $y - b \to 0$. Thus $x + y - a - b \to 0$ $[\![$Theorem 4-2-2$]\!]$ and so $x + y \to a + b$. $[\![$Let $a + b + U$ be a neighborhood of $a + b$, $U \in \mathscr{B}$. Then U is a neighborhood of 0 so $x + y - a - b \in U$ eventually; hence $x + y \in a + b + U$ eventually.$]\!]$

For continuity of multiplication, let $t \in \mathscr{K}$, $x \in X$. Let $tx + U$ be a neighborhood of tx, $U \in \mathscr{B}$. There exists $V \in \mathscr{B}$ with $V + V + V + V \subset U$ and $W \in \mathscr{B}$ with $tW \subset V$, $W \subset V$. $[\![$Let n be a positive integer with $2^n > |t|$. Choose W with $W + W + \cdots + W \subset V$ (2^n terms in the sum). Then $tW \subset 2^n W \subset V$ since W is balanced. Finally, to ensure $W \subset V$ simply replace W by a subset of $W \cap V$ chosen from $\mathscr{B}$.$]\!]$ Since V is absorbing we may find ε, $0 < \varepsilon < 1$, with $ax \in V$ for $|a| < \varepsilon$. Let N be the ε-neighborhood of t in $\mathscr{K}$. We shall show that $N(x + W) \subset tx + U$, and this will show that multiplication is continuous at $(t, x) \in \mathscr{K} \times X$.

To prove the inclusion let $s \in N$, $y \in x + W$. Then $sy = tx + t(y - x) + (s - t)x + (s - t)(y - x)$. Now $t(y - x) \in tW \subset V$; $(s - t)x \in V$ since $|s - t| < \varepsilon$; $(s - t)(y - x) \in (s - t)W \subset W \subset V$ since $|s - t| < 1$ and W is balanced. Thus $sy \in tx + V + V + V \subset tx + U$. $[\![V \subset V + V$ since $0 \in V.]\!]$

Lastly, the topology is unique since $\mathscr{B}$ generates a unique filter and this is $\mathscr{N}$.

PROBLEMS

1 Let $\mathscr{F}$ be a filter on a set X, $x \in X$. Show that $\mathscr{F}$ is the discrete filter at x if and only if $\{x\} \in \mathscr{F}$.

2 For a filter $\mathscr{F}$ in a topological space X, say that $\mathscr{F} \to x \in X$ if $\mathscr{F} \supset \mathscr{N}_x$. Show that $\mathscr{N}_x$ and the discrete filter at x both converge to x.

---

101 Show that the set of all balanced absorbing sets in a vector space does not satisfy the conditions of Theorem 4-3-5. Compare Prob. 4-1-102 $[\![$Prob. 1-5-101$]\!]$.

102 Let X be the set of real functions on a set S. Let $U_r = \{f \in X : \|f\|_\infty < r\}$ and $\mathscr{B} = \{U_r : r > 0\}$. Show that $\mathscr{B}$ does not satisfy the conditions of Theorem 4-3-5. Which operations would fail to be continuous if a topology is constructed from $\mathscr{B}$ by the procedure of Theorem 4-3-5?

103 Let $\mathscr{F}$ be a filter in a topological space X with $\mathscr{F} \to x$. Choose $x_\delta \in \delta$ for each $\delta \in \mathscr{F}$. Show that the net $(x_\delta : \mathscr{F})$ converges to x (order $\mathscr{F}$ by inclusion).

104 Let x be a net in a topological space X. Let $\mathscr{F} = \{A \subset X : x \in A \text{ eventually}\}$. Show that $\mathscr{F}$ is a filter and $\mathscr{F} \to a \in X$ if and only if $x_\delta \to a$.

105 For filters $\mathscr{F}$, $\mathscr{G}$ in a vector space, let $\mathscr{F} + \mathscr{G} = \{A + B : A \in \mathscr{F}, B \in \mathscr{G}\}$. Show that $\mathscr{F} + \mathscr{F} \supset \mathscr{F}$ if and only if $\mathscr{F}$ is additive and deduce Theorem 4-2-2. (See Prob. 4-3-2.)

---

201 If T is a vector topology for X_R, must it be a vector topology for X?

202 Invent a vector topology T for ω such that $T | c_0$ is induced by $\| \cdot \|_\infty$.

4-4 BOUNDED SETS

A set in a TVS is called *bounded* if it is absorbed by every neighborhood of 0; this means that S is bounded if and only if for every $U \in \mathcal{N}$ there exists $\varepsilon > 0$ such that $tS \subset U$ whenever $|t| < \varepsilon$. Some easily proved facts are spelled out in Probs. 4-4-2 to 4-4-5. The reader should consult these at once.

That *each singleton in a TVS is bounded* is just because neighborhoods of 0 are absorbing. *The union of two bounded sets is bounded.* [If $tA \subset U$ for $|t| < \varepsilon$ and $tB \subset U$ for $|t| < \varepsilon'$, we have $t[A \cup B] \subset U$ for $|t| < \min(\varepsilon, \varepsilon')$.] From these two facts it follows that *each finite set is bounded.*

It is a little surprising that there is a sequential test for boundedness:

1. Theorem The following are equivalent for a set S in a TVS:

(a) S is bounded.
(b) For every sequence $\{x_n\} \subset S$ and every sequence $\{\varepsilon_n\}$ of scalars with $\varepsilon_n \to 0$, we have $\varepsilon_n x_n \to 0$.
(c) For every sequence $\{x_n\} \subset S$, $(1/n)x_n \to 0$.

PROOF (a) implies (b). Let $U \in \mathcal{N}$, $tS \subset U$ for $|t| < \varepsilon$. Then $|\varepsilon_n| < \varepsilon$ eventually and so $\varepsilon_n x_n \in \varepsilon_n S \subset U$ eventually. That (b) implies (c) is trivial. (c) implies (a). Suppose that S is not bounded. There exists balanced $U \in \mathcal{N}$ such that $tS \not\subset U$ for some t with $|t| < \varepsilon$, no matter what $\varepsilon > 0$ is. Since U is balanced, $\varepsilon S \not\subset U$ for all $\varepsilon > 0$. In particular, $(1/n)S \not\subset U$ and so we may choose $x_n \in S \backslash nU$ for $n = 1, 2, \ldots$. Then $(1/n)x_n \notin U$ and so $(1/n)x_n \not\to 0$.

2. Definition A map between TVSs which preserves bounded sets is called bounded.

The content of Theorem 2-3-1 is that a linear map between seminormed spaces is continuous if and only if it is bounded. The situation in the general case is shown in Theorems 4-4-3 and 4-4-9 and Example 4-4-11.

3. Theorem A continuous linear map is bounded. Indeed, a sequentially continuous linear map is bounded.

PROOF Let $u: X \to Y$, S a bounded set in X. Let $\{y_n\} \subset u[S]$. Then $(1/n)y_n = (1/n)u(x_n) = u[(1/n)x_n] \to 0$ by Theorem 4-4-1. Thus, again by Theorem 4-4-1, $u[S]$ is bounded.

4. Corollary If T, T' are vector topologies for a vector space X and $T' \supset T$, then each set which is bounded in (X, T') is bounded in (X, T).

PROOF Apply Theorem 4-4-3 to the identity map.

5. Theorem Let Φ be a collection of vector topologies for a vector space X and $S \subset X$. Then S is bounded in $(X, \vee \Phi)$ if and only if it is bounded in (X, T) for each $T \in \Phi$.

PROOF $\Rightarrow$: This is trivial from Corollary 4-4-4. $\Leftarrow$: Let $\{x_n\} \subset S$. Then $(1/n)x_n \to 0$ in (X, T) for each T by Theorem 4-4-1 and so $(1/n)x_n \to 0$ in $\vee \Phi$ [Theorem 1-6-8]. By Theorem 4-4-1, S is $\vee \Phi$ bounded.

The familiar corollaries are given in Probs. 4-4-11 to 4-4-13.

We now take a quick dip into bornology—this assumes a more satisfactory form in a locally convex setting (Sec. 8-4).

6. Definition A bornivore is a set which absorbs all bounded sets. Thus, to say that B is a bornivore means that for every bounded set S there exists $\varepsilon > 0$ with $tS \subset B$ for $|t| < \varepsilon$.

Certainly each neighborhood of 0 is a bornivore by definition of bounded. In certain situations there are no others.

7. Example *Let X be a pseudometric vector space. Then every bornivore is a neighborhood of* 0. Let B be not a neighborhood of 0. Then $nB \notin \mathcal{N}$ for each positive integer n, so we may choose x_n with $d(x_n, 0) < 1/n$, $x_n \notin nB$. Let $S = \{x_n\}$. Then S is bounded [Prob. 4-4-9], but B does not absorb it. [$(1/n)S \not\subset B$ for each n, since $x_n \notin nB$.]

8. Lemma Let X, Y be TVSs, $u: X \to Y$ a bounded linear map, and B a bornivore in Y. Then $u^{-1}[B]$ is a bornivore in X.

PROOF Let $S \subset X$ be bounded. Then $u[S]$ is bounded, hence absorbed by B. It follows that S is absorbed by $u^{-1}[B]$.

The following converse to Theorem 4-4-3, which generalizes the equivalence of boundedness and continuity in Theorem 2-3-1, covers a large class of spaces. By Example 4-4-7 it includes all pseudometric vector spaces. Example 4-4-11 shows that the hypothesis cannot be dropped.

9. Theorem Let X be a TVS in which every bornivore is a neighborhood of 0. Then every bounded linear map $u: X \to Y$, Y a TVS, is continuous.

PROOF Let $U \in \mathcal{N}(Y)$. By Lemma 4-4-8 and the hypothesis, $u^{-1}[U] \in \mathcal{N}(X)$. Thus u is continuous at 0, hence everywhere.

10. Example *The norm and weak topologies on a normed space X have the same bounded sets.* The *weak topology* for any TVS X is $\sigma X'$; denote this by w and let n denote the norm topology. Since w is the smallest topology making each member of X' continuous [Prob. 1-6-7 and Definition 4-1-10], we have $n \supset w$ and so by Corollary 4-4-4 every n bounded set is w bounded. Conversely, let S be $\sigma X'$ bounded. Then each $f \in X'$ is bounded on S [Theorem 4-4-3] and so S is n bounded [Theorem 3-3-11].

11. Example *Two different vector topologies with the same bounded sets and a bounded linear map which is not continuous.* Let (X, n) be an infinite-dimensional normed space, $w = \sigma X'$. By Remark 4-1-14, $n \neq w$; by Example 4-4-10 they have the same bounded sets. The identity map from (X, w) to (X, n) is thus bounded and not continuous.

PROBLEMS

1 Let p be a continuous paranorm on a TVS. Show that p is bounded on bounded sets. [Use Theorem 4-4-1 and $(1/n)p(x_n) \leq p[(1/n)x_n]$, Prob. 2-1-5.] Hence a bounded set in a paranormed space is metrically bounded.

2 Show that a set S in a seminormed space is bounded if and only if it is norm bounded. [Apply Prob. 4-4-1.]

3 Show that if a set is absorbed by every member of a local base of neighborhoods of 0 it is bounded.

4 Show that a bounded vector subspace S of a TVS must be included in $\overline{\{0\}}$. Hence $\{0\}$ is the only bounded subspace of a separated TVS. [For $x \in S$, $(1/n)nx \to 0$ since $nx \in S$.]

5 Show that every subset of a bounded set is bounded.

6 Show that a set is bounded if and only if every countable subset is bounded [Theorem 4-4-1].

7 Let Y be a vector subspace of a TVS X and $S \subset Y$. Show that S is bounded in Y if and only if it is bounded in X [Theorem 4-4-1].

8 Show that the balanced closure of a bounded set is bounded [Prob. 4-4-3 and Theorem 4-2-7].

9 Show that every compact set is bounded. [Let $U \in \mathcal{N}$ be balanced and open. Then $\{nU : n = 1, 2, \ldots\}$ is an open cover. Apply Probs. 4-4-3 and 4-2-10.]

10 Show that every Cauchy sequence is bounded. [By Theorem 4-4-1 it is sufficient to show $x_n/n \to 0$ for any subsequence $\{x_n\}$. If $m \geq N, n \geq N$ implies $x_n - x_m \in U$, choose $N_1 > N$ such that $n > N_1$ implies $x_N/n \in U$. Then $n > N_1$ implies $x_n/n = x_N/n + (x_n - x_N)/n \in U + U$.]

11 Let $f : X \to Y$ be linear, where X is a vector space and Y a TVS. Let $S \subset X$. Show that S is σf bounded if and only if $f[S]$ is bounded [Theorems 4-4-1 and 1-6-10].

12 For a set F of linear maps from a vector space X to various TVSs, show that $S \subset X$ is σF bounded if and only if $f[S]$ is bounded for each $f \in F$ [Prob. 4-4-11 and Theorem 4-4-5].

13 Let (X_α) be a collection of TVSs. Show that S is bounded in πX_α if and only if each of its projections is bounded [Prob. 4-4-12]. Hence show that πS_α is bounded if S_α is bounded in X_α for each α.

14 If A, B are bounded sets and $t \in \mathcal{K}$, show that $tA + B$ is bounded [Prob. 4-4-13 and Theorem 4-4-3].

15 Let X be a TVS with $X' = X^*$. Show that every convergent sequence, hence every bounded set, is finite-dimensional [Example 3-3-14 and Theorem 4-4-1].

---

101 Show that in l^p and $L^p, 0 < p < 1$, a set is bounded if and only if it is metrically bounded. (In this respect they resemble normed spaces.)

102 Show that $\{ f : \| f \|_\infty < 1 \}$ is bounded in the topology of compact convergence (Prob. 4-1-105).

103 Let p, q be a seminorm and a paranorm respectively on a vector space X with p stronger than q. Show that $\{x : p(x) \leq 1\}$ is a bounded set in (X, q) [Corollary 4-4-4].

104 Show that the convex hull of a finite set is bounded.

105 Let $S = \{\delta^n\} \subset l^{1/2}$. Show that S is bounded [Prob. 4-4-101], but its convex hull is not.

106 Let (X, T) be a TVS. Let $b = \vee\{T' : T'$ is a vector topology whose bounded sets are exactly the T bounded sets$\}$. Show that a set is bounded in (X, b) if and only if it is bounded in (X, T). Thus there is a largest vector topology with a given class of bounded sets [Theorem 4-4-5].

107 Let X be a TVS in which every bounded set is finite-dimensional. Show that every absorbing set is a bornivore.

108 Show that the two vector topologies have the same bornivores if and only if they have the same bounded sets.

109 Let X be a TVS and X^b the set of bounded linear functionals. Show that $X^b = X^*$ if and only if every bounded set is finite-dimensional. (For $X^b \neq X'$ see Prob. 8-4-109.)

110 Let $\{X_\alpha : \alpha \in A\}$ be an infinite collection of nonindiscrete TVSs. Show that πX_α contains no bounded absorbing sets. Compare Theorem 4-1-12. [If S is bounded, let $B = \{\alpha(n)\} \subset A$; choose $a_n \in X_{\alpha(n)} \backslash nS_{\alpha(n)}$ using Prob. 4-4-4. Let $x_{\alpha(n)} = a_n$, $x_\alpha = 0$ for $\alpha \notin B$. For each n, $(1/n)x \notin S$ so S is not absorbing.]

111 Show that there exists no norm for ω which is comparable with its natural paranorm. (Problem 12-4-111 extends this.) [No stronger one by Probs. 4-4-103 and 4-4-110; no weaker one by Remark 4-1-13.]

---

201 Show that the convex hull of the unit disc in $L^p, 0 < p < 1$, is L^p. [Apply Hahn–Banach to the convex hull; see Prob. 2-3-120.]

202 Let X, Y be normed spaces and $u : X \to Y$ linear, continuous, open, onto. Let $B \subset Y$ be bounded. Must there exist a bounded set $A \subset X$ with $u[A] = B$?

---

301 Let X be a vector space and $\mathscr{B}$ a collection of subsets of X. Discuss necessary and sufficient conditions that X can be given a vector topology which makes $\mathscr{B}$ the family of bounded sets of X. [See [115].]

4-5 METRIZATION

It turns out that rather mild assumptions are enough to imply that a TVS is pseudometrizable. We begin with a criterion for generation by a seminorm (Theorem 4-5-2).

1. Lemma Let (X, T) be a TVS, B an absolutely convex absorbing set, and p the gauge of B. Then:

    (a) If B is bounded, $\sigma p \supset T$.

    (b) If B is a neighborhood of 0, $\sigma p \subset T$.

PROOF (a) Let $U \in \mathscr{N}(T)$. Since B is bounded, for some $\varepsilon > 0$, $U \supset \varepsilon B \supset \varepsilon\{x : p(x) < 1\} = \{x : p(x) < \varepsilon\}$. Thus U is a σp neighborhood of 0.

    (b) Let U be a σp neighborhood of 0. Then for some $\varepsilon > 0$, $U \supset \{x : p(x) \leq \varepsilon\} = \varepsilon\{x : p(x) \leq 1\} \supset \varepsilon B$. Thus $U \in \mathscr{N}(T)$.

2. Theorem Let X be a TVS. Then X is a seminormed space if and only if X has a bounded convex neighborhood U of 0.

PROOF ⇒: Take U to be the unit disc. ⇐: If U is absolutely convex its gauge gives the topology of X by Lemma 1. In any case, U certainly includes a balanced neighborhood V of 0 [Lemma 4-2-5], and the convex hull H of V is a neighborhood of 0 [$H \supset V$], is bounded [$H \subset U$], and is absolutely convex [Prob. 1-5-16].

The criterion for pseudometrizability turns out to be the obviously necessary condition: first countability (Theorem 4-5-5). The same result holds for topological groups, and in a modified form (countable base) for uniform spaces in general. We refer to [156] for details. For readers interested only in locally convex spaces, a very simple proof of Theorem 4-5-5 follows from the result of Prob. 7-2-6 since half of its proof uses only first countability.

A topological space X is called *first countable* if, for each $x \in X$, X has a countable local base of neighborhoods of x. For a TVS, this specializes, by the localization principle, to the assumption that $\mathcal{N}$ is generated by a countable filterbase; i.e., there exists $\{U_n\} \subset \mathcal{N}$ such for each $U \in \mathcal{N}$, $U \supset U_n$ for some n.

We shall isolate two of the steps in the proof. The first shows how to obtain a function satisfying many of the properties of a paranorm starting from a fairly arbitrary function.

3. Lemma Let X be a vector space and q a nonnegative real function defined on X. For $x \in X$ set

$$\| x \| = \inf \left\{ \sum_{k=1}^{n} q(x_k - x_{k-1}) : x_0 = 0, x_n = x \right\}$$

Then $\| x \| \geq 0$ and $\| x + y \| \leq \| x \| + \| y \|$ for $x, y \in X$. If $q(0) = 0$, then $\| 0 \| = 0$. If $q(-x) = q(x)$ for all x, then $\| -x \| = \| x \|$ for all x.

PROOF We shall refer to the set $(0, x_1, x_2, \ldots, x_{n-1}, x)$ as a chain ending at x. Let $\varepsilon > 0$ and choose chains $(x_i), (y_i)$ ending at x, y respectively with $\sum_{k=1}^{n} q(x_k - x_{k-1}) < \| x \| + \varepsilon$, $\sum_{k=1}^{m} q(y_k - y_{k-1}) < \| y \| + \varepsilon$. Let (z_i) be the chain $(0, x_1, \ldots, x_{n-1}, x, x + y_1, x + y_2, \ldots, x + y_{n-1}, x + y)$. Then $\| x + y \| \leq \sum_{k=1}^{m+n} q(z_k - z_{k-1}) = \sum_{k=1}^{n} q(x_k - x_{k-1}) + \sum_{k=1}^{m} q(y_k - y_{k-1}) < \| x \| + \| y \| + 2\varepsilon$.

The second result follows by taking $n = 1$ in the definition. To prove the third, if $(0, x_1, \ldots, x_n)$ is a chain ending at x, let $y_k = -x_k$. Then $\sum q(y_k - y_{k-1}) = \sum q(x_k - x_{k-1})$.

4. Lemma Let X be a vector space and q a nonnegative real function on X satisfying $q(0) = 0$, $q(x + y + z) \leq 2 \max \{q(x), q(y), q(z)\}$. Then for any $x_1, x_2, \ldots, x_n \in X$ we have $q(\sum_{i=1}^{n} x_i) \leq 2 \sum_{i=1}^{n} q(x_i)$.

PROOF Let $u = \sum_{i=1}^{n} q(x_i)$; we may assume $u > 0$. For $n = 1, 2, 3$, the result is trivial. Proceeding by induction, let $n > 3$. Let m be the largest integer such that $\sum_{i=1}^{m} q(x_i) \leq u/2$. (If this is false for $m = 1$ simply take $m = 0$ and ignore this inequality.) Then $0 \leq m < n$ and $\sum_{i=1}^{m+1} q(x_i) > u/2$, which implies that $\sum_{i=m+2}^{n} q(x_i) \leq u/2$. (If $m = n - 1$, the left-hand side is taken to be 0.) The sums on the left-hand sides of the following list have fewer than n terms (or are 0, the first if $m = 0$, the second if $m = n - 1$); hence the induction hypothesis gives

$$q\left(\sum_{i=1}^{m} x_i\right) \leq 2 \sum_{i=1}^{m} q(x_i) \leq u$$

$$q\left(\sum_{i=m+2}^{n} x_i\right) \leq 2 \sum_{i=m+2}^{n} q(x_i) \leq u$$

$$q(x_{m+1}) \qquad\qquad\qquad\qquad \leq u$$

Applying the hypothesis to these three inequalities yields the result.

5. Theorem The topology of a first countable TVS is given by a paranorm.

PROOF We set $U_0 = X$ and by induction construct a base for $\mathcal{N}$ which is a sequence $\{U_n\}$ of balanced neighborhoods of 0 satisfying for $n = 1, 2, \ldots$:

$$U_n + U_n + U_n \subset U_{n-1} \tag{4-5-1}$$

Now for $x \in \overline{\{0\}}$ we set $q(x) = 0$, while for $x \notin \overline{\{0\}}$ we set $q(x) = 2^{-k}$ where $k = k(x)$ is the largest integer such that $x \in U_k$. (Thus if $k = k(x)$, $x \in U_k \backslash U_{k+1}$.) We shall prove that for $\{x_n\} \subset X$

$$x_n \to 0 \quad \text{if and only if} \quad q(x_n) \to 0 \tag{4-5-2}$$

Let $x_n \to 0$ and let m be a positive integer. Then $x_n \in U_m$ eventually, and for such x_n either $x_n \in \overline{\{0\}}$, in which case $q(x_n) = 0$, or $k(x_n) \geq m$ and so $q(x_n) \leq 2^{-m}$. Thus $q(x_n) \to 0$. Conversely, if $q(x_n) \to 0$ let m be a positive integer. Then for sufficiently large n, $q(x_n) < 2^{-m}$. Now if $x_n \in \overline{\{0\}}$ we have $x_n \in U_m$ while if $x_n \notin \overline{\{0\}}$ we have, with $k = k(x_n)$, $q(x_n) = 2^{-k} < 2^{-m}$, which implies $k > m$ and so $x_n \in U_k \subset U_m$. So in any case $x_n \in U_m$ eventually. Thus $x_n \to 0$. This establishes (4-5-2).

Next we prove

$$q(x + y + z) \leq 2 \max \{q(x), q(y), q(z)\}$$

We may assume not all $x, y, z \in \overline{\{0\}}$. Suppose, for definiteness, that $q(x) = 2^{-k} \geq q(y)$ and $q(z)$. Then x, y, z all belong to U_k and so, by (4-5-1), $x + y + z \in U_{k-1}$. Thus $q(x + y + z) \leq 2^{-(k-1)} = 2 \cdot 2^{-k} = 2q(x)$.

We make a formal note that the conclusion of Lemma 4-5-4 is now established. Let us define $\|x\|$ by the formula of Lemma 4-5-3 and we shall prove

$$\tfrac{1}{2}q(x) \leq \|x\| \leq q(x) \qquad \text{for all } x \tag{4-5-3}$$

The second inequality follows by taking $n = 1$ in the definition of $\|x\|$. Let $(0, x_1, \ldots, x_n)$ be a chain ending at x. By Lemma 4-5-4 we have $\sum q(x_k - x_{k-1}) \geq \tfrac{1}{2}q[\sum(x_k - x_{k-1})] = \tfrac{1}{2}q(x)$. Taking inf yields the first inequality of (4-5-3).

By (4-5-2) and (4-5-3) it follows that $x_n \to 0$ in X if and only if $\| x_n \| \to 0$. Now set $d(x, y) = \| x - y \|$. It follows from Lemma 4-5-3 that d is a pseudometric for X. [That $g(-x) = q(x)$ follows from the fact that each U_n is balanced.] We have just proved that $x_n \to x$ in X if and only if $d(x_n, x) \to 0$ and this implies that d induces the topology of X. [They are both first countable. (Actually everything we said applies to nets.)] This supplies the last missing fact in the demonstration that $\| \cdot \|$ is a paranorm, namely that multiplication is continuous, for it is continuous in the original topology of X.

6. Remark Included in Theorem 4-5-5 is the remarkable fact that *a pseudometric vector space must be a paranormed space*, i.e., a paranorm can be defined which induces an equivalent pseudometric. Thus there is a sense in which there is no loss of generality when, in discussing pseudometric vector spaces, one restricts oneself to paranormed spaces. There may be a loss of generality in discussing isometric or uniform concepts such as completeness. The situation with respect to completeness is this. Suppose that d is a pseudometric and d' is an equivalent pseudometric derived from a paranorm: (a) d' may be complete and d not complete [Prob. 2-1-10]; (b) if d is complete, d' must be. If d is also derived from a paranorm this is quite trivial [Prob. 2-1-9]. In the general case, we have outlined a proof in Prob. 4-5-104.

The preceding theorem enables us to prove a topological result which, while it seems to have no applications, is nevertheless quite interesting. It may be compared with the fact that a TVS need not be normal [Prob. 4-5-302].

7. Theorem Every TVS X is a completely regular topological space.

PROOF Let F be a closed set and $x \notin F$. Let $\{U_n\}$ be a sequence of balanced neighborhoods of 0 with $x + U_n \not\subset F$ and $U_n + U_n \subset U_{n-1}$ for each n [Theorems 4-2-2 and 4-2-7]. Then $\{U_n\}$ is an additive filterbase of balanced absorbing sets and so, by Theorem 4-3-5, it is a local base of neighborhoods of 0 for a vector topology T'. Since T' is first countable it is pseudometrizable and hence completely regular. Thus there exists a T' continuous real function separating x from F. This function is also T continuous, where T is the original topology of X. [$T \supset T'$ since each U_n is a T neighborhood of 0.]

Theorem 4-5-7 shows that if we wish to distinguish between TVSs in a purely topological way we shall have to seek properties other than complete regularity. That such a purely topological classification may not be fruitful is indicated by results of the type given in Prob. 5-1-301 which, in spite of the enormous difficulty of its proof, must be regarded as a negative result.

We now define a *locally bounded* TVS to be one which has a bounded neighborhood of 0. Such spaces are pseudometrizable [Theorem 4-5-8], but not conversely [ω is not locally bounded by Example 4-1-13]. Thus the pseudometric vector spaces fall into two classes: those which are locally bounded and those which are not. For a *locally convex* TVS X (that is, X has a local base of convex neighborhoods of 0), we no longer have this distinction: *a locally convex TVS is locally bounded if and only if it is a seminormed space* [Theorem 4-5-2].

8. Theorem A locally bounded TVS is a paranormed space.

PROOF By Theorem 4-5-5 we have merely to check that it is first countable. Let U be a bounded neighborhood of 0. For any $V \in \mathcal{N}$ there exists an integer n such that $V \supset (1/n)U$. Thus $\{(1/n)U\}$ is a countable local base of neighborhoods of 0. [Each $(1/n)U$ *is* a neighborhood of 0.]

Locally bounded spaces have an additional noteworthy property. It was shown in Sec. 4-4 (Theorems 4-4-3 and 4-4-9 and Example 4-4-11) that a continuous linear map preserves bounded sets and that a linear map preserving bounded sets may or may not be continuous. We now obtain an "adjoint" statement, viz. that a linear map which is bounded on a neighborhood of 0 must be continuous (Theorem 4-5-9) and that a continuous linear map may or may not be bounded on a neighborhood of 0 (Prob. 4-5-3). The extra condition, compared to the one in Theorem 4-4-9, shifts from the domain to the range. (Another example of this "adjointness" of bounded sets and neighborhoods of 0 occurs in Lemma 4-5-1.)

9. Theorem Let $f : X \to Y$ be a linear map. If f takes some neighborhood U of 0 into a bounded set, f is continuous. If Y is locally bounded and f is continuous, then f takes some neighborhood of 0 into a bounded set.

PROOF Let $V \in \mathcal{N}(Y)$. Since $f[U]$ is bounded, $tf[U] \subset V$ for some $t \neq 0$. Thus $f^{-1}[V] \supset tU$, a neighborhood of 0, and so f is continuous. Next, choose a bounded $V \in \mathcal{N}(Y)$. Then f is bounded on $f^{-1}[V] \in \mathcal{N}(X)$.

We conclude with a useful criterion for continuity.

10. Theorem Let X be a TVS, $f \in X^{\#}$, and assume $f^{\perp}$ is closed. Then f is continuous.

PROOF Suppose f is not continuous; we shall prove $f^{\perp}$ dense (not surprisingly, by Prob. 4-2-5). Let $x \in X$, U a balanced neighborhood of 0. Now f is unbounded on U [Theorem 4-5-9] so $f[U]$ is an unbounded balanced set in $\mathcal{K}$; hence $f[U] = \mathcal{K}$. In particular, $f(u) = -f(x)$ for some $u \in U$. Then $x + u \in f^{\perp} \cap (x + U)$ so that $f^{\perp}$ meets every open set.

11. Corollary Let X be a separated TVS, $0 \neq y \in X$, $f \in X^{\#}$. Let $g(x) = f(x)y$. Then if $g : X \to X$ is continuous, so is f.

PROOF This is by Theorem 4-5-10 since $f^{\perp} = g^{\perp}$.

PROBLEMS

In this list X, Y are TVSs.

1 Let p be a seminorm on X. Prove the equivalence of
    (a) p is continuous,
    (b) $\sigma p \subset T$,
    (c) $\{x : p(x) < 1\}$ is a neighborhood of 0.
[See Lemma 4-5-1(b).]

2 Prove that a linear functional is continuous if and only if it is bounded on some neighborhood of 0 [Theorem 4-5-9].

3 Show that every continuous linear map to Y is bounded on a neighborhood of 0 if and only if Y is locally bounded.

---

101 Show that the following are locally bounded: any seminormed space, l^p, $0 < p < 1$, L^p, $0 < p < 1$ [Prob. 4-4-101].

102 Show that $\mathcal{M}$ is not locally bounded. [If $U(\varepsilon) = \{f : \|f\| < \varepsilon\}$, $U(\delta)$ does not absorb $U(\varepsilon)$ for $\varepsilon > \delta$.]

103 Prove the converses of (a) and (b) in Lemma 4-5-1.

104 Let d be a complete pseudometric and p a paranorm for a vector space Z. Show that (Z, p) is complete (i.e., a *topologically complete* paranormed space is complete). [By Prob. 1-6-301, Z is a G_δ in the completion of (Z, p). By Prob. 2-1-114, it is closed.]

105 Let X be of second category. Show that the following are equivalent:
    (a) X is locally bounded.
    (b) X is the union of a sequence of bounded sets.
    (c) The collection of bounded sets in X has a countable cobase.

106 Give φ the topology $\sigma\varphi^*$. Show that in Prob. 4-5-105, condition (c) holds but condition (a) does not [Prob. 4-4-15, Theorem 4-5-8, and Example 3-3-14].

107 For a TVS X show that $X' = X^*$ if and only if every vector subspace is closed [Theorem 4-5-10 and Prob. 1-5-20]. Compare Prob. 4-4-109.

108 A set S is called t *convex* if $S + S \subset tS$. (Thus a convex set must be 2 convex.) Show that a bounded neighborhood U of 0 in a TVS must be t convex for some $t \geq 2$. [Let $V + V \subset U$. Then $U \subset tV$ since U is bounded.]

109 A *quasinorm* satisfies $\|x\| \geq 0$, $\|tx\| = |t| \|x\|$, $\|x + y\| \leq k(\|x\| + \|y\|)$ for some fixed $k \geq 1$. (Thus $\|x - y\|$ need not define a pseudometric.) With $U_n = \{x : \|x\| < 1/n\}$, show that $\mathscr{B} = \{U_n\}$ satisfies the conditions of Theorem 4-3-5 [for example $U_{2kn} + U_{2kn} \subset U_n$]. The resulting TVS is called a *quasinormed space*.

110 A TVS is a quasinormed space (Prob. 4-5-109) if and only if it is locally bounded. [$\Rightarrow$: In Prob. 4-5-109, $U_1 \subset nU_n$ so U_1 is bounded. $\Leftarrow$: Let $U \in \mathcal{N}$ be bounded and balanced. Using Prob. 4-5-108, the gauge of U is a quasinorm as in the proof of Theorem 2-2-1. For the rest, imitate Lemma 4-5-1.]

111 Show that every vector topology T is $\bigvee\{\sigma p : p$ is a paranorm and $\sigma p \subset T\}$. [For each $U \in \mathcal{N}(X)$ follow the procedure of Theorem 4-5-7.]

---

201 If $\mathcal{N}(X)$ has a cofinal chain, X must be a paranormed space. (Here $\mathcal{N}$ is a poset, ordered by inclusion.) [See [153], Prob. 10.4.9.]

---

301 Let $0 < p \leq 1$. A p *norm* satisfies $\|x\| \geq 0$, $\|tx\| = |t|^p \|x\|$, and the triangle inequality. A TVS is p-normable if and only if it is locally bounded. Compare Prob. 4-5-110. [See [88], §15, sec. 10.]

302 In contrast with Theorem 4-5-7, a TVS need not be normal. Examples are R^R (see [156], Prob. 6.7.203) and l^∞ with its weak topology (see [23]).

FIVE

OPEN MAPPING AND CLOSED GRAPH THEOREMS

The third fundamental tool for the application of functional analysis is the closed graph theorem. Even more than the Hahn–Banach and uniform boundedness theorems, it has been subject to searching investigation and generalization. (See the Index.) The one advantage which the primitive form (Theorem 5-3-1) has over the ones given in Chap. 12 is that it makes no convexity assumptions.

5-1 FRÉCHET SPACE

The setting of the first version of the closed graph theorem is Fréchet space. A Fréchet space is a complete metric vector space. We know that the metric can be given by a (total) paranorm p ⟦Remark 4-5-6⟧, and those who have chased down the details of Prob. 4-5-104 know that p is complete. In any case, we define a *Fréchet space* to be a separated complete paranormed space, i.e., the paranorm is total and complete.

Every Banach space is a Fréchet space, but not conversely.

    1. Example ω *is a Fréchet space.* The proof is similar to but easier than those of Examples 3-1-3 and 3-1-4. (It also follows from Theorem 6-1-7 and Prob. 4-1-3.) *It is not a Banach space* ⟦Example 4-1-13⟧. The same is true of $\mathcal{M}$ and L^p, $0 < p < 1$ ⟦Probs. 2-3-119 and 2-3-120⟧. This will be proved for l^p, $0 < p < 1$, in Example 7-1-1.

    The following elementary criterion for completeness will be useful in simplifying a few computations. A series $\sum x_n$ in a paranormed space is said to be *absolutely convergent* if $\sum \| x_n \| < \infty$.

2. Theorem A paranormed space is complete if and only if every absolutely convergent series is convergent.

PROOF $\Rightarrow$: If $\sum x_n$ is absolutely convergent, $\left\| \sum_{i=m}^n x_i \right\| \leq \sum_{i=m}^n \| x_i \|$ so $\{\sum_{i=1}^n x_i\}$ is Cauchy, hence convergent. $\Leftarrow$: Suppose that $\{y_n\}$ is a non-convergent Cauchy sequence; let N_k have the property that $m > n \geq N_k$ implies $\| y_m - y_n \| < 2^{-k}$, $N_{k+1} > N_k$, $k = 1, 2, \ldots$. Let $x_k = y_{N_{k+1}} - y_{N_k}$. Then $\sum x_k$ is absolutely convergent but not convergent. [[$\{y_n\}$ has no convergent subsequence.]]

PROBLEMS

101 Let $H = \{z : |z| < 1\} \subset \mathscr{C}$. Let $X = C(H)$ with the topology of compact convergence. Show that X is a Fréchet space.

102 Let A be the set of analytic functions in X (Prob. 5-1-101). Show that A is a Fréchet space.

103 Let $X = C^\infty$, the infinitely differentiable real functions on $[0, 1]$. Let $p_n(f) = \max\{| f^{(k)}(t)| : 0 \leq t \leq 1, 0 \leq k \leq n\}$ for $n = 1, 2, \ldots$. Show that $(X, \sigma\{p_n\})$ is a Fréchet space.

104 If X is a noncomplete paranormed space, show that it contains a sequence $\{x_n\}$ such that $\sum x_n$ is not convergent but $\| x_n \| < 4^{-n}$ for all n [[as in Theorem 5-1-2]].

201 A topological space is called a *k space* if a subset whose intersection with every compact set K is closed in K must itself be closed. Let H be a hemicompact k space. Show that $C(H)$ (Prob. 4-1-105) is a Fréchet space. [[See [156], Prob. 8-1-120.]]

301 All separable infinite-dimensional Fréchet spaces are homeomorphic. [[See [5].]]

302 Every infinite-dimensional Hilbert space, in particular l_2, is homeomorphic with its unit disc. [[See [83], theorem II.1.6.]] (This provides a nice example of a metric space homeomorphic with its completion, namely $\{x : \| x \| < 1\}$. There are easy examples in R^2.)

303 There exists a locally convex Fréchet space X and a dense subspace D such that X has a bounded set which is not included in the closure of a bounded set in D. [[See [88], 29.6.]]

5-2 OPEN MAPS

A function is called *open* if the image of each open set is open. For example, the map $(x, y) \to x$ which carries $R^2 \to R$ is an open map. *An invertible function*, i.e., a bijection, *is open if and only if its inverse is continuous. A linear map between TVSs is open if and only if it preserves neighborhoods of 0.* [[Consider $f : X \to Y$ and let G be open in X, $y = f(x) \in f[G]$. Then $G - x \in \mathscr{N}(X)$ so $f[G - x] = f[G] - y \in \mathscr{N}(Y)$, that is, $f[G]$ is a neighborhood of y.]] Thus the following definition generalizes open maps: a linear map $f : X \to Y$ is called *almost open* if for each $U \in \mathscr{N}(X)$, $\overline{f[U]} \in \mathscr{N}(Y)$.

An easy example of an almost open map which is not open is the inclusion map from a dense subspace, for example, $i : \varphi \to c_0$.

1. Remark *Let $f : X \to Y$ be linear and suppose that for each $U \in \mathscr{N}(X)$, $\overline{f[U]}$ has nonempty interior. Then f is almost open.* Let $U \in \mathscr{N}(X)$, let $V \in \mathscr{N}(X)$

with $V - V \subset U$, and let y be interior to $\overline{f[V]}$. Then $\overline{f[U]} \supset \overline{f[V]} - f[V]$ [Theorem 4-2-11] $\supset \overline{f[V]} - y \in \mathcal{N}(Y)$.

2. Lemma A linear map whose range is of second category is almost open. More precisely, let X, Y be TVSs, $f : X \to Y$ be linear, and assume that $f[X]$ is of second category in Y. Then f is almost open.

PROOF Let $U \in \mathcal{N}(X)$, $V = f[U]$. Now U is absorbing so $X = \cup\{nU : n = 1, 2, \ldots\}$ which implies $f[X] = \cup\{nV\}$. By hypothesis, some $\overline{nV}$, hence $\overline{V}$ itself, has nonempty interior. By Remark 5-2-1, f is almost open.

We see from Lemma 5-2-2 that many linear maps are automatically almost ⊃pen, and from Lemma 5-2-3 that this may force them to be open.

3. Lemma Let X be a Fréchet space and Y a separated paranormed space (= metric vector space). Let $f : X \to Y$ be linear continuous and almost open. Then f is open.

PROOF Let $U \in \mathcal{N}(X)$ and we wish to show $f[U] \in \mathcal{N}(Y)$. It is sufficient to find $V \in \mathcal{N}(X)$ such that

$$f[U] \supset \overline{f[V]} \qquad (5\text{-}2\text{-}1)$$

since the hypothesis ensures that the latter is a neighborhood of 0. Now $U \supset N_\varepsilon = \{x : \|x\| < \varepsilon\}$ for some $\varepsilon > 0$ and we shall prove (5-2-1) for $V = N_{\varepsilon/2}$. To this end, let $z \in \overline{f[V]}$. Since f is almost open there exists a decreasing null sequence $\{\delta_n\}$ (of positive real numbers) such that

$$\overline{f[N_{\varepsilon/2^n}]} \supset \{y \in Y : \|y\| < \delta_n\}, \qquad n = 1, 2, \ldots \qquad (5\text{-}2\text{-}2)$$

By definition of z there exists $x_1 \in V = N_{\varepsilon/2}$ such that $\|z - f(x_1)\| < \delta_2$. Thus $z - f(x_1) \in \overline{f[N_{\varepsilon/2^2}]}$ by (5-2-2). Hence there exists $x_2 \in N_{\varepsilon/2^2}$ with $\|z - f(x_1) - f(x_2)\| < \delta_3$. Repeating this argument yields, for $n = 1, 2, \ldots$, the existence of $x_n \in N_{\varepsilon/2^n}$ with $\|z - \sum_{i=1}^n f(x_i)\| < \delta_{n+1}$.

Since $\|x_n\| < \varepsilon/2^n$ we may, by Theorem 5-1-2, set $x = \sum x_n$. Since the paranorm is continuous, $\|x\| \le \sum \|x_n\| < \varepsilon$ and so $x \in U$. Thus $z = f(x) \in f[U]$. $[\![f(\sum_{i=1}^n x_i) \to z$ since $\delta_n \to 0$; now f is continuous and Y is a Hausdorff space.$]\!]$ This proves (5-2-1) and Lemma 5-2-3.

4. Theorem: Open mapping theorem Let X, Y be Fréchet spaces. Let $f : X \to Y$ be linear, continuous, and onto. Then f is open.

PROOF As f is almost open by Lemma 5-2-2 and the Baire category theorem, it is open by Lemma 5-2-3.

The same proof gives the following useful extension.

5. Theorem Let X be a Fréchet space, Y a separated paranormed space, and $f: X \to Y$ linear and continuous. If $f[X]$ is of second category in Y, f is an open map of X onto Y.

PROOF That f is onto follows from the fact that $f[X]$ is an open vector subspace and hence is Y.

It turns out (Corollary 6-2-14) that the conditions of Theorem 5-2-5 imply that Y is a Fréchet space.

6. Corollary Let X, Y be Fréchet spaces and $f: X \to Y$ be linear, continuous, one to one, onto. Then f is a homeomorphism.

PROOF By Theorem 5-2-4, f is open, that is, f^{-1} is continuous.

7. Corollary: Uniqueness of topology Two comparable Fréchet topologies for a vector space must be equal.

PROOF Apply Corollary 5-2-6 to the identity map from the larger one.

8. Example *In Corollary 5-2-7, neither Fréchet topology can be allowed to be noncomplete.*
(a) For $X = l$, let $p(x) = \|x\|_1$, $q(x) = \|x\|_\infty$. Then p is strictly larger than q. (A generic example is given in Prob. 3-1-105.)
(b) Let X be a Banach space and $f \in X^\#\backslash X'$ [Example 3-3-14]. Let $p(x) = \|x\| + |f(x)|$. Then p is strictly larger than $\|\cdot\|$ [$f \in (X, p)'$]. Thus it cannot be complete [Corollary 5-2-7].

9. Example *In Corollary 5-2-7, "comparable" may not be omitted.* Let X, Y be two Banach spaces which are isomorphic as vector spaces but not linearly homeomorphic [Probs. 1-5-17 and 3-2-4]. We define a new norm for X by $p(x) = \|u(x)\|$ where $u: X \to Y$ is the isomorphism. The two norms for X are not equivalent; hence, by Corollary 5-2-7, they are not comparable since both are complete. The *sum of the two norms is not complete* by Corollary 5-2-7. Note also that u is almost open by Lemma 5-2-2, so "continuous" cannot be omitted in Lemma 5-2-3.

PROBLEMS

1 Show that l is a first category subspace of c_0, and $l^{1/2}$ is a first category subspace of l. [Apply Theorem 5-2-5 to the inclusion map.]

101 Let X be a vector space, Y a TVS, $f: X \to Y$ linear and onto. Show that $f: (X, \sigma f) \to Y$ is open.
102 In Example 5-2-8(b), show that $(X, p)' = X' + [f]$.

103 Show that a maximal subspace of a Banach space need not allow a larger complete norm; compare Prob. 3-1-105.

104 Let X, Y be Banach spaces, $u \in B(X, Y)$. Suppose there exist numbers $M, r, 0 < r < 1$ such that for each y with $\|y\| \leq 1$ there exists x with $\|x\| \leq M$, $\|y - ux\| \leq r$. Show that u is onto. [With $z = y - ux$, $y = ux + z$ so $u[D_M] + D_r \supset D_1$. It follows from Prob. 4-2-202 that u is almost open. A proof by the methods of Chap. 11 is given in [6], theorem 1.2.]

---

201 Give an example of two noncomplete norms whose sum is complete.

---

301 The proof of Lemma 5-2-2 uses only this property of the range of f. (*) Every closed absorbing set has nonempty interior. However, a TVS has property (*) if and only if it is of second category. [See [124], theorem 1.]

302 Every infinite-dimensional Banach space allows a strictly larger and a strictly smaller norm. (By Corollary 5-2-7 they are not complete.) [See Example 5-2-8(b) and [151].]

5-3 CLOSED GRAPH

If X, Y are topological spaces and $f: X \to Y$, the *graph* of f is defined to be $\{(x, y): y = f(x)\}$. It is a subset of $X \times Y$, and f is said to have *closed graph* if its graph is a closed subset of $X \times Y$. The definition is often extended to functions whose domain is a subset of X; this will not be done in this book.

1. Theorem: Closed graph theorem Let X, Y be Fréchet spaces. Let $f: X \to Y$ be linear and with closed graph G. Then f is continuous.

PROOF The map $P_1: G \to X$ given by $P_1(x, y) = x$ is a homeomorphism, by Corollary 5-2-6. [G is a Fréchet space by Prob. 1-6-11.] The map $P_2: G \to Y$ given by $P_2(x, y) = y$ is continuous, and $f = P_2 \circ P_1^{-1}$.

2. Theorem Let X, Y be TVSs and $f: X \to Y$ be linear. Then f has closed graph if and only if for each net x in X such that $x \to 0$ and $f(x) \to y$ it follows that $y = 0$.

PROOF $\Rightarrow$: Since $[x, f(x)] \to (0, y)$ it follows that $(0, y) \in \bar{G} = G$ where G is the graph of f. Hence $y = f(0) = 0$. $\Leftarrow$: Let $(a, b) \in \bar{G}$. There is a net g in G with $g \to (a, b)$ in $X \times Y$, say $g = [x, f(x)]$ where x is a net in X, $x \to a$, $f(x) \to b$. Then $x - a \to 0$, $f(x - a) = f(x) - f(a) \to b - f(a)$. Hence $b - f(a) = 0$, $f(a) = b$, $(a, b) \in G$.

3. Corollary Let X, Y be TVSs with Y separated. Then a continuous linear map $f: X \to Y$ must have closed graph.

PROOF If $x \to 0$, $f(x) \to y$, we have $f(x) \to 0$ and so $y = 0$.

The condition on Y cannot be omitted [Prob. 5-3-2] and the converse of Corollary 5-3-3 is false [Probs. 5-3-4 and 5-3-102]. This adds a little spice to Theorem 5-3-1.

4. Remark The immense power of the closed graph theorem derives from the good behavior and abundance of closed graph functions. For example, the *inverse of a closed graph function also has closed graph.* ⟦It has the same graph!⟧ Thus (by Corollary 5-3-3) *the inverse of a continuous map between separated spaces has closed graph.* Because of this we can deduce Corollary 5-2-7 from Theorem 5-3-1 by considering the identity map from the smaller to the larger topology. Remarks 5-3-5, Lemma 5-5-1, and Probs. 5-5-101 and 12-5-103 give further evidence.

5. Remark (This may be omitted.) Certain results proved for separated spaces become false without this assumption but remain true if the relevant map is assumed closed graph rather than continuous. As an example see Prob. 5-3-103. The result is a generalization since, in the separated case, continuous implies closed graph. For further evidence of the value of closed graph maps see [107], p. 77, [155], and [156]. We do not insist on this aspect since if X, Y are TVSs and $f: X \to Y$ is linear, closed graph and onto, Y must be separated.

Close examination of the last step of the proof of Lemma 5-2-3 shows that with $u_n = \sum_{i=1}^{n} x_i$ we have $u_n \to x$, $f(u_n) \to z$, from which it was concluded that $z = f(x)$. We can draw the same conclusion merely by assuming f to have closed graph (even if Y is not separated). Thus we have this extension of Theorem 5-2-4.

6. Theorem: Open mapping theorem Let X, Y be Fréchet spaces. Let $f: X \to Y$ be linear, closed graph, and onto. Then f is continuous and open.

PROOF This is what was just proved. Another proof is to cite Theorem 5-3-1 and then apply Theorem 5-2-4.

As a first application, consider the subject of complementation. Let X be a TVS and let A, B be vector subspaces which are *algebraically complementary*, that is, $A + B = X$, $A \cap B = \{0\}$. Define $P: X \to A$ by $Px = a$ where $x = a + b$, $a \in A$, $b \in B$. This is called the *projection on A*.

7. Theorem With the above notation, A, B are closed if and only if P has closed graph.

PROOF $\Rightarrow$: Let $x \to 0$, $Px \to y$. Then $x - Px \in B$ and $x - Px \to -y$; thus $y \in B$. But $y \in A$ since $Px \in A$ and $Px \to y$. Thus $y = 0$. $\Leftarrow$: Let a be a net in A, $a \to x \in X$. Now $a - x \to 0$, $P(a - x) = a - Px \to x - Px$ and so $x - Px = 0$. Thus $x = Px \in A$ and A is closed. Let b be a net in B, $b \to y$. Then $b - y \to 0$, $P(b - y) = -P(y)$, so $Py = 0$ and so $y \in B$. Thus B is closed.

The results of the second half of this proof can also be deduced from Prob. 5-3-106. ⟦Take $K = \{0\}$, $f = P$, $f = I - P$.⟧

Continuing with the same notation, A, B are called *complementary* (sometimes, topologically complementary) if P is continuous; A is said to be *complemented* in X if it is one of a pair of complementary subspaces A, B in which case we write $X = A \oplus B$.

8. Theorem Algebraically complementary closed subspaces of a Fréchet space are complementary.

PROOF This follows from Theorem 5-3-7 and the closed graph theorem.

Problem 5-3-114 shows that the result is false for normed space in general.

9. Remark *Complementary subspaces of a separated space must be closed* [Corollary 5-3-3 and Theorem 5-3-7]. "Separated" cannot be omitted here. [Consider two complementary subspaces of an indiscrete space X, for example, $\{0\}$ and X.]

Examples of noncomplemented subspaces may be found by consulting the Index. An important result is that every real Banach space not isomorphic to a Hilbert space has a noncomplemented closed subspace (see [100]). Conversely, it is classical that every closed subspace of Hilbert space is complemented.

PROBLEMS

1 If f has closed graph and x is a net with $x \to a$, $f(x) \to b$, show that $b = f(a)$.

2 Let X be a TVS and $i: X \to X$ the identity map. Show that X is separated if and only if i has closed graph.

3 Deduce the open mapping theorem for a one-to-one function from Theorem 5-3-1 and Remark 5-3-4 applied to f^{-1}. (For functions which are not one to one see Prob. 6-2-106.)

4 Let T_1, T_2 be separated vector topologies for a vector space X such that T_1 is strictly larger. Show that $i: (X, T_2) \to (X, T_1)$ has closed graph and is not continuous [Corollary 5-3-3 and Remark 5-3-4].

5 Let S be a complemented subspace of a Banach space X. Let $\{f_n\} \subset S'$ with $f_n(x) \to 0$ for each $x \in S$. Show that each f_n can be extended to $g_n \in X'$ with $g_n(x) \to 0$ for all $x \in X$. Compare Example 14-4-9. [Let $g_n = f_n \circ P$ where P projects X onto S.]

6 Suppose that a Banach space X is linearly homeomorphic with the dual of a Banach space. Show that X is complemented in X'' [Prob. 3-2-5].

---

101 Deduce from Example 5-2-8 that neither space in Theorem 5-3-1 can be allowed to be noncomplete.

102 Let X be the continuously differentiable real functions on $[0, 1]$, with $\| f \|_\infty$, $Y = C([0, 1])$. Let $Df = f'$. Show that $D: X \to Y$ has closed graph and is not continuous. Is X complete?

103 Let X, Y be real TVSs and $f: X \to Y$ an additive map, that is, $f(x + y) = f(x) + f(y)$.

    (a) Show that $f(tx) = tf(x)$ for rational t.

    (b) If f has closed graph show that it must be linear.

    (c) If Y is separated and f is continuous, show that f must be linear.

    (d) Show that the assumption on Y in (c) cannot be omitted.

104 Prove analogues of Theorem 5-3-2 and Corollary 5-3-3 for general topological spaces and functions.

105 There exists no sequence u such that $\sum |t_n| < \infty$ if and only if $\{u_n t_n\}$ is bounded. Deduce this from the fact that this would make the map $x \to \{x_n/u_n\}$ a linear homeomorphism of l^∞ onto l (at least if $u_n \neq 0$ for all n). [Use Probs. 2-3-6 and 2-3-7.]

106 If $f: X \to Y$ has closed graph and K is a compact set in Y, show that $f^{-1}[K]$ is closed.

107 Show that a linear functional with closed graph must be continuous.

108 Show that the existence of a divergent Cauchy sequence in (X, p), Example 5-2-8(b), leads directly to a proof that f cannot have closed graph. (The result is a special case of Prob. 5-3-107.)

109 Let (X, T) and Y be separated TVSs which are complete (Sec. 6-1) and $f: X \to Y$ be a linear map. Let $T_1 = T \vee \sigma f$. Show that (X, T_1) is complete if and only if f has closed graph.

110 Deduce the closed graph theorem from Prob. 5-3-109 and Corollary 5-2-7.

111 Vector subspaces A, B of a TVS X are called *quasicomplementary* if A, B are closed, $A \cap B = \{0\}$, and $A + B$ is a dense proper subspace of X. Let $A = \{x \in cs : x_n = 0$ for n odd$\}$, $B = \{x \in cs : x_n = 0$ for n even$\}$. Show that A, B are quasicomplementary in cs. [$A + B \supset \varphi$, $\{(-1)^n/n\} \notin A + B$.]

112 Let A, B be algebraically complementary subspaces of a TVS X. Define $f: A \times B \to X$ by $f(a, b) = a + b$. Show that f is a continuous linear bijection (one to one and onto).

113 Show that in Prob. 5-3-112, f is a homeomorphism if and only if A, B are complementary. [$\Rightarrow: P = P_1 \circ f^{-1}$ where $P_1 : A \times B \to A$; $\Leftarrow: P_1 \circ f^{-1} = P$, $P_2 \circ f^{-1} = I - P$ are continuous. Now see Theorem 1-6-11.]

114 Let A, B be quasicomplementary subspaces of a Banach space Y (Prob. 5-3-111). Let $X = A + B$. Show that A, B are algebraically complementary in X but not complementary [Probs. 5-3-113 and 1-6-11].

115 Let X be a TVS. Show that any subspace which is algebraically complementary to $\overline{\{0\}}$ is complementary. [$P: X \to \overline{\{0\}}$ is continuous since $\overline{\{0\}}$ is indiscrete.] Hence every TVS is the product of a separated and an indiscrete space [Prob. 5-3-113].

116 Show that each X_β is a complemented subspace of $\pi X \alpha$.

117 Show that l^∞ is complemented in each normed space which includes it [Prob. 2-3-109]. Compare Prob. 2-3-303.

118 Let X, Y be TVSs and $f: X \to Y$ be linear with closed graph. Show that for each $y \in Y$, $y \neq 0$, there exists $U \in \mathcal{N}(X)$ with $y \notin \overline{f[U]}$.

119 Let X, Y, Z be topological spaces, $u: X \to Y$, $f: Y \to Z$. Define $h: X \times Z \to Y \times Z$ by $(x, z) \to (ux, z)$. Show that h^{-1} [graph of f] is the graph of $f \circ u$.

120 Deduce from Prob. 5-3-119, or directly, that a closed graph function of a continuous function must have closed graph.

121 Let R^2 have the topology which is the euclidean topology on the axes and is discrete elsewhere. Show that addition is not continuous but has closed graph.

---

201 Let X, Y be TVSs and $f: X \to Y$ be continuous and linear. Show that f is right invertible if and only if it is an open map and $f^\perp$ is complemented in X. Show that f is left invertible if and only if it is a linear homeomorphism (into) and $f[X]$ is complemented in Y. [See [58], p. 123.]

202 Let X be a complete metric space, Y a T_2 space, $f: X \to Y$ be one to one, continuous, almost open. Show that f is a homeomorphism. [See [156], Prob. 9.1.118. The open mapping theorem can be deduced from this. See [156], Prob. 12.4.123.]

203 Let X be a locally compact, separable group, Y a separated group of second category, $f: X \to Y$ a continuous homeomorphism. Prove that f is open. [See [156], Prob. 12.2.117. For further developments see [156], Prob. 12.2.123.]

---

301 Let $X = c_0$ or $l^p, p \geq 1$. Then every infinite-dimensional complemented subspace of X is linearly homeomorphic with X. [See [16].]

302 Let $X = C[0, 1]$ or L^p, $1 < p < \infty$, and $X = Y \oplus Z$. Then either Y or Z is linearly homeomorphic with X. ⟦See [16].⟧

303 An infinite-dimensional subspace of l^∞ is complemented if and only if it is linearly homeomorphic with l^∞. ⟦See [98].⟧

304 The spaces $C[0, 1]$, c_0, L^p, l^p, $p > 1$, $p \neq 2$, have noncomplemented closed subspaces. ⟦See [88], sec. 31.3, and [97].⟧

5-4 THE BASIS

The main result, Theorem 5-4-4, is that a basis for a Fréchet space must be a Schauder basis. The part of this section beginning with Lemma 5-4-3 is not used in the remainder of the book.

A *basis* for a TVS X is a sequence $\{b^n\}$ such that every $x \in X$ has a unique representation $x = \sum t_n b^n$. For example, c_0 has $\{\delta^n\}$ as basis. ⟦Example 2-3-4. If, also, $x = \sum t_n \delta^n$ we have $t_k = x_k$ by taking the kth coordinate.⟧ Bases for c, l^p were shown in Sec. 2-3. A basis is rarely a Hamel basis. A space with this property would have a countable Hamel basis and would thus be φ with a suitable topology. Further information is contained in Probs. 6-3-107, 6-3-108, and 9-4-111.

If X has a basis $\{b^n\}$ the functionals l_n, given by $l_n(x) = t_n$ when $x = \sum t_n b^n$, are linear. They have the property of forming with $\{b^n\}$ a *biorthogonal system*, that is, $l_n(b^k) = 0$ if $n \neq k$, $l_n(b^n) = 1$. They are called the *coordinate functionals* and $\{b^n\}$ is called a *Schauder basis* if each $l_n \in X'$.

1. Example Let X be a sequence space with basis $B = \{\delta^n\}$. Then X is a K space (Definition 3-1-2) if and only if B is a Schauder basis.

2. Remark *Let X be a TVS with basis $\{b^n\}$. Then X is linearly homeomorphic with a sequence space Y with basis $\{\delta^n\}$. Further, Y is a K space if and only if $\{b^n\}$ is a Schauder basis for X.* Namely, let $Y = \{t : \sum t_n b^n$ converges in $X\} = \{t : \text{for some } x \in X, t_n = l_n(x) \text{ for all } n\}$. The map $u : X \to Y$ given by $u(x) = t$, where $x = \sum t_n b^n$, is an isomorphism onto and becomes a homeomorphism when Y is identified with X, that is, the open sets of Y are defined to be $u[G]$ for G open in X. (This is the quotient topology, Sec. 6-2.) Also, $u(b^n) = \delta^n$ and l_n is identified with $P_n \in Y^*$, $P_n(t) = t_n$. Thus $P_n \in Y'$ if and only if $l_n \in X'$; so the result follows by Example 5-4-1.

The rest of this section may be omitted.

3. Lemma Let (X, p) be a paranormed sequence space with basis $\{\delta^n\}$. Let $q(x) = \sup \{p(\sum_{i=1}^n x_i \delta^i) : n = 1, 2, \ldots\}$. Then q is a paranorm for X and $q \geq p$.

PROOF A conjectured easy proof is disposed of by Prob. 2-1-302. For $x \in X$, let $u^n = u^n(x) = \sum_{i=1}^n x_i \delta^i = (x_1, x_2, \ldots, x_n, 0, 0, \ldots)$. Then $q(x) = \sup p(u^n)$. Only continuity of multiplication presents difficulty. We first prove

$$r_k \in \mathcal{K}, r_k \to 0 \quad \text{implies} \quad q(r_k x) \to 0 \quad \text{for each } x \in X \quad (5\text{-}4\text{-}1)$$

Let $\varepsilon > 0$. Now $\{u^n\}$ is a convergent sequence in (X, p); hence it is bounded, so there exists $\delta > 0$ such that $|t| < \delta$ implies $p(tu^n) < \varepsilon$ for all n. Choose N so that $k > N$ implies $|r_k| < \delta$. For $k > N$ we have $p(r_k u^n) < \varepsilon$ for all n and so $q(r_k x) \le \varepsilon$.

We next prove

$$q(x^k) \to 0 \quad \text{implies} \quad q(tx^k) \to 0 \quad \text{for all } t \in \mathcal{K} \quad (5\text{-}4\text{-}2)$$

Fix $t \in \mathcal{K}$. Now $x^k = \{x_n^k\}$ and $u^{k,n} = (x_1^k, x_2^k, \ldots, x_n^k, 0, 0, \ldots)$. Let $\varepsilon > 0$. Let U be a neighborhood of 0 in (X, p) with $p(tu) < \varepsilon$ for all $u \in U$. There exists N such that $k > N$ implies $u^{k,n} \in U$ for all n. $[\![p(u^{k,n}) \le q(x^k) \to 0.]\!]$ For $k > N$ we have $p(tu^{k,n}) < \varepsilon$ for all n and so $q(tx^k) \le \varepsilon$.

Next

$$r_k \in \mathcal{K}, x^k \in X, r_k \to 0, q(x^k) \to 0 \quad \text{imply} \quad q(r_k x^k) \to 0 \quad (5\text{-}4\text{-}3)$$

Let $\varepsilon > 0$. Let U be a balanced neighborhood of 0 in (X, p) with $p(u) < \varepsilon$ for all $u \in U$. There exists N such that $k > N$ implies $|r_k| < 1$ and $u^{k,n} \in U$ for all n. $[\![\text{See the proof of (5-4-2).}]\!]$ For $k > N$ we have $r_k u^{k,n} \in r_k U \subset U$ for all n, and so $p(r_k u^{k,n}) < \varepsilon$ for all n, from which $q(r_k x^k) \le \varepsilon$.

Finally, $t_k \to t \in \mathcal{K}$, $q(x^k - x) \to 0$ implies $q(t_k x^k - tx) \to 0$, for with $r_k = t_k - t$, $y^k = x^k - x$ it follows that $q(t_k x^k - tx) \le q(r_k y^k) + q(r_k y) + q(ty^k) \to 0$ by (5-4-3), (5-4-1), and (5-4-2).

4. Theorem A basis for a Fréchet space must be a Schauder basis, i.e., the coordinate functionals are continuous.

PROOF By Remark 5-4-2 we may take X to be a sequence space with basis $\{\delta^n\}$. Form q as in Lemma 5-4-3. Each $u^n: X \to X$ defined in the proof of Lemma 5-4-3 is continuous on (X, q). $[\![q[u^n(x)] \le q(x).]\!]$ Thus $u^n - u^{n-1}$ is continuous for each n. But $u^n(x) - u^{n-1}(x) = x_n \delta^n = P_n(x)\delta^n$ and so each P_n is continuous on $(X, .q)$ $[\![\text{Corollary 4-5-11}]\!]$. Since, also, $q \ge p$ it will follow from the closed graph theorem, in the form of Corollary 5-2-7, that q, p are equivalent when we show that (X, q) is complete. In particular, Theorem 5-4-4 will be proved. To this end, let $\{x^n\}$ be a q-Cauchy sequence. Since coordinates are continuous, $\{x_k^n\}$ is a Cauchy sequence in $\mathcal{K}$ for each fixed k. Let $x_k = \lim \{x_k^n: n \to \infty\}$. We shall show that

$$\sum x_k \delta^k \quad \text{converges in} \quad (X, p) \quad (5\text{-}4\text{-}4)$$

Clearly,

$$p\left(\sum_{k=u}^{v} x_k \delta^k\right) = \lim_{n \to \infty} p\left(\sum_{k=u}^{v} x_k^n \delta^k\right) \quad (5\text{-}4\text{-}5)$$

Given $\varepsilon > 0$, choose N so that $n > m \geq N$ implies $q(x^m - x^n) < \varepsilon$. Then for $n > N$ we have $p(\sum_{k=u}^{v} x_k^n \delta^k) \leq p[\sum_{k=u}^{v} (x_k^n - x_k^N) \delta^k] + p(\sum_{k=u}^{v} x_k^N \delta^k) \leq 2\varepsilon + p(\sum_{k=u}^{v} x_k^N \delta^k) < 3\varepsilon$ for sufficiently large u, v (depending on ε, N, and, hence, only on ε). From (5-4-5), $p(\sum_{k=u}^{v} x_k \delta^k) \leq 3\varepsilon$ for sufficiently large u, v. This proves (5-4-4) since (X, p) is complete. Let $a = \sum x_k \delta^k \in (X, p)$. (We could prove $a = x$ but do not need to.) The proof is concluded by showing that $q(x^n - a) \to 0$. Let $\varepsilon > 0$ and choose N as before. If $n > m \geq N$ we have $p[\sum_{k=1}^{r} (x_k^n - x_k^m) \delta^k] \leq q(x^n - x^m) < \varepsilon$ for any r. Thus for any such r, $p[\sum_{k=1}^{r} (x_k - x_k^m) \delta^k] \leq \varepsilon$ whenever $m \geq N$. Thus $q(a - x^m) \leq \varepsilon$ for $m \geq N$.

The book [134] is recommended for further study of bases.

PROBLEMS

101 Let X be the polynomials with $\| f \| = \max \{| f(z)| : |z| \leq 1\}$, $b_n(z) = z^n$. Show that $\{b_n\}$ is a basis for X but not a Schauder basis. Deduce that completeness cannot be omitted in Theorem 5-4-4 and that Remark 5-4-2 leads to a sequence space which is not a K space.

102 Deduce that c is linearly homeomorphic with c_0 by the map given in Remark 5-4-2 and the basis $(1, \delta^1, \delta^2, \ldots)$ given for c in Prob. 2-3-3. (How does this fit in with Prob. 3-1-301?)

103 Let $y = \{1/n\}$, $B = \{y, \delta^1, \delta^2, \ldots\}$, $X = $ span of B in c_0. Show that B is a Hamel basis but not a basis for X.

104 Are the coordinate functionals relative to B continuous in Prob. 5-4-103?

105 Let $\{b^n\}$ be a basis for a Banach space. Show that for some K, $1/\| b^n \| \leq \| l_n \| \leq K/\| b^n \|$ for all n.

106 By considering $\cos n\pi x$ and Fourier coefficients, construct a Schauder basis for a dense subspace, the trigonometric polynomials, of $C[0, 1]$ which is not a basis for the whole space.

107 We call a sequence *basic* if it is a basis for its linear closure. Show that any subsequence of a Schauder basis for a TVS is basic.

108 In cs show that $\{\delta^n\}$ is a Schauder basis. Find $x \in$ cs which cannot be written $y + z$ with $y \in$ linear closure of $\{\delta^{2n}\}$, $z \in$ linear closure of $\{\delta^{2n-1}\}$ [Prob. 5-3-111].

109 Show that $\mathcal{M}$, L^p, $0 < p < 1$ are separable Fréchet spaces with no basis [Probs. 2-3-119 and 2-3-120].

110 Give an example of a separable Banach space with no basis. [Let $X = R$ with scalar field the rationals. For any $\{b^n\}$ there exist rationals r_n with $\sum r_n b^n = 0$. For a "legitimate" example see Prob. 5-4-301.]

111 Suppose that a set X has defined on it a separating sequence of maps into $\mathcal{K}$. Show that $|X| \leq \mathfrak{c}$. ($|X|$ is the cardinal number of X, $\mathfrak{c} = |R|$.) [Map $X \to \omega$ by $x \to \{f_n(x)\}$.]

112 If X is a TVS with $|X| > \mathfrak{c}$, show that X has no basis [Prob. 5-4-111]. In Prob. 9-6-113 this is used to construct a separable locally convex space with no basis.

201 Show that c is not linearly isometric with c_0. (See Prob. 5-4-102.)

202 Show that Schauder cannot be omitted in Prob. 5-4-107.

203 Does X, Prob. 5-4-103, have a basis which is also a Hamel basis?

301 (P. Enflo). There exists a separable Banach space with no basis. [See [101], pp. 12 and 48.]

302 A weak basis (the series converges weakly) for a locally convex Fréchet space is a basis. [See [10].] Compare Prob. 9-3-140.

5-5 *FH* SPACES

The subject matter of this section has important applications to analysis, some of which are shown. However, no reference is made to this material until Sec. 15-2, and no reference at all is made to the material beginning with Lemma 5-5-10.

1. Lemma : Closed graph lemma Let X, Y be topological spaces and $f : X \rightarrow Y$ a function with closed graph. If X, Y are both given larger topologies, f still has closed graph.

PROOF The topology of $X \times Y$ has been made larger, so the graph of f remains closed.

We now present a framework in which applications of the closed graph theorem are natural and easy. Assume given a fixed vector space H which has a (not necessarily vector) Hausdorff topology. An *FH space* is a vector subspace X of H which is a Fréchet space and is continuously embedded in H, that is, the topology of X is larger than the relative topology of H. The phrase "let X, Y be *FH* spaces" includes the assumption that H is the same for both. A *BH space* is a normed *FH* space.

2. Example Suppose we take $H = \omega = \mathscr{K}^N$. This has the product topology, i.e., the smallest topology such that the coordinates P_n are continuous, where $P_n(x) = x_n$. Thus an *FH* space (with this H) is simply a Fréchet K space in the sense of Definition 3-1-2. [Continuity of coordinates is preserved when the topology is enlarged.] For this reason an *FH* space with $H = \omega$ is called an *FK* space. Examples 3-1-3 and 3-1-4 list the familiar ones, the first five of which are *BK* spaces (Banach instead of Fréchet).

3. Example Let S be a set and $H = \mathscr{K}^S$, the scalar-valued functions on S, with the product topology, i.e., the smallest topology such that the points P_s are continuous, where $P_s(f) = f(s)$ for $f \in H$, $s \in S$. Thus an *FH* space (with this H) is a Fréchet K-function space in the sense of Definition 3-1-5. As in Example 3-1-6, $C_0(T)$ is a *BH* space with $H = \mathscr{K}^T$, the disc algebra is a *BH* space with $H = \mathscr{C}^T$, $T = \{z : |z| \leq 1\}$.

4. Example Let $H = \mathscr{M}$. Then L^p, $p > 0$, is an *FH* space and a *BH* space if $p \geq 1$.

5. Example *Let X be a Fréchet space with a basis. Then X is linearly homeomorphic with an FK space. This is the content of Remark 5-4-2 and Theorem 5-4-4.*

The fundamental result is the following continuity criterion. Its necessity is trivial so we omit that part of the statement.

6. Theorem Let Y be an *FH* space, X a Fréchet space, and $f : X \to Y$ linear. If $f : X \to H$ is continuous, then $f : X \to Y$ is continuous.

PROOF By Lemma 5-5-1, $f : X \to Y$ has closed graph. ⟦Let the original topology on Y be the topology of H and the new larger topology, the Fréchet topology of Y.⟧ So f is continuous by the closed graph theorem.

7. Remark If Y is an *FK* space in Theorem 5-5-6, the condition for continuity is $f : X \to \omega$ is continuous. This, in turn, is true if and only if $P_n \circ f$ is continuous for each n ⟦Theorem 1-6-11⟧.

8. Corollary Let X, Y be *FH* spaces with $X \subset Y$. Then the inclusion map is continuous, that is, X has a larger topology than that of Y (on X). In particular, the topology of an *FH* space is unique, i.e., there is at most one way to make a vector subspace of H into an *FH* space.

PROOF Apply Theorem 5-5-6 to the inclusion map.

The reader may have noticed that the topologies of the increasing list $l^{1/2} \subset l \subset l^2 \subset l^3 \subset c_0 \subset l^\infty \subset \omega$ as given in Sec. 1-2 are decreasing in size. For example, $l \subset c_0$ and the given norms, $p(x) = \|x\|_1$, $q(x) = \|x\|_\infty$ respectively, have p strictly stronger than q. It follows from Corollary 5-5-8 and Prob. 5-5-103 that this situation is inevitable under the circumstances and that the choice of paranorms in Sec. 1-2 is forced (to within equivalence) by the uniqueness part of Corollary 5-5-8.

9. Theorem The intersection X of countably many *FH* spaces (X_n, p_n) is an *FH* space with $\bigvee \{p_n\}$.

PROOF Here p_n stands for both the paranorm and the topology of X_n. Let p be the Fréchet combination of $\{p_n\}$ so that $p = \bigvee \{p_n\}$ ⟦Theorem 2-1-2⟧. If x is a Cauchy sequence in (X, p), it is a Cauchy sequence in each (X_n, p_n), say $x \to t_n \in X_n$. Then $x \to t_n$ in H so all the t_n are the same, say $t_n = t$. Then $t \in X$ and $x \to t$ in (X, p) ⟦Theorem 2-1-2⟧.

The rest of this section may be omitted. Lemma 5-5-10 describes a key procedure in the construction of *FK* spaces in summability. In the statement, p denotes both the paranorm and the topology of X, and σu and $\bigvee$ are discussed in Definition 4-1-10 and Theorem 1-6-8.

10. Lemma Let (X, p) and (Y, q) be *FH* spaces and $u : X \to H$ be a continuous linear map. Let $S = u^{-1}[Y]$. Then $(S, p \bigvee \sigma u)$ is an *FH* space and $u : S \to Y$ is continuous.

PROOF This topology for S is metrizable since it is given by $p + q \circ u$. It is also larger than p and hence larger than the relative topology of H. Now let x be a Cauchy sequence in S. Then x is Cauchy in (X, p), say $x \to a$ in (X, p). Also, x is Cauchy in $(X, \sigma u)$ so ux is Cauchy in Y, say $ux \to b \in Y$. Now $ux \to ua$ in H [$u: X \to H$ is continuous] and $ux \to b$ in H [Y is an *FH* space]. Since H is Hausdorff, $b = ua$ so $a \in S$ and $x \to a$ in S. [$x \to a$ in p and in σu since $ux \to ua$, so $x \to a$ in $p \vee \sigma u$ by Theorem 1-6-8.]

11. Lemma If u, Lemma 5-5-10, is one to one and $u[X] \supset Y$, $(S, \sigma u)$ is an *FH* space.

PROOF Now $u: (S, \sigma u) \to Y$ is an isometry onto so $(S, \sigma u)$ is a Fréchet space. Since $(S, p \vee \sigma u)$ is also a Fréchet space it follows from the open mapping theorem, Corollary 5-2-7, that the two topologies are the same; so Lemma 5-5-10 gives the result.

12. Example *Let A be a row-finite matrix (each row $\in \varphi$). Then $c_A = \{x : Ax \in c\}$ is an FK space with $\sigma\{p_n\}$ where $p_0(x) = \| Ax \|_\infty$, $p_n(x) = |x_n|$ for $n = 1, 2, \ldots$.* In Lemma 5-5-10, take $X = \omega$, $Y = c$, $u = A$. That A is continuous is pointed out in Prob. 5-5-102.

13. Example *Let $a \in \omega$. Then $a^\beta = \{x : \sum a_k x_k \text{ converges}\}$ is an FK space with $\sigma\{p_n\}$ where $p_0(x) = \sup \{|\sum_{k=1}^r a_k x_k| : r = 1, 2, \ldots\}$, $p_n(x) = |x_n|$ for $n = 1, 2, \ldots$,* by Example 5-5-12, since $a^\beta = c_A$ if the nth row of A is $\sum_{k=1}^n a_k \delta^k$. Alternatively, use Lemma 5-5-10 with $X = \omega$, $Y = \text{cs}$.

If $a_n \neq 0$ for a certain n, p_n may be omitted in Example 5-5-13 since $p_n(x) \leq 2 p_0(x)/|a_n|$. In particular, $1^\beta = \text{cs}$ is a *BK* space with p_0.

14. Example *Let A be a matrix. Then $\omega_A = \{x : Ax \text{ exists}\}$ is an FK space with $\sigma\{p_n\}$ where $p_{2n}(x) = |x_n|$ for $n = 1, 2, \ldots, p_{2n-1}(x) = \sup \{|\sum_{k=1}^r a_{nk} x_k| : r = 1, 2, \ldots\}$ for $n = 1, 2, \ldots$,* by Example 5-5-13 and Theorem 5-5-9.

15. Example *Let A be a matrix. Then $c_A = \{x : Ax \in c\}$ is an FK space with $\sigma\{p_n\}$ where p_n is given in Example 5-5-14 and p_0 in Example 5-5-12.* In Lemma 5-5-10 take $x = \omega_A$ (Example 5-5-14), $Y = c$, $u = A$. Of course, if A is row finite the odd-numbered p_n may be omitted [Example 5-5-12].

A matrix A is called *reversible* if for each $y \in c$ there exists unique x such that $y = Ax$.

16. Theorem Let A be a reversible matrix. Then (c_A, p) is a *BK* space with $p(x) = \| Ax \|_\infty$.

PROOF Since (c_A, p) is equivalent with c it is a Banach space. It is an *FK* space with a larger topology, by Example 5-5-15. These must be the same by the open mapping theorem.

17. Remark If A is reversible, it follows by the equivalence of c_A with c and Prob. 2-3-3 that each $f \in c_A'$ is given by $f(x) = \alpha \lim Ax + \sum t_n(Ax)_n$, with $t \in l$.

18. Theorem Let A be a reversible matrix. Then there exists a matrix T whose rows are in l and a sequence α such that for each $y \in c$, the equation $y = Ax$ has the unique solution $x = \alpha \lim y + Ty$.

PROOF Apply Remark 5-5-17 to each $P_r \in c_A'$ where $P_r(x) = x_r$.

Summability theory may be pursued further through the problems; the Index gives references. The treatise [162] is also recommended.

PROBLEMS

101 Show that the closed graph Lemma 5-5-1 becomes false if "closed graph" is replaced by "continuous" in the hypothesis and conclusion.

102 Show that an arbitrary matrix map between FK spaces must be continuous $[(P_n \circ A)(x) = \sum a_{nk} y_k$ is continuous by Theorem 3-3-13. Apply Theorem 5-5-6.]

103 Let X, Y be FH spaces with $X \subset Y$. Show that the topology of X is strictly larger if and only if X is not a closed subspace of Y.

104 Let X, Y be FH spaces with $X \neq Y$. Show that $X \cap Y$ is of first category in Y. A special case is $X \subset Y$ [Theorem 5-5-9, Corollary 5-5-8, and Theorem 5-2-5].

105 Deduce that $\int |f_n|^p \to 0$ implies $\int |f_n|^q \to 0$ if $q > p > 0$ from Example 5-5-4 and Corollary 5-5-8.

106 Let X be a Fréchet space with each of two topologies T and T'. Suppose X has a Hausdorff topology smaller than both T and T'. Show that $T = T'$. (This improves Corollary 5-2-7.) [See Corollary 5-5-8.]

107 Let X be a Fréchet space with each of two unequal topologies T and T' [e.g., Example 5-2-9]. Show that $T \cap T'$ is not Hausdorff [Prob. 5-5-106]. Hence show that the intersection of two vector topologies need not be a vector topology [Theorem 4-2-8]. Compare Probs. 2-1-201, 2-2-201, and 6-2-112 and Example 13-1-12.

108 Let X be a Fréchet space with each of two topologies T and T'. Suppose there exists a total family of linear functionals on X, each of which is both T and T' continuous. Show that $T = T'$ [Prob. 5-5-106]. Deduce that c_0 with the new complete norm given in Example 5-2-9 cannot be a K space.

109 Let X be an FH space with $H = \mathscr{H}^S$ for some set S. Suppose that $h \in H$ is a *multiplier* for X, that is, $fh \in X$ for all $f \in X$. Show that the map $f \to fh$ is continuous [Theorem 5-5-6].

110 Let $H = R^{[0,1]}$ and X be an FH space of differentiable functions. Show that the derivative map $f \to f'$ is continuous: $X \to Y$ where Y is any FH space for which the map is defined. [Let $u_n(f) = n[f(t + 1/n) - f(t)]$. Apply Theorems 5-5-6 and 3-3-13.]

111 Let $\{f_n\}$ be a sequence of real differentiable functions on $[0, 1]$ such that f_n, but not f_n', tends uniformly to 0. Show that there exists a function which is not continuously differentiable but can be uniformly approximated by finite linear combinations of $\{f_n\}$ [Prob. 5-5-110].

112 Let (X, T) be an FH space and p a lower semicontinuous extended seminorm [$p(x) = \infty$ is allowed]. Show that $\{x : p(x) < \infty\}$ is an FH space with $T \vee p$.

113 Deduce Theorem 5-5-9 for two spaces from Lemma 5-5-10 [$u = i$].

114 Show that $\{\delta^n\}$ is a basis for ω_A, Example 5-5-14 [directly, or use $\omega_A = \cap\{r^\beta : r$ a row of $A\}$]. Deduce that for each $f \in \omega'_A$, $f(x) = \sum \beta_k x_k$. If A is row finite, $\beta \in \varphi$; if $\|A\| < \infty$, $\beta \in l$. [The series converges for all $x \in \omega_A$.]

115 Let u, v be sequences with $u_n \to 1$, $v_n \neq 0$ for all n. Let A be the matrix whose nth row is $u_n \delta^1 + \sum_{i=n+1}^{\infty} v_i \delta^i$. Show that A is reversible. Deduce that α, Theorem 5-5-18, may be unbounded. $[x_n = \alpha_n \lim y + (y_n - y_{n+1})/v_n$ with $\alpha_n = (u_{n+1} - u_n)/v_n$, $x_1 = \lim y.]$

116 Let A be the matrix whose nth row is $\delta^n + \frac{1}{2}\delta^{n+1}$. Show that A has an inverse [one way is by Prob. 5-5-201] but is not reversible. [It is not one to one.]

117 Show that a reversible matrix has a unique right inverse matrix but need have no left inverse matrix.

118 If in Theorem 5-5-18, $A(T1) = (AT)1$ and $\alpha \in \omega_A$, show that $\alpha = 0$. Deduce that $\alpha = 0$ if A is row finite.

119 Let $\{X_n\}$ be a decreasing sequence of *FH* spaces such that $X = \cap X_n$ is a dense proper subset of each X_k. Show that X is not a *BK* space. [The unit disc would involve only finitely many of the paranorms.] A corollary is $l \neq \cap l^{1+1/n}$.

---

201 A Banach algebra is a Banach space X and an algebra with 1 such that $\|xy\| \leq \|x\|\|y\|$, $\|1\| = 1$.

    (*a*) Show that $\|1 - x\| < 1$ implies $x \in X^{-1}$. [Let $y = \sum (1 - x)^n$.]

    (*b*) Show that every maximal subspace which is an ideal is closed [by (*a*) and Prob. 4-2-5].

    (*c*) Show that every scalar homomorphism is continuous [by (*b*) and Theorem 4-5-10].

Deduce that all Banach algebra norms are equivalent for an algebra which has a total family of scalar homomorphisms [by (*c*) and Prob. 5-5-108].

202 Let A be a Banach algebra of complex functions on a compact T_2 space T. Show that $\|f\| \geq \|f\|_\infty$ for all $f \in A$. [Each map $t \to f(t)$ is a scalar homomorphism and hence has norm ≤ 1.] Deduce that A is a *BH* space for $H = \mathscr{C}^T$.

203 Construct two reversible matrices whose product is not reversible. [See [1]. The product in the opposite order is reversible.]

---

301 If A is a conservative matrix with $c_A \neq c$, $(c_A \cap l^\infty, \|\cdot\|_\infty)$ must be nonseparable. [See [154], p. 98, or [2].]

302 If A is a conservative matrix, $c_A \cap l^\infty = c$ if and only if c is closed in c_A. [See [159].]

SIX

FIVE TOPICS

6-1 COMPLETENESS

There are many forms of completeness which play a role in the study of TVS. Perhaps the least important is the unmodified form given in the first part of Definition 6-1-2.

1. Definition A net $x = (x_\delta : D)$ in a TVS X is called a Cauchy net if for each $U \in \mathcal{N}$ there exists $\delta \in D$ such that $\alpha \geq \delta$, $\beta \geq \delta$ imply $x_\alpha - x_\beta \in U$.

The defining condition may be expressed as $x_\alpha - x_\beta \in U$ eventually.
A Cauchy sequence is a special case of a Cauchy net.

2. Definition Let X be a TVS and $S \subset X$. We call S complete if every Cauchy net in S converges to a point in S, sequentially complete if every Cauchy sequence in S converges to a point in S.

The net is Cauchy as a net in X.
It turns out that we have been using sequential completeness in discussing pseudometrizable spaces, but, no matter, the concepts are equivalent in this case.

3. Theorem A sequentially complete paranormed space is complete.

PROOF Let $x = (x_\delta : D)$ be a Cauchy net. For each integer $n > 0$, choose δ_n so that $\delta_n > \delta_{n-1}$ and $\| x_\alpha - x_\beta \| < 1/n$ for $\alpha, \beta \geq \delta_n$. Let $y_n = x_{\delta_n}$. Then $\{y_n\}$ is a Cauchy sequence. $[\![\| y_m - y_k \| \leq \| y_m - y_n \| + \| y_n - y_k \| < 2/n$ if $m > n$, $k > n.]\!]$ So $y_n \to y$, say. Then $x \to y$. $[\![$Let $\varepsilon > 0$. Choose $n > 2/\varepsilon$ so that $\| y_n - y \| < \varepsilon/2$. Then $\delta > \delta_n$ implies $\| x_\delta - y \| \leq \| x_\delta - y_n \| + \| y_n - y \| < 1/n + \varepsilon/2 < \varepsilon.]\!]$

A product of complete spaces is complete [Theorem 6-1-7], but inheritance of completeness by weak and sup topologies is erratic.

4. Example Let S be a nonclosed subspace of a complete separated TVS. Then S has the weak topology by the inclusion map but is not complete.

Some other results are given in Probs. 6-1-108 to 6-1-111.

5. Lemma Let X be a TVS and $P: X \to S$ a continuous projection onto a subspace S. Let A be a subset of S such that $P^{-1}[A]$ is complete. Then A is complete.

PROOF Let x be a Cauchy net in A. Then x is a Cauchy net in X and hence $x \to b \in P^{-1}[A]$. Then $x = Px \to Pb \in A$.

6. Remark A special case is that S is complete if X is. This would be trivial if X were separated since S would then be closed. See also Prob. 6-1-111.

7. Theorem Let X_α be a collection of TVSs and $A_\alpha \subset X_\alpha$ for each α. Then πA_α is complete if and only if each A_α is complete.

PROOF $\Leftarrow$: Let x be a Cauchy net in πA_α. Then each x_α is a Cauchy net in X_α [P_α is continuous and linear]; hence $x_\alpha \to a_\alpha \in A_\alpha$. Define $a \in \pi A_\alpha$ by $P_\alpha a = a_\alpha$ for each α. Then $x \to a$ [Theorem 1-6-11]. $\Rightarrow$: By Remark 6-1-6.

8. Definition A TVS is called *boundedly complete* if every bounded closed set is complete.

Some authors use *quasicomplete* for this concept.
A complete space must be boundedly complete [every closed subset is complete] and a *boundedly complete space must be sequentially complete.* [The closure of a Cauchy sequence is complete by Prob. 4-4-10.] Thus, by Theorem 6-1-3, these concepts coincide for a paranormed space. In general, the converse of each of these implications is false [Examples 9-3-12 and 9-3-14].
Theorems 6-1-13 and 6-1-16 give a useful criterion for completeness.

9. Definition Let T, T' be vector topologies for a vector space X. We say T' is *F linked* to T if T' has a local base of neighborhoods of 0, each of which is T closed.

10. Example *If $T' \subset T$ then T' is F linked to T.* [By Theorem 4-2-7, T' has a local base of neighborhood of 0 which are T' closed, hence T closed.] But T is not necessarily F linked to T' [Example 6-1-15], so *the relation F linked is not symmetric.*

The fundamental result now given shows the role played by F linked topologies. It is trivial in the case given in Example 6-1-10.

11. Lemma Let T, T' be vector topologies for a vector space X with T' being F linked to T and $x = (x_\delta : D)$ a net in X. If x is a Cauchy net in (X, T') and $x \to a$ in (X, T), then $x \to a$ in (X, T').

PROOF Let U be a T' neighborhood of 0 which is T closed. Choose $\delta \in D$ such that $\alpha \geq \delta$, $\beta \geq \delta$ implies $x_\alpha - x_\beta \in U$. Fix $\alpha \geq \delta$. Then $x_\alpha - x_\beta \in U$ for all $\beta \geq \delta$. Since U is T closed, $x_\alpha - a \in U$. Since this holds for all $\alpha \geq \delta$ and the set of such U is a base for $\mathcal{N}(X, T')$, we have $x \to a$ in (X, T').

Note. Remarks 6-1-12 and 6-1-14 may be omitted if A (Theorem 6-1-17) is also assumed closed. This will cover all applications.

12. Remark If the net in Lemma 6-1-11 is a sequence it is sufficient to assume that T' has a local base of neighborhoods of 0 each of which is T sequentially closed ⟦the same proof⟧.

13. Theorem Let T, T' be vector topologies for a vector space X such that $T' \supset T$ and T' is F linked to T. Let $S \subset (X, T)$ be complete (or sequentially complete). Then S is T' complete (sequentially complete).

PROOF Let x be a Cauchy net or sequence in (X, T'). Then x is Cauchy in (X, T) and hence convergent. Say $x \to a$ in (X, T). By Lemma 6-1-11, $x \to a$ in (X, T').

14. Remark For the second part of Theorem 6-1-13 it is sufficient that T' have a local base of neighborhoods of 0, each of which is T sequentially closed ⟦Remark 6-1-12⟧.

15. Example Let (X, T') be complete and $T \supset T'$ such that (X, T) is not complete ⟦Example 5-2-8(b)⟧. By Theorem 6-1-13, T is not F linked to T'.

16. Theorem Let (X, T) be a TVS which is complete, boundedly complete, or sequentially complete. Let T' be a larger vector topology which is F linked to T. Then (X, T') is, respectively, complete, boundedly complete, or sequentially complete.

PROOF The first and third are the special case $S = X$ of Theorem 6-1-13. Now let S be a bounded closed set in (X, T'). Let $\bar{S}$ be the T closure of S. Then $\bar{S}$ is a bounded closed set in (X, T) ⟦it is bounded by Corollary 4-4-4 and Prob. 4-4-8⟧ and hence is T complete. By Theorem 6-1-13, $\bar{S}$ is T'-complete, and so also is its T' closed subset S.

This result has an important application. We have seen that the gauge of a certain type of set is a seminorm (Theorem 2-2-1). Theorem 6-1-17 gives a sufficient condition that it be a complete norm. Sets with this property are studied in [28] and [139].

17. Theorem Let (X, T) be a separated TVS. Let A be an absolutely convex, bounded, sequentially complete set in X. Let Z be the span of A and p the gauge of A, defined on Z. Then (Z, p) is a Banach space.

PROOF First, note that A is absorbing in Z. [For $z \in Z$, $z \neq 0$, $z = \sum t_i a^i$, a finite sum! Let $t = \sum |t_i|$. Then $|b| < 1/t$ implies $bz \in A$ since $bz = \sum (bt_i) a^i$ and $\sum |bt_i| < 1$.] Now $p \supset T|Z$ [Lemma 4-5-1; we are writing p for σp] and so (Z, p) is separated, that is, p is a norm. Also, p is F linked to $T|Z$ in the sequential form of Remark 6-1-14. [$\{\varepsilon A : \varepsilon > 0\}$ is a local base of neighborhoods of 0 in (Z, p) and each εA is sequentially complete in X, hence in $(Z, T|Z)$, and so is sequentially closed in $(Z, T|Z)$.] By Theorem 6-1-13 and Remark 6-1-14, A is p sequentially complete. Thus (Z, p) is a normed space with a sequentially complete neighborhood of 0 [$A \supset \{x : p(x) < 1\}$]. This obviously makes (Z, p) sequentially complete. [Each Cauchy sequence is bounded and hence included in a multiple of A.]

The next result finds its most satisfactory form in Theorem 6-5-7.

18. Lemma A compact subset K of a TVS X is complete.

PROOF Let $x = (x_\delta : D)$ be a Cauchy net in K. Let $T_\alpha = \{x_\delta : \delta \geq \alpha\}$ for each $\alpha \in D$. Since D is directed, $\{T_\alpha : \alpha \in D\}$ has the finite intersection property (each finite intersection is nonempty); thus, by compactness, there exists $k \in K$ such that k belongs to the closure of each T_α. We shall prove that $x \to k$. Let $U \in \mathcal{N}$. Let $V \in \mathcal{N}$ with $V + V \subset U$. There exists α such that $\delta \geq \alpha$, $\delta' \geq \alpha$ implies $x_\delta - x_{\delta'} \in V$. Since $k \in \overline{T_\alpha}$, $k + V$ meets T_α, say $x_\delta \in k + V$ with $\delta \geq \alpha$. But then $\delta' \geq \alpha$ implies $x_{\delta'} - k = x_{\delta'} - x_\delta + x_\delta - k \in V + V \subset U$. Hence $x \to k$.

If X is not separated, a compact set need not be closed; for example, $\{0\}$ is not.

Some useful completeness results are given in [41], theorem 1, [42], and [154], p. 77. The converse of Lemma 6-1-11 is considered in [42].

PROBLEMS

1 Show that a closed subset of a complete TVS is complete.

2 Show that a complete subset of a separated TVS is closed and that "separated" cannot be omitted.

3 Suppose that, in Theorem 6-1-17, A is absorbing. Show that X gets a larger complete norm. If $A \in \mathcal{N}$, show that (X, T) is a Banach space.

4 In $\vee \Phi$ (Example 4-1-6) show that a net is Cauchy if and only if it is Cauchy in (X, T) for each T.

5 Show that a net is Cauchy in σF (Definition 4-1-10) if and only if it is Cauchy in σf for each $f \in F$. [Prove it for one map and use Prob. 6-1-4.]

6 Show that a net in a product of TVSs is Cauchy if and only if each projection is [Prob. 6-1-5].

---

101 Let $X = \omega$, $A = \{x : \|x\|_\infty \leq 1\}$ in Theorem 6-1-17. Show that $Z = l^\infty$. [A is compact since it is D^N, where D is the unit disc in $\mathcal{K}$.]

102 If T is F linked to T_1 and $T_2 \supset T_1$, show that T is F linked to T_2.

103 Let $x = (x_\delta : D)$ be a net in a TVS. Suppose that for each $U \in \mathcal{N}$ there exists δ such that $\alpha \geq \beta \geq \delta$ implies $x_\alpha - x_\beta \in U$. Must x be a Cauchy net?

104 Let $\bar{X}$ be the set of Cauchy sequences in a paranormed space X. For x, y in $\bar{X}$ let $\|x - y\| = \lim \|x_n - y_n\|$. Show that $\bar{X}$ is a complete paranormed space in which X is densely embedded via $a \to (a, a, a, \ldots)$. [Problem 1-6-111 should come in handy.]

105 Deduce from Prob. 6-1-104 that a totally paranormed space ($=$ metric vector space) has a Fréchet *completion* (a complete Fréchet space in which it is densely embedded). [Take the complement of $\overline{\{0\}}$ in $\bar{X}$ using Prob. 5-3-115.]

106 Let T, T' be vector topologies such that $T' \supset T$ and T' has a local base $\mathcal{B}$ of neighborhoods of 0 such that for each $U \in \mathcal{B}$, if x is a T' Cauchy net in U and $x \to a$ in T, then $a \in U$. Deduce the results of Theorem 6-1-16.

107 If a TVS has a complete neighborhood of 0 it is complete. [It has a closed complete $U \in \mathcal{N}$ since it is regular. If $x_\delta - x_{\delta'} \in U$, $x_\delta - x_{\delta'}$ is a convergent net for fixed δ'.]

108 Let X be a vector space, Y a TVS, $f : X \to Y$ linear, and $S \subset X$. Show that S is σf complete if and only if $f[S]$ is complete in Y.

109 Let S be a complete set in (X, T) and (X, T'). Show that S need not be complete in $(X, T \vee T')$ [Example 5-2-9].

110 If, in Prob. 6-1-109, there is a Hausdorff topology for X which is smaller than both T and T', show that S is complete in $(X, T \vee T')$. [If $x \to a, b$ in T, T', then $a = b$.]

111 Find a complete set in R^2 whose projection on the X axis is not complete and a noncomplete set both of whose projections are complete. (Compare Lemma 6-1-5.)

112 Let T, T' be vector topologies for a vector space X such that $T' \supset T$ and $\mathcal{N}(T')$ contains a T complete set. Show that every Cauchy net in (X, T') is convergent in (X, T). What does this say if $T' = T$?

113 In Prob. 6-1-112 assume also that T' is F linked to T. Show that (X, T') is complete.

114 Let $x^n = \sum_{k=1}^n \delta^k$. Show that $\{x^n\}$ is Cauchy in w, the weak topology of c (Example 4-4-10), $x^n \to 1$ in ω, but $x^n \not\to 1$ in (c, w). Draw a conclusion from Lemma 6-1-11.

115 Let $\{x^n\}$ be a bounded sequence in a TVS and $t \in l$. Show that $\{\sum_{i=1}^n t_i x^i\}$ is a Cauchy sequence.

---

201 A convex set with nonempty interior is called a *convex body*. Show that a sequentially closed convex body must be closed. [See [154], p. 58.] This is false for convex sets in general [Prob. 9-3-122].

202 Can the proof of Theorem 5-4-4 be simplified by using Prob. 6-1-106?

203 Let X be a subspace of two complete TVSs Y and Z with $X = Y \cap Z$. Suppose $Y \cup Z$ has a T_2 topology T which, on X, is smaller than both $T_Y | X$ and $T_Z | X$. Show that $(X, T_Y \wedge T_Z)$ is complete.

204 In Prob. 6-1-203, it is not sufficient that T is defined only on X and smaller than T_Y and T_Z there.

205 Let H be a k space (Prob. 5-1-201). Show that $C(H)$ is complete. [See [156], Prob. 8-1-120.]

6-2 QUOTIENTS

If we dualize the notion of weak topology, replacing "domain" by "range" and "smallest" by "largest," we arrive at a concept called the *quotient topology* because of its historic link with quotients of groups. Let X be a TVS, Y a vector space, and $f: X \to Y$ a linear map onto. The quotient topology Qf is the largest vector topology for Y which makes f continuous. Of course, we do not know a priori that such a topology exists. We could (and do, in Sec. 13-1) define Qf to be $\bigvee \{T: T$ is a vector topology for Y making f continuous$\}$ [meaningful since the indiscrete topology is a candidate] and then conclude that $f: X \to (Y, Qf)$ is continuous by Theorem 1-6-8. However, it is a little easier to do it this way.

1. Lemma Let X be a TVS, Y a vector space, and $f: X \to Y$ a linear map onto. Let $\mathscr{B} = \{f[U]: U$ a balanced neighborhood of 0 in $X\}$. Then $\mathscr{B}$ is an additive filterbase of balanced absorbing sets.

PROOF If $U, V \in \mathscr{N}(X)$ are balanced, $U \cap V \supset W \in \mathscr{N}$, also balanced, and $f[W] \subset f[U] \cap f[V]$, so $\mathscr{B}$ is a filterbase. Also $\mathscr{B}$ is additive since if $V + V \subset U$ then $f[V] + f[V] \subset f[U]$. The members of $\mathscr{B}$ are balanced and also absorbing. [If $y \in Y$, $y = f(x)$ for some x; since x is absorbed by each $U \in \mathscr{N}$, y is absorbed by $f[U]$.]

2. Definition The quotient topology Qf is the vector topology generated by $\mathscr{B}$ (Lemma 6-2-1) as in Theorem 4-3-5.

3. Theorem With this notation, $f: X \to (Y, Qf)$ is continuous and open; moreover, Qf is the only topology for Y for which this is true.

PROOF First f is continuous [let $V \in \mathscr{N}(Y, Qf)$; then by definition $V \supset f[U]$ for some $U \in \mathscr{N}(X)$ and so $f^{-1}[V] \supset U$] and open [let $U \in \mathscr{N}(X)$ be balanced; then $f[U] \in \mathscr{N}(Y, Qf)$ by definition].

Now let T be a topology for Y which makes f continuous; then $T \subset Qf$ by Prob. 1-6-12. On the other hand, if $f: X \to (Y, T)$ is open, Prob. 1-6-12 implies that $T \supset Qf$.

4. Remark Part of the proof of Theorem 6-2-3 shows that *Qf is the largest topology making f continuous, and the smallest making f open.* It is interesting that these are vector topologies: Problem 4-1-8 is a similar (dual) result for weak topologies.

A linear map $q: X \to Y$ onto is said to be a *quotient map* if Y has the quotient topology. By Theorem 6-2-3, "quotient map" is synonymous with "continuous open linear map." If Y has the quotient topology by some map, we call Y a *quotient of X.*

Note that a one-to-one quotient map is a homeomorphism. Thus the quotient concept is a generalization of homeomorphism to maps which are not one to one.

The expected criterion for continuity has the same dual character as was mentioned above (compared to that for weak topologies, Theorem 1-6-10), namely, we discuss maps *from* a quotient.

5. Theorem Let Y be a quotient of X by means of $q: X \to Y$ and let $f: Y \to Z$ be a linear map, Z a TVS. Then $f: Y \to Z$ is continuous if and only if $f \circ q: X \to Z$ is continuous.

PROOF $\Rightarrow$: Since q and f are continuous. $\Leftarrow$: Let $V \in \mathcal{N}(Z)$. There exists $U \in \mathcal{N}(X)$ such that $f[qU] \subset V$. Since q is open, $q[U] \in \mathcal{N}(Y)$ and so f is continuous.

6. Remark Quotients in the group context are usually formed out of subspaces. This is equivalent to the formulation given above. Let X be a TVS and S a vector subspace. Let $Y = \{x + S : x \in X\}$ and make Y a vector space by defining $x + S + x' + S = x + x' + S$, $t(x + S) = tx + S$ for $t \in \mathcal{K}$. (Note that S is the 0 of Y.) Define $q: X \to Y$ by $q(x) = x + S$. Then (Y, Qq) is written X/S and called the *quotient of X by S*. Conversely, *every quotient map is of this form*, i.e., if $q: X \to Y$ is a quotient map, there exists a subspace $S \subset X$ such that $Y = X/S$; namely, let $S = q^{\perp}$ and then define $g: Y \to X/S$ by $g(y) = x + S$ if $y = q(x)$ [g is well defined since q is onto and if $y = q(x) = q(x')$, we have $x - x' \in q^{\perp} = S$ and so $x + S = x' + S$]. Then g is linear and one to one [if $g(y) = 0 = S$, $y = q(x)$ with $x + S = S$, that is, $x \in S$ which implies $y = 0$] and onto [given $x + S \in X/S$, $g[q(x)] = x + S$]. Finally, we show that g is a homeomorphism. Let $Q: X \to X/S$ be $Qx = x + S$. Then $Q = g \circ q$ is continuous; hence g is continuous by Theorem 6-2-5. On the other hand, $q = g^{-1} \circ Q$ is continuous; hence g^{-1} is continuous.

7. Remark The equation $Y = X/S$ in Remark 6-2-6 is to be interpreted in the sense of linear homeomorphism.

8. Theorem Let X be a TVS and S a subspace. Then X/S is separated if and only if S is closed. Equivalently, if $q: X \to Y$ is a quotient map, Y is separated if and only if $q^{\perp}$ is closed.

PROOF The equivalence is by Remark 6-2-6. We prove the second formulation. $\Rightarrow$: $q^{\perp} = q^{-1}[\{0\}]$ and $\{0\}$ is closed. $\Leftarrow$: Let $0 \neq y \in Y$. Then $y = q(x)$ with $x \in X \backslash q^{\perp}$. There exists $U \in \mathcal{N}(X)$ with $x - U \notin q^{\perp}$. Thus $y \notin q[U]$. [If $y = q(u)$ with $u \in U$ we would have $q(x - u) = 0$, that is, $x - u \in q^{\perp}$, contradicting the choice of U.] Since q is open, $q[U] \in \mathcal{N}(Y)$ and so Y is separated, by Theorem 4-2-8.

There is an important special case in which we can conclude that any onto map is a quotient map.

9. Theorem Let X, Y be Fréchet spaces. Then any continuous linear map from X onto Y is a quotient map.

PROOF This is the open mapping theorem, Theorem 5-2-4.

10. Remark Thus the words "image" and "quotient" may be used interchangeably in discussing Fréchet spaces.

11. Theorem A quotient Y of a paranormed space X is a paranormed space.

PROOF It would be easy to prove this by observing that Y is first countable and applying Theorem 4-5-5. However, it is useful to have an explicit representation for the paranorm. For $y \in Y$, let $\| y \| = \inf \{ \| x \| : y = q(x) \}$ where $q : X \to Y$ is the quotient map. (Note that $\{ x : y = q(x) \} = x + S$ where $S = q^{\perp}$; thus $\| y \|$ is the distance in X from $x + S$ to 0.) Clearly $\| 0 \| = 0$, $\| y \| \geq 0$ for all y. Next, $\| y + z \| \leq \| y \| + \| z \|$. [Let $\varepsilon > 0$ and let $w, x \in X$ with $y = q(w)$, $z = q(x)$, $\| w \| < \| y \| + \varepsilon$, $\| x \| < \| z \| + \varepsilon$. Then $y + z = q(w + x)$ so $\| y + z \| \leq \| w + x \| \leq \| w \| + \| x \| < \| y \| + \| z \| + 2\varepsilon$.]

The formula $d(y, z) = \| y - z \|$ defines a translation invariant pseudometric d on Y. We shall show that d induces the quotient topology. Since that is a vector topology, it follows that $\| \cdot \|$ is a paranorm. First, $q : X \to (Y, d)$ is continuous. [Let $x \to a$ in X. Then $d[q(x), q(a)] = \| q(x - a) \| \leq \| x - a \| \to 0$.] Also, $q : X \to (Y, d)$ is open. [Let G be an open set in X, $b \in q[G]$, $b = q(a)$, $a \in G$. There exists $\varepsilon > 0$ such that $\| x - a \| < \varepsilon$ implies $x \in G$. Now let $y \in Y$ with $\| y - b \| < \varepsilon/2$. Let $w \in X$ with $y - b = q(w)$, $\| w \| < \| y - b \| + \varepsilon/2$. Then $w + a \in G$ since $\| w + a - a \| = \| w \| < \varepsilon$, and $y = q(w) + b = q(w + a) \in q[G]$. Thus $\| y - b \| < \varepsilon/2$ implies $y \in q[G]$, so b is interior to $q[G]$; thus $q[G]$ is open.] By Theorem 6-2-3, d induces the quotient topology.

12. Theorem A quotient of a seminormed space is a seminormed space.

PROOF We merely observe that the paranorm given in the proof of Theorem 6-2-11 is a seminorm since for $t \neq 0$, $\| ty \| = \inf \{ \| x \| : ty = q(x) \} = \inf \{ \| x \| : y = q(x/t) \} = | t | \inf \{ \| x/t \| : y = q(x/t) \} = | t | \| y \|$.

13. Remark By Theorems 6-2-8 and 6-2-12 it follows that X/S is a normed space if X is a seminormed space and S is a closed subspace. A similar result follows from Theorems 6-2-8 and 6-2-11.

The next result may be compared with the fact that a quotient of a complete TVS need not be complete [Prob. 12-4-301]. For some special kinds of quotient, completeness is preserved. (See Prob. 6-2-114 and also Prob. 6-2-4 with Remark 6-1-6.)

14. Theorem A quotient of a complete paranormed space is complete.

PROOF Let $q: X \to Y$ be the quotient map and let $\sum y_n$ be an absolutely convergent series in Y. For each n choose $x_n \in X$ with $y_n = q(x_n)$, $\| x_n \| < \| y_n \| + 2^{-n}$. (We are using the paranorm for Y defined in the proof of Theorem 6-2-11.) Then $\sum x_n$ is absolutely convergent and hence convergent in X. Since q is continuous it follows that $\sum y_n = \sum q(x_n)$ converges to $q(\sum x_n)$. The result follows from Theorem 5-1-2.

15. Corollary Every separated quotient of a Fréchet (or Banach) space is a Fréchet (Banach) space.

This means that a separated TVS which is a quotient of a Fréchet space is a Fréchet space. As in Remark 6-2-6, an equivalent formulation is: every quotient of a Fréchet space by a closed subspace is a Fréchet space. Similar remarks hold for Banach space.

16. Remark It is possible that a quotient of X may not be linearly homeomorphic with any subspace of X [Probs. 6-2-201, 8-1-109, and 14-1-103]; equivalently, that a subspace of X may not be a quotient of X [Example 14-4-9]. However, if A is a complemented subspace of X, say $X = A \oplus B$, then A is a quotient of X; indeed, the projection $a + b \to a$ is linear continuous and open and A is linearly homeomorphic with X/B. Thus a complemented subspace is a special case of a quotient. Since $A \oplus B = A \times B$ [Prob. 5-3-113] these remarks are extended by Prob. 6-2-4.

PROBLEMS

1 Let X be a normed space, S a subspace, and $q: X \to X/S$ the quotient map. Show that $\| q \| = 1$.

2 Let X, Y be normed spaces and S a closed subspace of X. Let $S^{\perp} = \{ f \in B(X, Y): f = 0 \text{ on } S \}$. Show that $S^{\perp}$ is equivalent with $B(X/S, Y)$. [For $f \in S^{\perp}$, set $g(x + S) = f(x)$ so that $f = g \circ q$. If $0 < t < \| g \|$ let $\| x + S \| < 1$, $\| g(x + S) \| > t$; make $\| x \| < 1$. Then $\| f \| > t$. Thus $\| f \| \geq \| g \|$.]

3 Let X be a TVS, A a subspace, and B a subset of X. Let $q: X \to X/A$ be the quotient map. Show that $q^{-1}\{q[B]\} = A + B$. [For example, $q(x) \in q(B)$ implies $x - b \in q^{\perp} = A$ for some $b \in B$.]

4 Show that each X_{β} is a quotient of πX_{α} [Prob. 1-6-13].

---

101 Follow the suggestion made before Lemma 6-2-1 to define Qf as $\vee \{ T: T \text{ is a vector topology making } f \text{ continuous} \}$. Show that $f: X \to (Y, Qf)$ is open.

102 Show that X/S is indiscrete if and only if S is dense.

103 Describe the largest vector topology for Y which makes $f: X \to Y$ continuous if f is not onto.

104 Let X be a seminormed space, S a subspace, $Y = X/S$, and $q: X \to Y$ the quotient map. Let $D_X = \{ x: \| x \| \leq 1 \}$, $N_X = \{ x: \| x \| < 1 \}$, show that $q[N_X] = N_Y$, $q[D_X] \subset D_Y$, and give an example in which equality does not hold [Prob. 2-3-104].

105 Show that each seminormed space is the product of a normed space and an indiscrete space [Remark 6-2-13 and Prob. 5-3-115].

106 Deduce the open mapping theorem from Prob. 5-3-3 using the hint to Prob. 6-2-2 with $S = f^\perp$. [g is one to one.]

107 Can we deduce from Remark 6-2-6 that if $f: X \to Y$, $g: X \to Y$, and $f^\perp = g^\perp$, then $Qf = Qg$ since $Y = X/S$?

108 In contrast with Theorem 6-2-3 there may be several vector topologies for X which make $f: X \to Y$ continuous and open. [See p, q in Prob. 2-1-1. Take f to be a projection.]

109 Let $q: X \to Y$ be a quotient map. Show that X need not have the topology σq [Prob. 6-2-108].

110 Let $f: X \to Y$ be linear and onto and suppose that X has the topology σf. Show that f must be a quotient map.

111 Let X be a TVS and S a sequentially closed but dense vector subspace (e.g., Prob. 8-1-12). Let $x \in X\backslash S$ and set $u_n = x + S$ for all n. Show that $u_n \to 0$ in X/S but no choice of $x_n \in u_n$ can result in $x_n \to 0$ in X. [See Prob. 6-2-102.] (This is why no net characterization for quotients similar to Theorem 1-6-10 was given.)

112 Let T and T' be vector topologies for a vector space X. Let $f: [(X, T) \times (X, T')] \to X$ be defined by $f(x, y) = x + y$. Show that $Qf = T \wedge T'$. (Not $T \cap T'$; see Prob. 5-5-107.)

113 If X is a vector space, Y a normed space, and $f: X \to Y$ linear, then $p(x) = \| f(x) \|$ defines a seminorm on X. Show that conversely every seminorm p on X can be obtained in this way. [Take $Y = X/p^\perp$, $f = $ quotient map.]

114 Let X be a complete TVS. Show that $X/\overline{\{0\}}$ is complete [Probs. 6-2-4 and 5-3-115].

115 Show that the restriction of a quotient map to a closed subspace need not be a quotient map. (See Prob. 6-2-118.) [In Prob. 5-3-111, let $q: X \to X/A$. Then $q^{-1}(q[A]) = A + B$ is not closed so $q[A]$ is not closed. Now see Theorem 6-2-14.]

116 Let X, Y be vector spaces, $u: X \to Y$ linear. A subspace S of X is called *saturated* if $S = u^{-1}(u[S])$. Show that S is saturated if and only if $S \supset u^\perp$.

117 Let S be saturated as in Prob. 6-2-116. Show that $u[A \cap S] = u[A] \cap u[S]$ for all $A \subset X$.

118 If $q: X \to X/A$ and $S \supset A$, show that $q|S$ is a quotient map [Probs. 6-2-116 and 6-2-117].

---

201 Show that every separable Banach space is a quotient of l. [Map $\delta^n \to x_n$ where $\{x_n\}$ is dense.]

202 Show that every Banach space is a quotient of $l(I)$ for some I. [Take I to be dense and imitate Prob. 6-2-201.]

203 Show that every separated quotient of l^2 is $l^2(I)$ with I a set of integers. (Compare Prob. 6-2-201.)

---

301 Every separated quotient of c_0 is linearly homeomorphic with a subspace of c_0. [See [101], p. 53.]

6-3 FINITE-DIMENSIONAL SPACE

There is only one n-dimensional separated TVS and that is $\mathscr{K}^n$, euclidean n space. This is the content of Theorem 6-3-2. A special case was worked out in Prob. 3-2-110.

1. Remark $\mathscr{K}^n$ is the space of n-tuples of scalars, thus is $\mathscr{K} \times \mathscr{K} \times \cdots \times \mathscr{K}$. Moreover, the euclidean topology is the product topology since the following statements are equivalent: $x \to 0$ in the product topology, $x_i \to 0$ for $i = 1, 2, \ldots, n$, $\sum |x_i|^2 \to 0$.

2. Theorem Let X be an n-dimensional separated TVS, $n < \infty$. Then X is linearly homeomorphic with $\mathcal{K}^n$.

PROOF For $n = 0$, $X = \{0\}$ and the result is true. Proceeding by induction, let the dimension be $n > 0$. We first note that $X' = X^\#$, that is, every linear functional is continuous. [Let $f \in X^\#$. Then $f^\perp$ is $(n - 1)$ dimensional. By the induction hypothesis $f^\perp$ is linearly homeomorphic with $\mathcal{K}^{n-1}$; thus it is complete, hence closed, and the result follows by Theorem 4-5-10.] Now let $(b^1, b^2, \ldots, b^n)$ be a basis for X. Define $u: \mathcal{K}^n \to X$ by $u(a_1, a_2, \ldots, a_n) = \sum a_i b^i$. Then u is a continuous isomorphism onto X and it remains to prove that u^{-1} is continuous. Now for each i, let $P_i(a) = a_i$ for $a \in \mathcal{K}^n$; then $P_i \circ u^{-1}$ is a linear functional on X, hence, as just proved, is continuous. It follows from Theorem 1-6-11 that u^{-1} is continuous.

An example was given in Prob. 5-3-111 of two closed subspaces whose sum is not closed. The next result is a useful commentary. It is interesting, too, because the finite-dimensional subspace need not be closed.

3. Theorem The sum of a closed and a finite-dimensional subspace of a TVS must be closed.

PROOF Let A be a closed and B a finite-dimensional subspace of X. Then X/A is separated [Theorem 6-2-8] and so $q[B]$ is a closed subspace where $q: X \to X/A$ is the quotient map [Prob. 6-3-1]. Since q is continuous, $q^{-1}\{q[B]\}$ is closed; but this is $A + B$ [Prob. 6-2-3].

It is possible that a one-dimensional subspace of a separated TVS may be noncomplemented. This is the case, for example, if $X' = \{0\}$ (e.g., Prob. 2-3-119), since no maximal subspace can be closed [Theorem 4-5-10]. However, if there is a closed algebraic complement, it is indeed a complement.

4. Theorem Let X be a TVS. Then each closed subspace A of finite co-dimension is complemented. Any algebraic complement B will do.

PROOF Let $(b^1, b^2, \ldots, b^n)$ be a basis for B. Each $x \in X$ can be written uniquely as $a + \sum t_i b^i$, $a \in A$. Let $f_i(x) = t_i$. Then $f_i^\perp$ is closed for each i. [It is $A + \text{span} \{b^j : j \neq i\}$; by Theorem 6-3-3, this is closed.] Thus f_i is continuous [Theorem 4-5-10]. It follows that the projection on B is continuous since this is the map $x \to \sum f_i(x) b^i$ [Prob. 6-3-2]. By definition A, B are complemented.

5. Remark The reader may have become gradually aware that some theorems require separation as a hypothesis while others do not. A rule of thumb is that if some set in the theorem is assumed closed, or some map is assumed to have closed graph, then the spaces need not be assumed to be separated. Problem 6-3-110 shows a variation.

PROBLEMS

1 Show that every finite-dimensional subspace of a separated TVS is closed. [It is complete by Theorem 6-3-2.]

2 Let X, Y be TVSs, $f \in X'$, $y \in Y$. Define $g : X \to Y$ by $g(x) = f(x)y$. Show that g is continuous.

3 Show that every linear map from a separated finite-dimensional TVS X into a TVS Y is continuous. [Take $X = \mathscr{K}^n$ and apply Prob. 6-3-2.]

4 Show that every finite-dimensional separated TVS has a Schauder basis [Theorem 6-3-2].

---

101 Let X be a TVS, $f \in X^*$, and suppose f vanishes on some closed linear subspace of finite codimension. Show that $f \in X'$.

102 What property of a point x determines whether or not its span is linearly homeomorphic with $\mathscr{K}$?

103 Let X be a vector space and S a subspace of finite codimension. Show that there is exactly one vector topology for X such that $\overline{\{0\}} = S$.

104 Give R^2 two different topologies in each of which $\overline{\{0\}}$ is one dimensional.

105 State and prove a version of Theorem 6-3-2 for nonseparated spaces [Prob. 6-3-103].

106 Give an example of a noncontinuous linear map from a normed space into a finite-dimensional TVS.

107 Show that a separated TVS which has a countably infinite Hamel basis must be of first category [Prob. 6-3-1].

108 Show that no Fréchet space can have a countably infinite Hamel basis [Prob. 6-3-107].

109 Show that every finite-dimensional subspace S of a TVS such that $S \supset \overline{\{0\}}$ must be closed. This extends Prob. 6-3-1.

110 Let X be a finite-dimensional TVS, $f \in X^*$, $f = 0$ on $\overline{\{0\}}$. Show that $f \in X'$.

111 Let (X, T) be a finite-dimensional TVS with dual X'. Show that $T = \sigma X'$ [Prob. 6-3-103].

6-4 TOTALLY BOUNDED SETS

A much stronger condition than boundedness is the following: a set S in a TVS is called *totally bounded* if for each neighborhood U of 0 there is a finite set F with $S \subset F + U$. Thus such a set is one which can be covered by finitely many translates of any given neighborhood. ("Can be patrolled by finitely many arbitrarily near-sighted policemen" is an image which was suggested some years ago.) Many authors use *precompact* for totally bounded; this is because of Theorem 6-5-7 and the customary use of pre-P for a set whose completion has property P. It appears that total boundedness is near compactness. In contrast, it is far from boundedness, as is shown by Theorem 6-4-2 and the existence of locally bounded spaces.

1. Lemma Every compact set is totally bounded, and every totally bounded set is bounded.

PROOF Let S be compact and U an open neighborhood of 0. Then $\{s + U : s \in S\}$ is an open cover of S and so can be reduced to a finite cover. Next, let S be totally bounded and U a balanced neighborhood of 0. Let V be a balanced neighborhood of 0 with $V + V \subset U$. Now $S \subset F + V$ with F finite; since F is bounded, $F \subset nV$ for some $n \geq 1$. Thus $S \subset nV + V \subset nV + nV = n(V + V) \subset nU$. It follows that $tS \subset U$ for $|t| < 1/n$, so S is bounded.

A totally bounded set need not be compact, e.g., an interval (a, b) of reals. A bounded set need not be totally bounded, as is shown by Theorem 6-4-2 applied to a locally bounded space; the difference is pointed up by Prob. 6-4-109.

2. Theorem Let X be a separated TVS which has a totally bounded neighborhood of 0. Then X is finite-dimensional.

PROOF A rather mild form of the hypothesis is sufficient, namely, there exists $U \in \mathcal{N}$ with $U \subset F_0 + \frac{1}{2}U$ for some finite F_0. (See also Prob. 6-4-109.) Let F be the span of F_0, a finite-dimensional subspace of X; then $U \subset F + \frac{1}{2}U \subset F + \frac{1}{2}(F + \frac{1}{2}U) = F + \frac{1}{4}U$. Continuing, $U \subset F + 2^{-n}U$ for $n = 1, 2, \ldots$. Now $\{2^{-n}U\}$ is a local base of neighborhoods of 0. [For $V \in \mathcal{N}$, $V \supset 2^{-n}U$ for some n since U is bounded.] So $U \subset \cap\{F + 2^{-n}U\} = \overline{F}$ [Prob. 4-2-7] $= F$ [Prob. 6-3-1]. Since F is a vector subspace and includes a neighborhood of 0, $F = X$.

3. Theorem A continuous linear image of a totally bounded set is totally bounded.

PROOF Say $S \subset X$, $f : X \to Y$. Let $U \in \mathcal{N}(Y)$. Then $f^{-1}[U] \in \mathcal{N}(X)$, so $S \subset F + f^{-1}[U]$ for some finite F. Then $f[S] \subset f[F] + U$.

4. Definition Let X be a TVS, $S \subset X$, $A \subset X$. We say that S is small of order A, if $S - S \subset A$.

5. Theorem A set S is totally bounded if and only if for each $U \in \mathcal{N}$, S is a finite union of sets which are small of order U.

PROOF $\Rightarrow$: Let $V \in \mathcal{N}$ with $V - V \subset U$. Then $S \subset F + V = \cup\{x + V : x \in F\}$ and $(x + V) - (x + V) = V - V \subset U$, so each $x + V$ is small of order U. Now set $S_x = (x + V) \cap S$. Then $S = \cup S_x$, and each $S_x \subset x + V$ and hence is small of order U. $\Leftarrow$: Let $U \in \mathcal{N}$. Then $S = \cup\{S_i : i = 1, 2, \ldots, n\}$ where S_i is small of order U for each i. We may assume each S_i to be nonempty. Let $x_i \in S_i$ for each i, $F = (x_1, x_2, \ldots, x_n)$. Then $S_i - x_i \subset S_i - S_i \subset U$, so $S_i \subset x_i + U$. Then $S = \cup S_i \subset \cup\{x_i + U\} = F + U$. So S is totally bounded.

6. Theorem Let X be a vector space, Y a TVS, $f : X \to Y$ linear, and $S \subset X$. Then S is σf totally bounded if and only if $f[S]$ is totally bounded.

PROOF $\Rightarrow$: By Theorem 6-4-3. $\Leftarrow$: Let $U \in \mathcal{N}(X, \sigma f)$. There exists $V \in \mathcal{N}(Y)$ such that $U \supset f^{-1}[V]$. Now $f[S] = \cup A_i$ with each A_i small of order V [Theorem 6-4-5]. Let $S_i = f^{-1}[A_i]$. Then S_i is small of order U. [For

$x, y \in S_i, f(x - y) = f(x) - f(y) \in A_i - A_i \subset V$, so $x - y \in f^{-1}[V] \subset U$.] Also, $S \subset \cup S_i$. [If $x \in S$, $f(x) \in$ some A_i, so $x \in S_i$.] Thus S is totally bounded.

7. Theorem Let Φ be a collection of vector topologies for a vector space X, $S \subset X$. Then S is $\vee \Phi$ totally bounded if and only if it is T totally bounded for each $T \in \Phi$.

PROOF $\Rightarrow$: By Theorem 6-4-3, since $T \subset \vee \Phi$. $\Leftarrow$: Let $U \in \mathcal{N}(X, \vee \Phi)$. Then there exist $T_1, T_2, \ldots, T_n \in \Phi$, and $V_j \in \mathcal{N}(X, T_j)$ for each j, such that $U \supset \cap V_j$. For each $j = 1, 2, \ldots, n$, let $S = \cup \{S_{ij} : i = 1, 2, \ldots, m_j\}$ with each S_{ij} small of order V_j. For each j, $1 \leq j \leq n$, select an integer $i(j)$, $1 \leq i(j) \leq m_j$, and set $A = \cap \{S_{i(j)j} : j = 1, 2, \ldots, n\}$. Then A is small of order U. [Let $x, y \in A$. For each j; $x, y \in S_{i(j)j}$ and so $x - y \in V_j$. Thus $x - y \in \cap V_j \subset U$.] Now S is included in the union of all such A. [Let $x \in S$. For each j, $x \in \cup \{S_{ij} : i = 1, 2, \ldots, m_j\}$, so $x \in S_{ij}$ for some i. Call it $i(j)$.] The set of all such A is finite. Thus S is totally bounded by Theorem 6-4-5.

Theorems 6-4-6 and 6-4-7 have the two obvious corollaries, as follows.

8. Corollary Let X be a vector space and F a set of linear maps $f : X \to Y_f$ where each Y_f is a TVS. Then $S \subset (X, \sigma F)$ is totally bounded if and only if $f[S]$ is totally bounded for each $f \in F$.

9. Corollary Let X_α be a set of TVSs, $S \subset \pi X_\alpha$. Then S is totally bounded if and only if each of its projections is totally bounded. In particular, if $S_\alpha \subset X_\alpha$ is totally bounded for each α, πS_α is totally bounded in πX_α.

There is also an easy direct proof of Corollary 6-4-9. If each projection of S is totally bounded let $U \in \mathcal{N}(\pi X_\alpha)$ and we may assume $U = \pi U_\alpha$ with $U_\alpha \in \mathcal{N}(X_\alpha)$, $U_\alpha = X_\alpha$ for all but finitely many α [Prob. 1-6-105]. Now for each α, $P_\alpha[S] \subset F_\alpha + U_\alpha$ where each F_α is finite and $F_\alpha = \{0\}$ for all but finitely many α. Then $S \subset \pi F_\alpha + U$.

10. Definition A BTB space is a TVS in which every bounded set is totally bounded.

For example, R (the reals) is a BTB space.

The next three results are immediate from Theorems 6-4-7 to 6-4-9 and 4-4-5 and Probs. 4-4-12 and 4-4-13.

11. Theorem Let Φ be a collection of vector topologies for a vector space X such that (X, T) is a BTB space for each $T \in \Phi$. Then $(X, \vee \Phi)$ is a BTB space.

12. Theorem Let F be a collection of linear maps from a vector space X to various BTB spaces. Then $(X, \sigma F)$ is a BTB space.

13. Theorem A product of BTB spaces is a BTB space.

14. Example $\omega = \mathcal{K}^N$ is a BTB space. No locally bounded space, in particular no normed space, can be a BTB space unless it is finite dimensional, by Theorem 6-4-2.

PROBLEMS

1 Show that the following are totally bounded (TB):
   (a) any subset of a TB set,
   (b) any finite set,
   (c) the sum of two TB sets,
   (d) the union of two TB sets.

2 Let X be a TVS, Y a subspace, and $S \subset Y$. Show that S is totally bounded in Y if and only if it is totally bounded in X [Theorem 6-4-5].

3 Let U be a closed neighborhood of 0 and S a set which is small of order U. Show that $\bar{S}$ is small of order U [Theorem 4-2-11].

4 Show that the closure of a totally bounded set is totally bounded [Prob. 6-4-3 and Theorem 6-4-5].

5 Show that the balanced hull of a totally bounded set S is totally bounded. [It is the image of $\{z : |z| \le 1\} \times S$ under $(z, x) \to zx$; Corollary 6-4-9.]

6 Show that in the definition of totally bounded, the set F may be chosen to be a subset of S [Theorem 6-4-5].

---

101 Show that "separated" cannot be omitted from Theorem 6-4-2.

102 Let $S = \{n^{-1/2}\delta^n\} \subset l^{1/2}$. Show that S is totally bounded but its convex hull is not even bounded.

103 Is the set $\{\delta^n\}$ a totally bounded subset of c_0?

104 Show that a set is totally bounded if and only if all its countable subsets are totally bounded.

105 Let S be a nonconvergent Cauchy sequence. Show that S is closed and totally bounded but not compact.

106 A countably compact set must be totally bounded. [If S is not TB, $S \not\subset F + U$ for some U. Let $x^n \in S \setminus \cup \{x^i + U : i = 1, 2, \ldots, n - 1\}$. Then $\{x^n\}$ has no cluster point.]

107 A locally compact separated TVS must be finite-dimensional [Lemma 6-4-1 and Theorem 6-4-2].

108 Let X be a separated TVS for which there exists $U \in \mathcal{N}$, and $t, 0 < t < 1$, such that $U \subset F + tU$ for some finite set F. Show that X is finite dimensional.

109 Let D be the unit disc in an infinite-dimensional normed space. Show that D cannot be covered by finitely many translates of tD if $0 < t < 1$ [Prob. 6-4-108].

---

201 Let $\mathcal{B} \subset \mathcal{N}$ have the property that every $U \in \mathcal{N}$ includes a finite intersection of members of $\mathcal{B}$. Suppose a set S has the property that for every $U \in \mathcal{B}$, S is a finite union of sets which are small of order U. Show that S is totally bounded.

202 With $\mathcal{B}$ as in Prob. 6-4-201, suppose that for every $U \in \mathcal{B}$, $S \subset F + U$ for some finite F. Show that S need not be totally bounded. [An example with $S \subset R$ may be given.]

203 Must a quotient of a BTB space be a BTB space? (Compare Prob. 4-4-202.)

---

301 Let X be an infinite-dimensional Banach space, $\{x^n\} \subset X$, $\varepsilon_n \to 0$ and $B_n = \{x : \|x - x^n\| \le \varepsilon_n\}$. Then $X \ne \cup B_n$. [See [87].]

302 A linear functional is bounded if and only if it is bounded on each totally bounded set. If two vector topologies have the same totally bounded sets, they have the same bounded sets. [See [146], Lemma 2.3.]

6-5 COMPACT SETS

The main result (Theorem 6-5-7) shows that total boundedness falls short of compactness just by possible lack of completeness.

Filters are the handiest tool for proving compactness. They will be used to prove Theorem 6-5-7 and will not appear again until Sec. 13-3.

A filterbase $\mathscr{F}$ on a TVS is called *Cauchy* if for each $U \in \mathscr{N}$, $\mathscr{F}$ contains a set which is small of order U. It is called *convergent to* x if $\mathscr{F}' \supset N_x$, where $\mathscr{F}'$ is the filter generated by $\mathscr{F}$. In symbols, $\mathscr{F} \to x$. Thus $\mathscr{F} \to x$ if and only if for each $U \in \mathscr{N}$, $x + U \supset A$ for some $A \in \mathscr{F}$.

If S is a subset of a TVS X and $\mathscr{F}$ is a filterbase on S, we say $\mathscr{F}$ is Cauchy if it is Cauchy as a filterbase on X.

1. Lemma Let $\mathscr{F}$ be a Cauchy filterbase on a TVS and suppose $x \in \bar{A}$ for each $A \in \mathscr{F}$. Then $\mathscr{F} \to x$.

PROOF Let $U \in \mathscr{N}$ be closed. Choose $A \in \mathscr{F}$, small of order U. Then $A - x \subset \overline{A} - \overline{A} \subset \overline{A - A} \subset U$; hence $A \subset x + U$.

2. Lemma Let S be a complete subset of a TVS X. Then each Cauchy filterbase $\mathscr{F}$ on S converges to a point of S.

PROOF The converse, which is not needed, is left as a problem. For each $A \in \mathscr{F}$ choose $x_A \in A$. Then $x = \{x_A : \mathscr{F}\}$ is a net when we observe that $\mathscr{F}$ is directed under inclusion. Moreover, x is Cauchy. [Let $U \in \mathscr{N}(X)$. Let $B \in \mathscr{F}$ be small of order U. Then $A \geq B$, $A' \geq B$ means $A \subset B$, $A' \subset B$ and so implies that $x_A - x_{A'} \in A - A' \subset B - B \subset U$.] So $x \to s \in S$. But then $\mathscr{F} \to s$. [By Lemma 6-5-1 it is sufficient to show that $s \in \bar{A}$ for each $A \in \mathscr{F}$. Let $U \in \mathscr{N}$. There exists $B \in \mathscr{F}$ such that $B' \subset B$ implies $x_{B'} \in s + U$. Then $A \cap B$ includes some $C \in \mathscr{F}$ and $C \subset B$, hence $x_C \in s + U$. But $x_C \in C \subset A$, so A meets $s + U$. Thus $s \in \bar{A}$.]]

An *ultrafilter* is a maximal filter, that is, $\mathscr{F}$ is an ultrafilter means that whenever $\mathscr{F}'$ is a filter with $\mathscr{F}' \supset \mathscr{F}$ it follows that $\mathscr{F}' = \mathscr{F}$. The simplest type of ultrafilter is the discrete filter at a point x, that is, $\{S : x \in S\}$. (We are speaking of filters on a fixed set X.)

3. Lemma Let S be a set and $\mathscr{F}$ an ultrafilter on S. Let $A \subset S$. Then either $A \in \mathscr{F}$ or $S \backslash A \in \mathscr{F}$.

PROOF If $S \backslash A \notin \mathscr{F}$ then $S \backslash A$ includes no member of $\mathscr{F}$; hence A meets every member of $\mathscr{F}$. Thus the set of finite intersections of A with members of $\mathscr{F}$ is a filterbase $\mathscr{F}' \supset \mathscr{F}$. But then $\mathscr{F}' = \mathscr{F}$ and so $A \in \mathscr{F}$ [$A = A \cap S \in \mathscr{F}'$].

4. Lemma With $S, \mathscr{F}$ as in Lemma 6-5-3, suppose $S = \cup S_i$, a finite union. Then at least one $S_i \in \mathscr{F}$.

PROOF Since $\cap\{S\backslash S_i\}$ is empty, at least one $S\backslash S_i \notin \mathscr{F}$. By Lemma 6-5-3, $S_i \in \mathscr{F}$.

5. Lemma Let S be a totally bounded set in a TVS X and $\mathscr{F}$ an ultrafilter on S. Then $\mathscr{F}$ is Cauchy.

PROOF The converse, which is not needed, is left as a problem. Let $U \in \mathscr{N}(X)$. Then $S = \cup S_i$, a finite union, where each S_i is small of order U [Theorem 6-4-5]. The result follows by Lemma 6-5-4.

6. Lemma Let S be a set and $\sum$ a collection of subsets of S with the finite intersection property. Then $\sum$ is included in some ultrafilter on S.

PROOF The hypothesis means that every finite subcollection of $\sum$ has non-empty intersection. Let P be the collection of all filters $\mathscr{F}$ on S such that $\mathscr{F} \supset \sum$. Then P is not empty since the set of finite intersections of members of $\sum$ is a filterbase which generates a member of P. Partially order P by containment and let C be a maximal chain. Let $\mathscr{F}' = \cup\{\mathscr{F} : \mathscr{F} \in C\}$. Then $\mathscr{F}' \supset \sum$ and $\mathscr{F}'$ is a filter. [For example, if $A, B \in \mathscr{F}'$ then $A \in \mathscr{F}_1$, $B \in \mathscr{F}_2$ for some $\mathscr{F}_1, \mathscr{F}_2 \in C$. We may assume $\mathscr{F}_1 \supset \mathscr{F}_2$. Then $A \cap B \in \mathscr{F}_1 \subset \mathscr{F}'$.] It is also an ultrafilter since any larger filter could be adjoined to C, contradicting its maximality.

7. Theorem A set K in a TVS X is compact if and only if it is totally bounded and complete.

PROOF $\Rightarrow$: By Lemmas 6-1-18 and 6-4-1. $\Leftarrow$: Let C be a collection of closed sets in K with the finite intersection property. By Lemma 6-5-6, there is an ultrafilter $\mathscr{F}$ on K with $\mathscr{F} \supset C$. By Lemmas 6-5-5 and 6-5-2, $\mathscr{F} \to k \in K$. We shall show that $k \in \cap C$ and this will imply that K is compact. Let $S \in C$, $U \in \mathscr{N}(X)$. Then $k + U \supset A$ for some $A \in \mathscr{F}$. [$\mathscr{F}$ is only a filterbase on X.] Since, also, $S \in \mathscr{F}$ it follows that $k + U$ meets S. Thus $k \in \bar{S}$. But S is closed in K so $k \in S$.

Thus a complete BTB space (Definition 6-4-10) has the classical property that its closed bounded sets are compact. (See Probs. 6-5-111, 6-5-301, and 6-4-105.)

The behavior of compactness in sup, weak, and product topologies is spelled out in Probs. 6-5-1 and 6-5-103 to 6-5-105. Of course, a quotient map preserves compact sets, as does any continuous map, linear or not.

8. Lemma Let A, B be absolutely convex compact sets in a TVS X. Then the absolutely convex hull H of $A \cup B$ is compact.

PROOF Let $D = \{(z, w) \in \mathscr{K}^2 : |z| + |w| \leq 1\}$. Define $f : D \times A \times B \to X$ by $f(z, w, a, b) = za + wb$. The range of f is H. [$\supset$: If $h \in H$, $h = \sum t_i x^i$,

$\sum |t_i| \leq 1$, $x^i \in A \cup B$. Arrange $x^1, x^2, \ldots, x^k \in A$; $x^{k+1}, x^{k+2}, \ldots, x^n \in B$. We may assume that $n > k$ and that no $t_i = 0$ by taking $x^i = 0$ if necessary, and dropping those x^i for which $t_i = 0$. Let $z = \sum_{i=1}^{k} |t_i|$, $w = \sum_{i=k+1}^{n} |t_i|$. Let $a = \sum_{i=1}^{k} (t_i/z)x^i$, $b = \sum_{i=k+1}^{n} (t_i/w)x^i$. Then $a \in A$, $b \in B$, and $h = f(z, w, a, b)$.] Since f is continuous and $D \times A \times B$ is compact [Prob. 6-5-1], the result follows.

For the reader's information, we list a few results to be proved later: in a locally convex space the absolutely convex closure of a compact set must be totally bounded [Theorem 7-1-5] but need not be compact [Example 9-2-9].

Analysts are fond of saying that compact sets behave in many ways like finite sets. Lemma 6-5-11 is a useful example of this. Compare Theorem 6-3-3 in which finite-dimensional replaces compact.

9. Remark We are going to give an elementary proof of Lemma 6-5-11. If one knows subnets a very short proof runs as follows. Let $k + f \to x$. Then k has a subnet $k' \to a \in K$; also, $x - f' \to a$ and hence $x - a = \lim f' \in F$. So $x \in K + F$.

10. Lemma Let F, K be disjoint sets in a TVS X with F closed, K compact. Then there exists $U \in \mathcal{N}(X)$ with $F + U \not\phi K$.

PROOF For each $k \in K$ choose open $U_k \in \mathcal{N}(X)$ with $k - U_k \not\phi F$. Reduce the open cover $\{k - U_k : k \in K\}$ to a finite cover $\{k(i) - U_{k(i)} : i = 1, 2, \ldots, n\}$ of K. Let $U = \cap U_{k(i)}$. Then $K - U \not\phi F$.

11. Lemma The sum of a compact set K and a closed set F in a TVS X is closed.

PROOF Let $x \notin K + F$. Then $x - F$ is closed and does not meet K. By Lemma 6-5-10, $x - F + U \not\phi K$ for some $U \in \mathcal{N}(X)$. Thus $x + U \not\phi K + F$.

PROBLEMS

1 Let X_α be a collection of TVSs, $A_\alpha \subset X_\alpha$ for each α. Show that πA_α is compact if and only if each A_α is compact. (This is a special case of Tychonoff's theorem.) [See Theorem 6-5-7, Corollary 6-4-9, and Theorem 6-1-7.]

2 Show that the balanced hull of a compact set is compact [like Prob. 6-4-5].

3 Prove Lemma 6-5-8 for any finite collection. [Imitate the proof using $\mathcal{K}^n$.]

---

101 If every Cauchy filter on a set S in a TVS is convergent to a point in S, show that S is complete [Prob. 4-3-103].

102 If every ultrafilter on a set S in a TVS is Cauchy, show that S is totally bounded.

103 Let X be a vector space, Y a TVS, $f: X \to Y$ linear and onto, $S \subset X$. Show that S is σf compact if and only if $f[S]$ is compact [Prob. 6-1-108 and Theorem 6-4-6].

104 Let p, q be noncomparable norms for a vector space X [Example 5-2-9]. Let $x_n \to a \in (X, p)$, $x_n \to b$ in (X, q), $a \neq b$. Let $S = (a, b, x_1, x_2, \ldots)$. Show that S is compact in (X, p), (X, q) but not in $(X, p + q) = (X, p \vee q)$.

105 State and prove an analogue of Prob. 6-1-110 for compactness [Theorems 6-5-7 and 6-4-7].

106 Show that the unit disc in l^∞ is a compact subset of ω [Prob. 6-5-1].

107 A TVS is called N *complete* (or *von Neumann complete*) if each closed totally bounded set is compact. Show that X is N complete if and only if each closed totally bounded set is complete.

108 Show that a boundedly complete space is N complete and an N-complete space is sequentially complete.

109 State and prove a result about N completeness in F-linked topologies like that of Theorem 6-1-16.

110 Show that a Cauchy filter which contains a complete set must be convergent.

111 A boundedly complete BTB space is called a *semi-Montel space*. (See Prob. 6-5-301.) Discuss preservation of this property under sup, weak, product, and quotient topologies. Show that a TVS is semi-Montel if and only if closed bounded sets are compact.

---

301 (Montel's theorem). Let 0 be an open set of complex numbers and X the analytic functions on 0 endowed with the topology of compact convergence. Then X is a Fréchet space [Prob. 5-1-201] and a BTB space. [See [138], Theorem 14.6 or [58], Theorem 3.9.] Hence closed bounded sets in X are compact.

LOCAL CONVEXITY

A locally convex (lc) space is a TVS such that each neighborhood of 0 includes a convex neighborhood of 0. The initial motivation for studying lc spaces is the "largeness" of the dual space. This permits a fruitful duality theory, i.e., study of X by examination of X'. In recent years spaces with "large" duals have been studied and many results have thus been extended to spaces which, while not necessarily locally convex, have this property. We shall develop the convex theory and give indications of the more general theory in the problems. (See Non-lc in the Index.)

7-1 LOCALLY CONVEX SPACE

As we have said, a lc (locally convex) space is a TVS which has a local base of convex neighborhoods of 0. For example, every normed space is locally convex. Non-locally convex spaces are abundant. The next example shows one and the Index may be consulted for others.

1. Example $l^{1/2}$ *is not locally convex.* (This also follows from Probs. 7-1-1 and 4-4-105.) Let $U = \{x : \| x \| \le 1\}$. We shall show that U includes no convex neighborhood of 0. To this end let $V \in \mathscr{N}$ be convex. For some $\varepsilon > 0$, $V \supset \{x : \| x \| \le \varepsilon\}$. In particular, $\varepsilon^2 \delta^k \in V$ for each k. Choose an integer $n > 1/\varepsilon^2$ and set $x = (1/n) \sum_{k=1}^{n} \varepsilon^2 \delta^k = (\varepsilon^2/n, \varepsilon^2/n, \ldots, \varepsilon^2/n, 0, 0, \ldots)$. Then $x \in V$ and $\| x \| = \sum_{k=1}^{n} (\varepsilon^2/n)^{1/2} = n^{1/2} \varepsilon > 1$. So $V \not\subset U$.

2. Theorem A lc space has a local base of neighborhoods of 0 which are barrels (Definition 3-3-1).

PROOF Let $U \in \mathcal{N}$ be closed. There exists absolutely convex $V \in \mathcal{N}$ with $V \subset U$. [See the proof of Theorem 4-5-2.] Then $\bar{V} \subset U$ and $\bar{V}$ is also absolutely convex [Theorem 4-2-12].

In defining lc topologies the procedure of Theorem 4-3-5 may be simplified.

3. Lemma Let X be a vector space and $\mathcal{B}$ a filterbase of convex sets such that for each $U \in \mathcal{B}$, $\frac{1}{2}U$ includes a member of $\mathcal{B}$. Then $\mathcal{B}$ is additive.

PROOF If $\frac{1}{2}U \supset V$, then $V + V \subset \frac{1}{2}U + \frac{1}{2}U = U$ since U is convex.

4. Theorem Let X be a vector space and $\mathcal{B}$ an additive filterbase (see Lemma 7-1-3) of absolutely convex absorbing sets. Then there is a unique lc topology for X for which $\mathcal{B}$ is a local base of neighborhoods of 0.

PROOF This follows from Theorem 4-3-5.

Locally convex spaces have important permanence properties, as shown in Theorem 7-1-5 and Prob. 7-1-1. These are lacking in general [Probs. 4-4-105 and 6-4-102].

5. Theorem Let B be a totally bounded subset of a lc space. Then B_1, the absolutely convex closure of B is also totally bounded.

PROOF Since closure preserves total boundedness [Prob. 6-4-4], it is sufficient to prove this with B_1 taken to be the absolutely convex hull. Let $U \in \mathcal{N}$. Choose absolutely convex $V \in \mathcal{N}$ with $V + V \subset U$. Then $B \subset F + V$ with $F = (f^1, f^2, \ldots, f^n)$. Since F is bounded, $F \subset mV$ for some integer m. For $t \in \mathcal{K}^n$ define $\|t\|_1 = \sum |t_i|$ (this is equivalent to the euclidean norm, by Theorem 6-3-2) and let D be the unit disc in $\mathcal{K}^n$. Since D is totally bounded, it contains a finite subset A such that for each $t \in D$, $\|t - a\|_1 \leq 1/m$ for some $a \in A$. Let $G = \{\sum_{i=1}^{n} a_i f^i : a \in A\}$, a finite set. The proof is concluded by showing that $B_1 \subset G + U$. To this end, let $x \in B_1$. So $x = \sum \{r_i b^i : i = 1, 2, \ldots, s\}$ with $b^i \in B$, $\sum |r_i| \leq 1$. Since $B \subset F + V$, we have $b^i = f^{j(i)} + v^i$ for each i, $v^i \in V$, so $x = \sum r_i f^{j(i)} + \sum r_i v^i = w + v$, say. Now $v \in V$ since V is absolutely convex, and so we have only to prove that $w \in G + V$. We write $w = \sum q_i f^i$ and it is fairly easy to see that each q_i is 0 or is a sum of the r_j and each r_j appears exactly once. Thus $\sum |q_i| \leq \sum |r_i| \leq 1$ so $q \in D$. Choose $a \in A$ with $\|q - a\|_1 \leq 1/m$. Then $w = \sum a_i f^i + \sum (q_i - a_i) f^i = g + v_1$, say, with $g \in G$. Finally, $v_1 \in V$ since $v_1 = \sum m(q_i - a_i)[f^i/m]$, each $f^i/m \in V$, $\sum |m(q_i - a_i)| = m\|q - a\|_1 \leq 1$, and V is absolutely convex.

PROBLEMS

1 Show that in a lc space, the absolutely convex hull of a bounded set is bounded. ⟦Use an absolutely convex neighborhood of 0.⟧

2 Show that a quotient of a lc space is lc ⟦Lemma 6-2-1⟧.

3 Show that the interior of a convex set is convex. ⟦For $0 < t < 1$, $tA^i + (1 - t)A^i$ is an open set included in A and hence is included in A^i.⟧

4 Show that every neighborhood of 0 in a lc space includes an absolutely convex open neighborhood of 0 ⟦Theorem 7-1-2 and Probs. 7-1-3 and 4-2-10⟧.

5 Let X be a vector space. Let $\mathscr{B}$ be the collection of absolutely convex absorbing sets in X. Show that it satisfies Theorem 7-1-4 and that the resulting topology is the largest lc topology for X. (Compare Prob. 4-3-101.)

6 Show that the largest lc topology is separated ⟦Prob. 2-2-8⟧.

7 Let a vector space X have the largest lc topology. Show that all seminorms are continuous and that every linear map from X to a lc space is continuous; in particular, $X' = X^*$. Deduce that every bounded set is finite dimensional ⟦Prob. 4-4-15⟧.

8 Show that an infinite-dimensional vector space with the largest lc topology cannot be metrizable ⟦Prob. 7-1-7 and Example 3-3-14⟧.

9 Let a vector space X have the largest lc topology. Show that every subspace is closed. ⟦See Prob. 7-1-7, Theorem 4-5-10, and Prob. 1-5-20. Alternatively, note that for each $f \in X^*$, $f = 0$ on S implies $f = 0$ on $\bar{S}$ since f is continuous.⟧

10 An infinite-dimensional vector space with the largest lc topology must be of first category. ⟦By Prob. 7-1-9, since it is the union of a sequence of subspaces as in Prob. 3-1-4.⟧

---

101 Prove the converse of Theorem 7-1-2.

102 Must every convex set be closed in the largest lc topology?

103 Given a TVS (X, T), let T_c be the vector topology generated by the set of absolutely convex neighborhoods of 0. Verify its legitimacy by Theorem 7-1-4; show that T_c is the largest lc topology smaller than T and that $(X, T_c)' = (X, T)'$.

104 Let (X, T) be a TVS and $\mathscr{B}$ a local base for $\mathcal{N}(T)$. Show that the set of absolutely convex hulls of the members of $\mathscr{B}$ is a local base for $\mathcal{N}(T_c)$ ⟦Prob. 7-1-103⟧.

105 If (X, T) is a paranormed space, so also is (X, T_c) ⟦Prob. 7-1-104 and Theorem 4-5-5⟧.

106 Let X be a vector space with an uncountable Hamel basis. Show that the largest vector topology for X is not lc ⟦Prob. 2-1-202⟧.

107 Show that the largest lc topology on any vector space is boundedly complete ⟦Prob. 7-1-7 and Theorem 6-3-2⟧. (It is actually complete, see Example 13-2-13.)

108 Let X be a non-lc TVS. Show that the set of nonconvex neighborhoods of 0 is a local base of neighborhoods of 0.

109 Let X be a lc space and $\{x^n\}$ a sequence in X with $x^n \to 0$. Show that $(1/n) \sum_{k=1}^{n} x^k \to 0$.

110 Let X be a vector space with σX^*. Show that every seminorm on X is sequentially continuous ⟦Prob. 4-4-15 and Theorem 6-3-2⟧.

111 Let X be infinite dimensional in Prob. 7-1-110. Show that every norm on X is sequentially continuous but not continuous ⟦Theorem 4-1-12⟧.

112 Show that $\sigma(X, X^*)$ and $\tau(X, X^*)$ have the same convergent sequences ⟦Prob. 7-1-110⟧.

---

201 Let X be a lc metric space which has a countable cobase for its bounded sets. Show that X is a normed space. Thus the equivalences of Prob. 4-5-105 hold for lc metric space. ⟦See [88], 29.1 (2).⟧

7-2 SEMINORMS

The topology of a lc space is given by a set P of seminorms (Theorem 7-2-2); thus its properties ought to be reflected by properties of the seminorms. This is the content of Theorems 7-2-4 to 7-2-7 and 7-2-14 and Probs. 7-2-6, 7-2-8, 7-2-102, and 7-2-106, and the obvious fact that $x \to 0$ if and only if $p(x) \to 0$ for each $p \in P$.

1. Lemma Let p be a seminorm on a TVS. Then p is continuous if and only if $\{x : p(x) \leq 1\} \in \mathcal{N}$.

PROOF $\Leftarrow$: Let $U_r = \{x : p(x) \leq r\}$. Suppose $x \to 0$. For any $\varepsilon > 0$, $U_\varepsilon = \varepsilon U_1 \in \mathcal{N}$; thus $x \in U_\varepsilon$ eventually, that is, $p(x) \leq \varepsilon$ eventually. This shows that $p(x) \to 0$, hence that p is continuous at 0. Continuity elsewhere follows from $|p(x) - p(y)| \leq p(x - y)$.

2. Theorem Let (X, T) be a lc space. There exists a set P of seminorms such that $T = \sigma P$.

PROOF Let P be the set of continuous seminorms. It is not empty since $0 \in P$. First $T \supset \sigma P$. $[\![$If $x \to 0$ in T, $p(x) \to 0$ for each $p \in P$; thus, by Example 4-1-8, $x \to 0$ in σP.$]\!]$ Also $T \subset \sigma P$. $[\![$Let $x \to 0$ in σP. Let $U \in \mathcal{N}$ be absolutely convex and let p be the gauge of U; it is continuous by Lemma 7-2-1. Now $p(x) < 1$ eventually, and so $x \in U$ eventually.$]\!]$

3. Definition We write (X, P) for any lc space whose topology is σP, P being a set of seminorms.

We know that P may be taken to be the set of all continuous seminorms, but when (X, P) is written it is assumed that P is some fixed set of seminorms, not necessarily all the continuous ones, such that σP is the topology of X.

We pause to show that σP is always locally convex. This, with Theorem 7-2-2 justifies Definition 7-2-3.

4. Theorem Let P be a family of seminorms on a vector space X. Then for each $U \in \mathcal{N}(X, \sigma P)$ there exist $\varepsilon > 0$ and $p_1, p_2, \ldots, p_n \in P$ such that $U \supset \cap_{i=1}^{n} \{x : p_i(x) < \varepsilon\}$.

PROOF We know that $\sigma P = \vee \{\sigma p : p \in P\}$ and the sets $\{x : p(x) < \varepsilon\}$ for various $\varepsilon > 0$ form a base of neighborhoods of 0 for $(X, \sigma P)$. As in Theorem 1-6-8, we take the set of finite intersections of such sets as a base for $\mathcal{N}(\sigma P)$. A typical one would be $\cap \{x : p_i(x) < \varepsilon_i\}$. This includes $\cap \{x : p_i(x) < \varepsilon\}$ if $\varepsilon = \min \varepsilon_i$.

5. Theorem Let (X, P) be a lc space and $f \in X^{\#}$. Then $f \in X'$ if and only if there exist M and $p_1, p_2, \ldots, p_n \in P$ such that $|f(x)| \leq M \sum p_i(x)$ for all x. If P

is the set of all continuous seminorms, this condition may be written $|f| \leq p$ for some $p \in P$.

PROOF $\Rightarrow$: The set $\{x : |f(x)| \leq 1\} \in \mathcal{N}$ so it includes a finite intersection $\cap \{x : p_i(x) < \varepsilon\}$ by Theorem 7-2-4. Let $p = \sum p_i$. Then $p(x) < \varepsilon$ implies $|f(x)| \leq 1$ and so $|f(x)| \leq (1/\varepsilon) \sum p_i(x)$ for all x [Prob. 2-2-4]. $\Leftarrow$: f is continuous at 0 and hence everywhere.

6. Theorem A set S in a lc space (X, P) is bounded if and only if $p[S]$ is bounded for each $p \in P$.

PROOF $\Rightarrow$: For each p, $U = \{x : p(x) < 1\} \in \mathcal{N}$; hence $mU \supset S$ for some m. But then $p(x) < m$ for all $x \in S$. $\Leftarrow$: Let $x_n \in S$. Then for each p, $p(x_n/n) = p(x_n)/n \to 0$. Thus $x_n/n \to 0$ in σP and so S is bounded, by Theorem 4-4-1.

7. Theorem The lc space (X, P) is separated if and only if P is total.

PROOF $\Leftarrow$: Let $y \neq 0$. There exists p with $p(y) > 0$. Then $\{x : p(x) < \frac{1}{2}p(y)\}$ is a neighborhood of 0 not containing y. $\Rightarrow$: Let P be not total. There exists $x \neq 0$ with $p(x) = 0$ for all p. By Theorem 7-2-4, $x \in U$ for all $U \in \mathcal{N}$, so the topology is not separated.

8. Corollary Let X be a lc separated space. Then X' is total over X.

PROOF Let $0 \neq x \in X$. By Theorem 7-2-7, P is total so there exists $p \in P$ with $p(x) \neq 0$. Let $Y = (X, p)$, a seminormed space. By the Hahn–Banach Theorem 2-3-9, there exists $f \in Y'$ with $f(x) \neq 0$. Since $\sigma P \supset \sigma p$, it follows that $f \in X'$.

9. Example The Fréchet spaces with trivial duals (Probs. 2-3-119 and 2-3-120) are not locally convex. This follows from Corollary 7-2-8 and illustrates the role of local convexity.

10. Remark *Let S be a vector subspace of a lc space (X, P). Then $S = (S, Q)$ where $Q = \{p | S : p \in P\}$.* [These statements are equivalent for a net x in $S : x \to 0$ in S, $x \to 0$ in X; $p(x) \to 0$ for all $p \in P$, $q(x) \to 0$ for all $q \in Q$.]

A corollary is the Hahn–Banach extension theorem for lc spaces. More precise forms of Theorem 7-2-11 and Corollary 7-2-12 for seminormed spaces were given in Theorems 2-3-7 and 2-3-9.

11. Theorem Let (X, P) be a lc space and $f \in S'$, where S is a vector subspace. Then f may be extended to $F \in X'$.

PROOF Let Q be as in Remark 7-2-10. By Theorem 7-2-5, $|f(x)| \leq M \sum q_i(x)$ for all $x \in S$ where $q_1, q_2, \ldots, q_n \in Q$. Now $q_i = p_i | S$ with $p_i \in P$; let $p(x) =$

$M \sum p_i(x)$ for $x \in X$. By the Hahn–Banach Theorem 2-2-2, f can be extended to $F \in X^*$ with $|F(x)| \leq p(x)$ for $x \in X$. By Theorem 7-2-5, F is continuous.

12. Corollary Let X be a lc space, S a vector subspace and $x \in X\backslash \bar{S}$. Then there exists $f \in X'$ with $f(x) = 1$, $f = 0$ on S.

PROOF Define $f(s + tx) = t$ for $s \in S$, $t \in \mathcal{K}$. This defines $f \in (S + [x])'$ with the required properties. $[\![f$ is continuous because $f^\perp = S$ is not dense in $S + [x]$. This uses Prob. 4-2-5 and Theorem 4-5-10.$]\!]$ Now Theorem 7-2-11 extends f to all of X.

13. Corollary A set S in a lc space X is fundamental if and only if $f \in X'$, $f = 0$ on S implies $f = 0$.

PROOF This is immediate from Corollary 7-2-12.

14. Theorem Let (Y, P) be a lc space and X a TVS. A linear map $f: X \to Y$ is continuous if and only if $p \circ f$ is continuous for each $p \in P$.

PROOF $\Rightarrow$: $p \circ f$ is the composition of continuous maps. $\Leftarrow$: Let $x \to 0$ in X. Then $p[f(x)] \to 0$ for each $p \in P$, so $f(x) \to 0$ [Example 4-1-8]. Thus f is continuous at 0 and hence everywhere.

In the lc setting, the dual of $(X, \vee T_a)$ may be found. This is the content of Theorem 7-2-16. The result fails for vector topologies in general (Prob. 7-2-301). It is extremely easy to see that $[(X, T_1) \times (X, T_2)]' = (X, T_1)' + (X, T_2)'$ since $f(x, y) = f[(x, 0) + (0, y)] = f(x, 0) + f(0, y) = g_1(x) + g_2(y)$ with $g_i \in (X, T_i)'$. The corresponding result is false for $T_1 \vee T_2$ [Prob. 7-2-301], but is true in the lc case. We begin with an application of the Hahn–Banach theorem.

15. Lemma Let p, q be seminorms on a vector space X. Let $f \in X^*$ satisfy $|f(x)| \leq p(x) + q(x)$ for all x. Then there exist $g, h \in X^*$ with $|g(x)| \leq p(x)$, $|h(x)| \leq q(x)$, $f = g + h$.

PROOF Let $Y = X \times X$, $r(x, y) = p(x) + q(y)$; r is a seminorm on Y. Define u on the diagonal $D = \{(x, x): x \in X\}$ by $u(x, x) = f(x)$. Then $|u(x, x)| \leq r(x, x)$ so, by the Hahn–Banach Theorem 2-2-2, we can extend u to be defined on Y with $|u(x, y)| \leq r(x, y)$. Then $f(x) = u(x, 0) + u(0, x)$, $|u(x, 0)| \leq r(x, 0) = p(x)$, $|u(0, x)| \leq r(0, x) = q(x)$.

16. Theorem Let Φ be a collection of lc topologies for a vector space X. Then $f \in (X, \vee \Phi)'$ if and only if there exist $T_1, T_2, \ldots, T_n \in \Phi$; $g_1, g_2, \ldots, g_n \in X^*$ with each $g_i \in (X, T_i)'$ and $f = \sum g_i$.

PROOF $\Leftarrow$: Each $g_i \in (X, \vee \Phi)'$ since $\vee \Phi \supset T_i$. $\Rightarrow$: Let P_T be the set of all continuous seminorms on (X, T) for each $T \in \Phi$. Then $(X, T) = (X, P_T)$ for each $T \in \Phi$, by Theorem 7-2-2; moreover, $\vee \Phi = \sigma(\cup\{P_T : T \in \Phi\})$. (See Prob. 7-2-2.) By Theorem 7-2-5, there exist $p_1, p_2, \ldots, p_n$ selected from $\cup P_T$ and M such that $|f(x)| \leq M \sum p_i(x)$ for all x. The result follows from Lemma 7-2-15.

This result may be expressed as follows: the dual of the sup is the direct sum (Sec. 13-2) of the separate duals. The dual of a space with a special sort of weak topology is given in Theorems 8-1-7 and 8-2-12; that of a product is given in Theorem 13-2-15.

PROBLEMS

1 If P is a set of seminorms on a vector space, show that σP is locally convex [Theorem 7-2-4]. Hence ω, $C(H)$, etc., are locally convex.

2 Let P_α be a set of seminorms on a vector space X for each $\alpha \in A$ (A is some set). Show that $\sigma(\cup P_\alpha) = \vee \{\sigma P_\alpha : \alpha \in A\}$ [Theorem 1-6-8 and Example 4-1-8].

3 Let (T_α) be a set of lc topologies for a vector space. Show that $\vee T_\alpha$ is locally convex [Probs. 7-2-1 and 7-2-2].

4 Show that the weak topology by linear maps to lc spaces is locally convex [Prob. 7-2-3].

5 Show that a product of lc spaces is locally convex [Prob. 7-2-4].

6 A lc space (X, P) is pseudometrizable if and only if P has a countable subset Q with $\sigma Q = \sigma P$. [See Theorem 2-1-2. If $\{U_n\}$ is a local base for $\mathcal{N}(\sigma P)$, apply Theorem 7-2-4 to each U_n.]

7 A lc space is a seminormed space if and only if it is locally bounded [Theorem 4-5-2].

8 A lc space (X, P) is a seminormed space if and only if P has a finite subset Q with $\sigma Q = \sigma P$. [Apply Theorem 7-2-4 to the unit disc. This can also be deduced from Prob. 7-2-105.]

9 If X is indiscrete, show that P (Theorem 7-2-2) has only one member.

10 With X, S, x as in Corollary 7-2-12 and $g \in S'$, g can be extended to $G \in X'$ with $G(x) = t$, where t is any preassigned scalar. [Extend g in any way, choose f as in Corollary 7-2-12, and let $G = g + [t - g(x)]f$.]

---

101 On c_0 let $p(x) = \|x\|_\infty$, $p_n(x) = |x_n|$. Is $\sigma p = \sigma\{p_n\}$?

102 Give a necessary and sufficient condition in terms of P for a net to be Cauchy in (X, P).

103 Let P, Q be sets of seminorms on a vector space. Show that $\sigma P \supset \sigma Q$ if and only if each $q \in Q$ is σP continuous.

104 Does Theorem 7-2-6 hold for total boundedness?

105 Let q be a seminorm on a lc space (X, P). Show that q is continuous if and only if there exist $M, p_1, p_2, \ldots, p_n \in P$ such that $q \leq M \sum p_i$. [Imitate the proof of Theorem 7-2-5.]

106 Let $S \subset (X, P)$. Show that $x \in \bar{S}$ if and only if for all $\varepsilon > 0$ and $p_1, p_2, \ldots, p_n \in P$, there exists $s \in S$ with $\sum p_i(x - s) < \varepsilon$ [Theorem 7-2-4].

107 Show that the largest lc topology (Prob. 7-1-5) is σP where P is the set of all seminorms.

108 Show that Theorem 7-2-11 fails without lc, even if X is separated. [Take $X' = 0$, dim $S = 1$.]

109 Criticize this proof of Theorem 4-5-10 for a lc space. Let $f(a) = 1$. Let $g \in X'$ with $g = 0$ on $f^\perp$, $g(a) = 1$ [Corollary 7-2-12]. By Theorem 1-5-1, $f = g$. Thus $f \in X'$.

110 Use Prob. 7-2-3 to give alternate definitions for T_c (Prob. 7-1-103) and the largest lc topology.

111 Suppose X is separated in Theorem 7-2-2. Can we make P a set of norms? [Prob. 4-4-111].

112 Every lc space is linearly homeomorphic into a product of seminormed spaces. [Map (X, P) into $\pi\{(X, p) : p \in P\}$ by the diagonal map $(ux)_p = x$ for each p.]

113 Every separated lc space is linearly homeomorphic into a product of Banach spaces (See Prob.

7-2-304). ⟦In Prob. 7-2-112 continue into $(X, p)/p^{\perp}$. Use Theorem 7-2-7. Finally, take the completion of each space, Remark 3-2-13.⟧

114 Let $H = (0, 1)$, $X = C(H)$ (Prob. 4-1-105). Show that X is not a normed space. ⟦Use Prob. 7-2-8.⟧

115 Let A be the set of analytic functions given in Prob. 5-1-102. Show that A is not a normed space ⟦Prob. 7-2-8⟧.

116 For $y \in l^{\infty}$, let $p_y(x) = |\sum y_n x_n| + \|x\|_{\infty}$. Let l have the topology $T = \sigma\{p_y : y \in l^{\infty}\}$. Show that T lies strictly between the weak and norm topologies of l.

117 Give an example of a non-lc space which obeys Corollary 7-2-12 ⟦Probs. 7-1-106 and 7-1-103⟧. But see Prob. 7-2-302.

118 Let (X, T) be a lc space and U an absolutely convex absorbing set. Show that there exists a lc topology T_1 for X such that $T_1 \supset T$, $U \in \mathcal{N}(T_1)$ and T_1 is the smallest lc topology with these two properties. ⟦Adjoin the gauge of U to P where $T = \sigma P$.⟧

119 Take (X, P), (Y, Q) to be locally convex in Lemma 5-5-10. Show that the topology of S is $\sigma(P \cup Q \circ u)$ where $Q \circ u = \{q \circ u : q \in Q\}$ ⟦Prob. 7-2-2⟧.

120 Let $f \in S'$ (Prob. 7-2-119). Show that $f = F + G \circ u$, $F \in X'$, $G \in Y'$. ⟦$f = F + g$, F, $g \in S'$, $|F| \leq p$, $|g| \leq q \circ u$, by Theorem 7-2-5 and Lemma 7-2-15. Define G on $u[X] \cap Y$ by $G(y) = g(x)$ if $y = u(x)$.⟧

121 Let A be a matrix and $f \in c'_A$ (Example 5-5-15). Show that $f(x) = \alpha \lim Ax + t(Ax) + \beta x$ with $t \in l$, where $t(Ax) = \sum t_n(Ax)_n$, $\beta x = \sum \beta_n x_n$. If A is row finite, $\beta \in \varphi$; if $\|A\| < \infty$, $\beta \in l$; and if A is reversible, $\beta = 0$. ⟦In Prob. 7-2-120, take $X = \omega_A$, $Y = c$. Use Probs. 5-5-114 and 2-3-3. The last part is by Remark 5-5-17.⟧

122 In Prob. 7-2-121, let A be conservative. Show that $\chi(f) = \alpha\chi(A)$ where $\chi(f) = y_1$ in Prob. 2-3-3 ⟦direct computation using $\sum\sum |t_n a_{nr}| \leq \|A\| \|t\|_1$⟧. Hence, if A is coregular α is uniquely determined by $f|_c$, in particular $\alpha = 0$ if $f = 0$ on c.

123 In Prob. 7-2-121, let A be conservative. Show that $f(x) = \alpha \lim Ax + \gamma x$ for $x \in c_A \cap l^{\infty}$. ⟦See the hint to Prob. 7-2-122.⟧ Deduce that if A is coregular, c is dense in $c_A \cap l^{\infty}$ (in c_A), and if A is conull, c_0 is dense in c ⟦Prob. 7-2-122⟧.

124 Let A be coregular and reversible. Show that c is dense in c_A if and only if A^t (the transposed matrix) is one to one on l. ⟦$\Rightarrow$: If $tA = 0$, $f(x) = t(Ax) = 0$ for $x \in c$ as in Prob. 7-2-123. $\Leftarrow$: $f = 0$ on c implies by Probs. 7-2-121 and 7-2-122 that $0 = f(x) = (tA)x$ for $x \in c$.⟧ (The two properties are not equivalent in general, depending crucially on $\beta = 0$. They are called, respectively, *perfect* and *type M*, in honor of S. Mazur.)

125 Let $\alpha \in \mathcal{K}$, $t \in l$; let C be the matrix whose nth row is $\sum_{i=1}^{n-1} t_i \delta^i + \alpha \delta^n$. Show that C is conservative and if $B = CA$, with A a matrix, then $c_B \supset c_A$ and $\lim_B x = \alpha \lim_A x + t(Ax)$ for $x \in c_A$.

---

301 There exists a vector space X with vector topologies T, T_1 such that $(X, T)' = (X, T_1)' = \{0\}$, but $(X, T \vee T_1)' \neq \{0\}$. ⟦See [84], p. 243.⟧

302 If a Fréchet space X obeys Corollary 7-2-12 it must be locally convex. ⟦See [77], corollary 5.3.⟧ Compare Prob. 7-2-117.

303 The converse of the last part of Prob. 7-2-123 is false: A may be coregular and have c_0 dense in c. ⟦See [152], p. 657.⟧

304 The Banach spaces in Prob. 7-2-113 cannot be taken to be reflexive even if X is a reflexive Fréchet space. ⟦See [39], Theorem 4.2.⟧

7-3 SEPARATION AND SUPPORT

The role of local convexity in duality (the study of a space by means of functions defined on it) is illustrated by the fact that a point and a distant convex set in

R^2 can be separated by a line, but a point and a nonconvex set cannot, in general. For example, any line through the center of a circle must pass through the circumference; thus the center cannot be distinguished from the circumference by means of linear functionals. Duality must also be useless for those spaces which have a meager supply of continuous linear functionals, for example, $\mathcal{M}$ and L^p, $0 < p < 1$ (Probs. 2-3-119 and 2-3-120). Local convexity turns out to be a sufficient assumption. We develop the tools in this section; duality is the subject of Chap. 8. (The reader might wonder about an indiscrete space which, although locally convex, has a trivial dual. The fact is that the dual faithfully reflects the properties of the space which are also trivial.)

We have already seen an analytic proof that points can be distinguished by members of the dual (Corollary 7-2-8). We shall now use geometric language, but the succeeding results can also be derived analytically. (See Prob. 7-3-104.)

Although topological assumptions are made, the discussion can be given a purely linear setting by using assumptions equivalent to openness in the largest lc topology (Prob. 7-1-5).

We first show that a certain type of convex set can be supported, i.e., a maximal subspace can be found which is disjoint from it. In the real case this implies that the set lies entirely on one side of the subspace. It is natural to begin with a way of recognizing maximal subspaces; this one is rather obvious.

1. Lemma Let X be a TVS and M a vector subspace which disconnects X, that is, $X \backslash M = G_1 \cup G_2$ with G_1, G_2 disjoint, open, and nonempty. Then M is a maximal subspace.

PROOF Let $y \notin M$, say $y \in G_1$. Let $M_1 = M + [y]$. We have to prove $M_1 = X$ and it is sufficient to show $M_1 \supset G_2$, since a proper subspace cannot have interior points. To this end let $x \in G_2$. Let $A_i = \{t \in R : tx + (1-t)y \in G_i\}$, $i = 1, 2$. Each A_i is open in R ⟦it is the inverse image of G_i under the continuous map $t \to tx + (1-t)y$⟧ and each A_i meets $[0, 1]$ ⟦$t = 0, 1$ respectively⟧. Since $[0, 1]$ is connected, there exists $t \notin A \cup B$, $0 < t < 1$. Let $m = tx + (1-t)y$. Then $m \notin G_1 \cup G_2$ so $m \in M$. Then $x = (1/t)m + (1 - 1/t)y \in M_1$.

Of course, Lemma 7-3-1 is vacuous unless X is a real TVS.

The next result is the geometrical form of the Hahn–Banach extension theorem. Its equivalence with the analytic form is shown in Remark 7-3-3 and Prob. 7-3-104.

2. Lemma Let X be a TVS, A a convex, open, nonempty subset, and S a vector subspace not meeting A. Then S is included in a closed maximal subspace which does not meet A.

PROOF It is sufficient to find a maximal subspace not meeting A; since it is not dense it must be closed ⟦Prob. 4-2-5⟧. Let P be the set of vector subspaces of X which include S and do not meet A. Order P by containment, let C

be a maximal chain, and set $M = \cup C$. An easy check shows M to be a vector subspace not meeting A. It remains only to show that M is a maximal subspace, i.e., has codimension 1.

Case (a). X is a real TVS.
As obvious candidates for an application of Lemma 7-3-1 we nominate $G_1 = M + \cup\{tA : t > 0\}$, $G_2 = -G_1$. Each G_i is nonempty and open [a union of translates of multiples of A]; they are disjoint from each other and from M. [If $x \in M \cap G_1$, $x = m + ta$ implies $a \in (x - m)/t \in M$, contradicting $A \pitchfork M$. If $x \in G_1 \cap G_2$, $x = m + ta = -(m' + t'a')$. Let $u = (ta + t'a')/(t + t')$. Then $u \in A$ since A is convex; but also $u = (-m - m')/(t + t') \in M$, contradicting $A \pitchfork M$.] Finally, $X \backslash M = G_1 \cup G_2$. [Let $x \notin M$. Then $M_1 = M + [x]$ properly includes M and so must meet A, since otherwise we could enlarge the maximal chain of subspaces. Let $a \in A \cap M_1$, $a = m + rx$. Surely $r \neq 0$ since $m \notin A$. If $r > 0$, $x = (-1/r)m + (1/r)a \in G_1$; if $r < 0$, $x = -[(1/r)m + (-1/r)a] \in G_2$. Thus $X \backslash M \subset G_1 \cup G_2$. The opposite inclusion is because M is disjoint from $G_1 \cup G_2$.] It now follows from Lemma 7-3-1 that M is maximal.

Case (b). X is a complex TVS.
Applying case (a) to X_R, which is X with real scalars only, we obtain a maximal subspace M_R of X_R which includes S and does not meet A. [S is a vector subspace and A a convex open set in X_R.] Let $g \in (X_R)^\#$ with $g^\perp = M_R$. Let $f \in X^\#$ with $g = Rf$ [Theorem 1-5-2] and set $M = f^\perp$. Then $M \subset M_R \pitchfork A$ and $M \supset S$. [Using Theorem 1-5-2, if $s \in S$, $f(s) = g(s) - ig(is) = 0$ since $is \in S$ also.]

The assumption that A is open cannot be dropped. See Probs. 7-3-101 and 7-3-102.

3. Remark *The original Hahn–Banach Theorem 2-2-2 can be deduced from Lemma 7-3-2.* We may assume $f \neq 0$. Choose $a \in S$ with $f(a) = 1$. Consider the seminormed space (X, p). Let $A = \{x : p(x - a) < 1\}$, a convex open set, and let $S_0 = f^\perp$ so $S_0 \subset S$. Now $S_0 \pitchfork A$. [$x \in A$ implies $|f(x)| = |1 + f(x - a)| \geq 1 - p(x - a) > 0$.] By Lemma 7-3-2, S_0 is included in a maximal subspace M not meeting A. Now $a \notin M$ since $a \in A$, so there exists $F \in X^\#$ with $F^\perp = M$, $F(a) = 1$. This is the required extension. First, $F \mid S = f$. [Since $F(a) = f(a) = 1$, it is sufficient, by Theorem 1-5-1, to prove that $f^\perp \subset (F \mid S)^\perp$, that is, $S_0 \subset M \cap S$. But this is obvious.] Finally, $|F(x)| \leq p(x)$ for all x. [If not, let $|F(x)| > p(x)$. Let $u = a - x/F(x)$. Then $F(u) = 0$ so $u \in M$. But $p(u - a) < 1$ so $u \in A$. This is a contradiction.]

4. Theorem Let X be a real lc space, A a convex subset, and $x \notin \bar{A}$. Then there exists $f \in X'$ with $f(x) > \sup \{f(a) : a \in A\}$.

PROOF Since A is not assumed to have interior we cannot apply Lemma 7-3-2 to it. However, let U be a convex open neighborhood of 0 [Prob. 7-1-4] with

$x + U \not\supset A$. Let $B = x + U - A$. Then B is open, convex, and does not contain 0. By Lemma 7-3-2 (with $S = \{0\}$, there exists a closed maximal subspace M not meeting B. Say $M = f^\perp$; f is continuous [Theorem 4-5-10] and f does not change sign in B. [If $f(a) = r > 0$, $f(b) = -s < 0$, let $u = (sa + rb)/(s + r)$. Then $u \in B$, since B is convex, and $f(u) = 0$.] We may assume $f(b) > 0$ for all $b \in B$. [If not, treat $-f$ instead.] Choose $u \in U$ such that $f(u) < 0$ [$f^\perp \not\supset U$ since a proper subspace has no interior]. Then for all $a \in A$, $f(x + u - a) > 0$ and so $f(x) > f(a) - f(u)$.

5. Theorem Let X be a lc space, A an absolutely convex subset, and $x \notin \bar{A}$. Then there exists $f \in X'$ with $|f(x)| > \sup\{|f(a)| : a \in A\}$.

PROOF Let X_R be X with real scalars. By Theorem 7-3-4, there exists $g \in (X_R)'$ with $g(x) > s = \sup\{g(a) : a \in A\}$. Let $f \in X'$ with $g = Rf$. [See Theorem 1-5-2; f is continuous since $f(y) = g(y) - ig(iy)$.] For $a \in A$, let $f(a) = re^{i\theta}$; then $f(ae^{-i\theta}) = r = g(ae^{-i\theta})$ since it is real. Since $ae^{-i\theta} \in A$, we have $r \leq s$. Thus $|f(x)| \geq g(x) > s \geq r = |f(a)|$.

Lemma 7-3-2 gives a weak sort of separation of a point and an open convex set. Theorems 7-3-4 and 7-3-5 give a strong separation of a point and a closed convex set. In the latter case the space had to be locally convex.

Every finite-codimensional subspace of a TVS is complemented [Theorem 6-3-4]. For lc spaces the same is true for a finite-dimensional subspace.

6. Theorem Each finite-dimensional subspace S of a lc space X is complemented.

PROOF We shall assume that X is separated. The general case, which is not important, is left to the reader. By Prob. 6-3-4, S has a Schauder basis $(b^1, b^2, \ldots, b^n)$ with coordinate functionals $l_i \in S'$. By Theorem 7-2-11, each l_i may be extended to all of X. Then $P(x) = \sum l_i(x)b^i$ clearly defines a continuous projection of X onto S.

PROBLEMS

101 Let A be the set of nonzero members of R^2 whose last nonzero coordinate is positive. (Thus A is the upper half-plane and the positive X axis.) Show that A is convex but no maximal subspace of R^2 is disjoint from A.

102 Let A be the set of nonzero members of (real) φ whose last nonzero coordinate is positive. Show that A is convex but that every linear function on φ changes sign on A. (This improves Prob. 7-3-101. Compare Prob. 2-1-112.)

103 Let X be a real vector space, $A, B \subset X$. Say that A, B are *separated by* $f \in X^\#$ if $f(a) \geq f(b)$ for $a \in A$, $b \in B$ [or $f(a) \leq f(b)$ for $a \in A$, $b \in B$] *strictly separated by* f if, for some real r, $f(a) > r > f(b)$ [or $f(a) < r < f(b)$]. Show that A, B can be (strictly) separated if and only if $A - B$ can be (strictly) separated from $\{0\}$.

104 Fill in the details of this proof of Lemma 7-3-2. We shall assume that $A = a + B$ with B absolutely convex. In the general case use Prob. 2-2-103. Let p be the gauge of B, $Y = S + [a]$, $f \in Y^*$ with $f^\perp = S$, $f(a) = 1$. Then $|f| \leq p$ on Y [Prob. 2-2-7]. Extend f to $F \in X^*$ with $|F| \leq p$. Then $F^\perp \not\pitchfork A$.

105 Let K, F be disjoint convex sets in a real lc space X with K compact, F closed. Show that they can be strictly separated by $f \in X'$ [Prob. 7-3-103, Lemma 6-5-11, and Theorem 7-3-4].

106 If $X' = \{0\}$, X has only one nonempty convex open subset [Lemma 7-3-2]; thus the convex hull of an open subset must be all of X [Prob. 7-1-3] and the associated lc topology is indiscrete.

107 Find a Fréchet space which allows no smaller separated lc topology [Probs. 7-3-106 and 2-3-119].

108 Find a Fréchet space which allows no larger lc topology [Prob. 2-1-202].

109 Prove Theorem 7-3-6 [Prob. 5-3-115].

110 Show that we cannot add to Theorem 7-3-6 (as we did in Theorem 6-3-4). "Any algebraic complement will do" [e.g., let dim $A = 1$].

111 Show that Theorem 7-3-6 becomes false if "locally convex" is omitted, even if X is a Fréchet space and dim $A = 1$.

EIGHT
DUALITY

Our study now takes a decisive turn with the introduction of duality—the study of a space by means of functions defined on it. In the case of a TVS X we try to derive properties of X from those of X'. Such properties are called *duality invariants*. Specifically, a property is called duality invariant if whenever it holds for a lc space (X, T) it also holds for (X, T_1) where T_1 is any other lc topology for X such that $(X, T_1)' = (X, T)'$.

8-1 COMPATIBLE TOPOLOGIES

1. Definition Let X be a vector space. Two vector topologies T and T_1 for X are called compatible with each other if they are both lc and $(X, T)' = (X, T_1)'$.

2. Definition Let X be a TVS. The weak topology for X is defined to be $\sigma X'$.

Recall that wX' was defined in Theorem 1-6-10 and it was decided to call it $\sigma X'$, as explained after Definition 4-1-10. One should keep in mind that for a net x in X, $x \to 0$ *weakly* (i.e., in the weak topology) *if and only if* $f(x) \to 0$ *for all* $f \in X'$ [Theorem 1-6-10]. It is clear that the weak topology of (X, T) is smaller than T since it is the smallest such that the members of X' are continuous.

3. Theorem Let (X, T) be a lc space. Then T and the weak topology are compatible with each other. The weak topology is the smallest topology for X which is compatible with T.

PROOF If f is T continuous and $x \to 0$ weakly, then $f(x) \to 0$, as just pointed out. Conversely, if f is weakly continuous it is T continuous since T is larger than the weak topology. Finally, if T_1 is any other compatible topology it is larger than the weak topology since the latter is the smallest with the given dual.

4. Example *If (X, T) is an infinite-dimensional normed space, the weak topology is strictly smaller than T.* This is part of Example 4-4-11.

5. Definition Let X be a vector space and Y a subspace of $X^{\#}$. The weak $*$ topology for Y is defined to be $\sigma \hat{X}$, where $\hat{X} = \{\hat{x} : x \in X\}$.

Recall that $\hat{x} \in Y^{\#}$ is defined by $\hat{x}(f) = f(x)$ for $f \in Y$.

6. Remark The weak $*$ topology is also called the *pointwise topology* since, for a net $f \in Y, f \to 0$ if and only if $f(x) \to 0$ for each $x \in X$. It is also a product topology [Prob. 8-1-4].

The next result gives the dual of $(Y, \text{weak} *)$. It is important that this is X. (See Remark 8-1-8.)

7. Theorem With X, Y as in Definition 8-1-5, let $g \in Y^{\#}$. Then g is weak $*$ continuous if and only if $g = \hat{x}$ for some $x \in X$.

PROOF $\Leftarrow$: By definition. $\Rightarrow$: Let $U = \{f \in Y : |g(f)| < 1\}$, a weak $*$ neighborhood of 0 in Y. As in the proof of Theorem 4-1-12, $U \supset S = \cap\{(\hat{x}_i)^{\perp} : i = 1, 2, \ldots, n\}$ for some $x_1, x_2, \ldots, x_n \in X$. Thus g is bounded on the vector subspace S, from which it follows that $g = 0$ on S. From Theorem 1-5-1 it follows that $g = \sum t_i \hat{x}_i = (\sum t_i x_i)\hat{\,}$.

8. Remark If, in Theorem 8-1-7, Y is total over X, the equation $g = \hat{x}$ determines x uniquely and so $(Y, \text{weak} *)'$ is isomorphic with X in a natural way. They are usually identified, so, for example, if X is a lc separated space, we have $(X', \text{weak} *)' = X$.

9. Example Let X be a normed space. The weak $*$ and norm topologies on X' are compatible with each other if and only if X is reflexive.

10. Theorem Let X be a vector space, $S \subset X^{\#}$. Then S is fundamental in $(X^{\#}, \text{weak} *)$ if and only if S is total over X. If S is a vector subspace of $X^{\#}$, it is dense in $(X^{\#}, \text{weak} *)$ if and only if it is total over X.

PROOF The second statement is a special case of the first. $\Rightarrow$: Let $f(x) = 0$ for all $f \in S$. Then $(\hat{x})^{\perp} = \{g \in X^{\#} : g(x) = 0\}$ is a weak $*$ closed vector subspace of $X^{\#}$ [$\hat{x}$ is weak $*$ continuous by definition] which includes S and hence is all of X^*. Thus $x = 0$. $\Leftarrow$: Let $Z = (X^*, \text{weak} *)$. Since Z is locally convex and separated [[$\hat{X}$ is total over $X^\#$]], it is sufficient to prove that $h \in Z'$, $h = 0$ on S implies $h = 0$ [[Corollary 7-2-13]]. Now such h must have the form $h = \hat{x}$ for some x by Theorem 8-1-7. The condition $h = 0$ on S means that $f(x) = 0$ for all $f \in S$. By hypothesis $x = 0$, so $h = 0$.

11. Corollary Let X be a lc space and S a weak $*$ closed proper subspace of X'. Then there exists $x \neq 0$ such that $f(x) = 0$ for all $f \in S$.

PROOF Since S is not weak $*$ dense in X' it is not total over X, by Theorem 8-1-10.

12. Abbreviation *A lcs space is a lc separated space.*

In the rest of this book, spaces will usually be assumed to be separated.

PROBLEMS

1 Let T and T_1 be topologies for a vector space X which are compatible with each other. Show that (X, T) and (X, T_1) have the same weak topology.

2 If T is separated so is any topology compatible with T. [[The weak topology is separated by Corollary 7-2-8.]]

3 Show that the weak $*$ topology is always separated.

4 Let X be a vector space and Y a subspace of X^*. Show that the weak $*$ topology for Y is the restriction to Y of the product topology of $\mathscr{K}^X$. [[Trivial; just check net convergence.]] Compare Prob. 8-1-106.

5 Let X be a vector space, $Y = X^*$ with the weak $*$ topology. Show that Y is complete. [[Using Prob. 8-1-4 and Theorem 6-1-7 show that Y is closed, i.e., if $f \to g \in \mathscr{K}^X$ then $g \in Y$; for example, $g(x + y) = \lim f(x + y) = \lim f(x) + \lim f(y) = g(x) + g(y)$. An easier proof follows from Prob. 8-1-106.]]

6 Let X be an infinite-dimensional normed space, $A = \{x : \|x\| \geq 1\}$. Show that A is closed in the norm topology but is dense in the weak topology [[Theorem 4-1-12]].

7 A normed space X is reflexive if and only if every $f \in X''$ is weak $*$ continuous [[Example 8-1-9]].

8 Show that weakly convergent sequences in l are norm convergent [[Theorem 1-3-2 and Example 2-3-5]].

9 Show that the result of Prob. 8-1-8 is false for c_0.

10 Suppose that two vector topologies have the same convergent sequences. Show that they have the same Cauchy sequences. [[$\{x^n\}$ is Cauchy if and only if $x^{u(n)} - x^{v(n)} \to 0$ for all sequences $\{u(n)\}, \{v(n)\}$ of integers.]]

11 Show that l with its weak topology is sequentially complete [[Probs. 8-1-8 and 8-1-10]] but not complete [[Theorem 8-1-10, $X = l^\infty$]].

12 Let X be a Banach space. Show that in $[X^*, \sigma(X^*, X)]$, X' is both dense and sequentially closed [[Theorems 8-1-10 and 3-3-13]].

101 Let X be a normed space. Show that on X the weak topology $\sigma X'$, the weak $*$ topology of X''

(that is, $\sigma X''\hat{}\,|\hat{x}$), and the weak topology of X'' (that is, $\sigma X'''|\hat{x}$) are all the same. [Check net convergence.]

102 Let X be a lcs space. Show that the weak $*$ topology on X' is metrizable if and only if X has a countable Hamel base. Indicate clearly where separation of X is used in your proof, e.g., the result is false if X is indiscrete [Prob. 7-2-6].

103 If X is an infinite-dimensional lc Fréchet space, the weak $*$ topology on X' cannot be metrizable [Probs. 8-1-102 and 6-3-107].

104 Let Φ be a collection of lc topologies for a vector space X. Let w be the weak topology of $(X, \vee\Phi)$ and $w(T)$ the weak topology of (X, T) for each $T \in \Phi$. Show that $w = \vee\{w(T): T \in \Phi\}$. [See Theorem 7-2-16. Check net convergence.]

105 Find a weakly compact set in c_0 which is not norm compact [Prob. 8-1-9].

106 Let H be a Hamel base for a vector space X. Show that $(X^{*}, \text{weak } *)$ is homeomonphic with $\mathcal{K}^{H}$.

107 Let $X = (l^{\infty}, \text{weak } *)$ considering $l^{\infty} = l'$, Example 2-3-5. Show that $\{\delta^k\}$ is a basis for X.

108 Show that c_0 is not weakly sequentially complete. [It is sequentially dense in $(l^{\infty}, \text{weak } *)$ by Prob. 8-1-107.] Compare Prob. 8-1-11.

109 Show that l has a noncomplemented closed subspace. [$q^{\perp}$ where $q: l \to c_0$. Use Probs. 8-1-8, 8-1-9, and 6-2-201.]

110 If T, T_1 are lc topologies for a vector space X with $T_1 \supset T$, show that $w(T_1) \supset w(T)$ where, for example, $w(T)$ is the weak topology of T.

111 Let A be a conservative matrix. Show that A is conull if and only if $\sum \delta^k = 1$, weakly, that is, $\sum f(\delta^k) = f(1)$ for all $f \in c'_A$. [$\Leftarrow$: Take $f(x) = \lim Ax$; $f \in c'_A$ by Prob. 5-5-102. $\Rightarrow$: Problem 7-2-122.] For this reason a *conservative* FK space, i.e., one which includes c, is called *conull* if $\sum \delta^k = 1$, weakly; otherwise it is *coregular*.

112 Let X, Y be conservative FK spaces with X conull and (a) $Y \supset X$ or (b) Y is a closed subspace of X. Show that Y is conull [(a) Probs. 8-1-110 and 8-1-111; (b) Prob. 5-5-103].

113 Let A, B be conservative matrices with A row finite. Show that $c_{AB} \supset c_B$.

114 Let Γ be the set of conservative matrices and fix a conull matrix B. Define $u \in \Gamma^{*}$ by $u(A) = \chi(AB)$. Show that u vanishes on the row-finite members of Γ [Probs. 8-1-113 and 8-1-112].

115 Show that AB is conull if $A, B \in \Gamma$ and B is conull. [See Prob. 8-1-114. Use a density argument in the matrix norm with Prob. 3-3-111.]

116 Show that $\chi(AB) = \chi(A)\chi(B)$. [$A[B - \chi(B)I]$ is conull by Prob. 8-1-115.]

117 Let A, B be matrices with $c_B \supset c_A$. Show that $\lim_B \in c'_A$. Here $\lim_B$ (Prob. 3-3-110) is to be restricted to c_A [Prob. 5-5-102].

118 For matrices A, B; A is called *consistent with* B if $\lim_A = \lim_B$ on $c_A \cap c_B$ and A is called *regular* if it is conservative and consistent with the identity matrix. Let A, B be regular and $c_B \supset c_A$. Show that $\lim_B x = \lim_A x$ for $x \in c_A \cap l^{\infty}$ [Probs. 8-1-117 and 7-2-123].

119 (S. Mazur's consistency theorem). Show that a reversible regular matrix A is consistent with every regular matrix B such that $c_B \supset c_A$ if and only if A is of type M [Probs. 8-1-117, 7-2-124, and 7-2-125 and Remark 5-5-17].

120 Show that a reversible coregular matrix A is consistent with every matrix B such that $c_B \supset c_A$, $\lim_B = \lim_A$ on c, if and only if A is of type M [like Prob. 8-1-119].

121 Give a coregular version of Prob. 8-1-118, like Prob. 8-1-120.

---

201 Let $\{x_n\}$ be a sequence of invertible members of a Banach algebra with identity such that $x_n \to 1$ weakly. Must $x_n^{-1} \to 1$ weakly? (See Prob. 5-5-201.)

202 Let A be a Banach algebra with identity. Is A^{-1} weakly open?

203 Show that two vector topologies with the same dense subsets must be equal. This improves Prob. 8-1-6.

8-2 DUAL PAIRS

The study is now switched to a symmetric setting which tries to ignore the precedence of a space over its dual and gives them equal status. Example 8-2-2 will show that the earlier discussions of X, X' are a special case.

1. Definition A dual pair is a triple (X, Y, b) where X, Y are vector spaces and b is a bilinear functional on $X \times Y$ written $b(x, y) = [x, y]$ such that (a) $[x, y] = 0$ for all $y \in Y$ implies $x = 0$ and (b) $[x, y] = 0$ for all $x \in X$ implies $y = 0$.

We are assuming that $[sx_1 + tx_2, y] = s[x_1, y] + t[x_2, y]$ and $[x, sy_1 + ty_2] = s[x, y_1] + t[x, y_2]$ for scalar s, t.

The separation conditions (a) and (b) on b are convenient but could be dispensed with.

2. Example *Let (X, T) be a lcs space, $Y = X'$, $[x, f] = f(x)$ for $x \in X$, $f \in Y$. Then (X, X') is a dual pair.* It is trivial that $[x, f] = 0$ for all x implies $f = 0$ but requires the Hahn–Banach theorem (in the form of Corollary 7-2-8) to conclude that $[x, f] = 0$ for all f implies $x = 0$.

3. Example A generalization of Example 8-2-2 (but really equivalent; see Example 8-2-13) is to let X be a vector space and Y any total subspace of $X^{\#}$, $[x, f] = f(x)$ for $x \in X$, $f \in Y$.

It is customary to abbreviate the presentation of dual pairs. The phrase "let (X, Y) be a dual pair" assumes that some particular bilinear form is specified. In the special case that $Y \subset X^{\#}$, the bilinear form is $[x, f] = f(x)$ for $x \in X$, $f \in Y$.

4. Remark *Every dual pair can be put in the form of Example 8-2-3 by means of the natural embedding.* With the notation of Definition 8-2-1, fix $y \in Y$. Let $\hat{y} \in X^{\#}$ be defined by $\hat{y}(x) = [x, y]$ for $x \in X$. The map $y \to \hat{y}$ is called the *natural embedding of Y into $X^{\#}$*. It is an isomorphism into; $\hat{Y} = \{\hat{y} : y \in Y\}$ is a total subspace of $X^{\#}$; and $(X, \hat{Y}, \hat{b})$ is a dual pair, where $\hat{b}(x, \hat{y}) = \hat{y}(x) = [x, y]$, of the form of Example 8-2-3. Similarly, the natural embedding of X into $Y^{\#}$ is defined by $\hat{x}(y) = [x, y]$.

5. Example In Example 8-2-3, for $f \in Y$, $\hat{f}(x) = [x, f] = f(x)$; thus $\hat{f} = f$, $\hat{Y} = Y$. Since $X^{\#}$ usually has total proper subspaces, this shows that *the natural embedding need not be onto $X^{\#}$*.

6. Example In Example 8-2-2, let X be a Banach space and fix $x \in X$. Then $\hat{x}(f) = f(x)$ for $f \in Y$. This is the natural embedding discussed in Sec. 3-2.

7. Example Let $b(x, y) = \sum x_i y_i$ for sequences x, y. Then (l, c, b) is a dual pair. If we set $b_1(x, y) = x_1 \lim y + \sum x_{i+1} y_i$, then again (l, c, b_1) is a dual pair. It is different from the first named one. The second dual pair has the property that the natural embedding of l into c^* is the equivalence of l with c' given in Prob. 2-3-3. The difference between these dual pairs is illustrated by Probs. 8-2-109, 8-2-110, 8-5-116, 8-5-117, 9-3-125, and 10-2-105.

8. Definition Let (X, Y) be a dual pair. Let T be a topology for X. Then T is called compatible with the dual pair if T is a lc (vector) topology such that $(X, T)' = \hat{Y}$. Dually, a compatible topology for Y is one such that $Y' = \hat{X}$.

Note that each compatible topology is separated. $[$If $x \to 0$ and $x \to a$, $[a, y] = \lim [x, y] = (0, y] = 0$ for each $y \in Y$; thus $a = 0.]$
Thus if T is compatible and $u \in X^*$ is T-continuous, there must exist $y \in Y$ such that $u(x) = [x, y]$ for all $x \in X$. Moreover, if $x \to 0$ in (X, T) it follows that $[x, y] \to 0$ for each $y \in Y$.

9. Example In Example 8-2-2, T is compatible with the dual pair since $(X, T)' = Y' = \hat{Y}$ as in Example 8-2-5.

10. Example In Example 8-2-6, the norm topology for Y is compatible with the dual pair if and only if X is reflexive.

We now repeat part of Sec. 8-1 in the present setting.

11. Definition Let (X, Y) be a dual pair. The topology $\sigma(X, Y)$ for X is defined to be $\sigma \hat{Y}$. If S is a vector subspace of X, $\sigma(S, Y)$ is defined to be $\sigma(X, Y)|_S$.

Recall that for a net $x \in [X, \sigma(X, Y)]$, $x \to 0$ if and only if $[x, y] \to 0$ for each $y \in Y$. $[x \to 0$ if and only if $\hat{y}(x) \to 0$ for each y, by Theorem 1-6-10.$]$ The extra definition for $\sigma(S, Y)$ was thrown in because, if $\hat{S}$ is not total over Y, (S, Y) is not a dual pair.
In Remark 8-2-4, the embedding of Y into X^* was given, and then that of X into Y^*. In Definition 8-2-8, mention of a compatible topology for X was followed by one for Y. Similarly, Definition 8-2-11 implies the dual definition of $\sigma(Y, X)$ to be $\sigma(\hat{X})$, a topology on Y. From now on the reader should supply the second (dual) half of each idea.

12. Theorem Let (X, Y) be a dual pair. Then $\sigma(X, Y)$ is compatible with the dual pair and is the smallest compatible topology for X.

PROOF It will be more convenient to prove this for $\sigma(Y, X)$. We identify $[Y, \sigma(Y, X)]$ with $[\hat{Y}, \sigma(\hat{Y}, X)] = (\hat{Y}, \text{weak} *)$ and compatibility follows from Theorem 8-1-7. If T is a compatibe topology for Y and $y \to 0$ in (Y, T) it follows that $[x, y] \to 0$ for each $x \in X$ and so $T \supset \sigma(Y, X)$.

13. Example In Example 8-2-3, give X the topology $\sigma(X, Y)$. Then $Y = X'$ by Theorem 8-2-12. Thus Example 8-2-3 is a special case of Example 8-2-2. An interesting possibility is to take $Y = X^\#$ where X is a vector space. Then $[X, \sigma(X, Y)]$ is an example of a space satisfying $X' = X^\#$. It cannot be metrizable (if infinite-dimensional) by Example 3-3-14.

We saw (Theorem 8-2-12) that there is a smallest compatible topology. Is there a largest one? To answer this question is simple: there is usually a largest topology with any given property P, namely, let $T = \vee \{T': T'$ has property $P\}$. (We did this for bounded sets in Prob. 4-4-106.) The only difficulty is that T might not have property P. In the present case it does.

14. Theorem Let (X, Y) be a dual pair and let $\tau(X, Y) = \vee \{T': T'$ is a topology for X compatible with the dual pair$\}$. Then $\tau(X, Y)$ is compatible and is the largest compatible topology.

PROOF This is immediate from Theorem 7-2-16.

I do not know why τ was chosen for the name. Some authors use m; this makes sense since m stands for maximal and since this topology is called the *Mackey topology* in honor of the rather difficult discovery of the form of the basic neighborhoods of 0 (Theorem 9-2-3) by G. W. Mackey.

15. Definition A lcs space (X, T) is called relatively strong (sometimes a Mackey space) if $T = \tau(X, X')$.

PROBLEMS

In this list (X, Y) is a dual pair.

1 Show that every lc topology lying between two compatible topologies is compatible.

2 Show that a lc topology for X is compatible if and only if it lies between $\sigma(X, Y)$ and $\tau(X, Y)$.

3 Let S be a subspace of X. Show that $\sigma(S, Y) = \sigma\{\hat{y} \mid S : y \in Y\} = \sigma(X, Y)|_S$. [Check net convergence.]

4 Consider the dual pair (ω, φ) with $[x, y] = \sum x_i y_i$. Show that $\sigma(\omega, \varphi)$ is the Fréchet topology T for ω. $[T = \sigma\{|P_n| : n = 1, 2, \ldots\}$ where $P_n(x) = x_n$; $\sigma \subset T$ by Theorem 8-2-12; and $\sigma \supset T$ since each $P_n \in \hat{\varphi}$; $P_n(x) = [x, \delta^n]$.] Thus the weak topology can be metrizable. (Compare Probs. 8-1-102, 8-1-103, and 8-4-1.)

5 Let $B \subset X$. Show that B is $\sigma(X, Y)$ bounded if and only if $\{[x, y] : x \in B\}$ is bounded for each $y \in Y$ [Prob. 4-4-12].

6 Show that $[X, \sigma(X, Y)]$ is a BTB space [Theorem 6-4-12; $\mathscr{K}$ is a BTB space] but an infinite-dimensional normed space cannot be a BTB space [Theorem 6-4-2].

---

101 Give examples of a lcs topology T for X which is not compatible with the dual pair (X, Y) and

   (a) $T \subset \sigma(X, Y)$ (b) $T \supset \sigma(X, Y)$ (c) neither (a) nor (b).

[Just pick $S \subset X^\#$, $T = \sigma(X, S)$. Make sure that Y allows such choices; see Prob. 8-2-106.]

102 Show that a vector space with its largest lc topology is relatively strong.

103 Find b_1, b_2 such that (l^∞, l, b_1), (l^∞, c_0, b_2) are dual pairs. ⟦Example, $b_1(x, y) = \sum x_n y_n$.⟧

104 Let X be a lc space, not necessarily separated. Discuss the dual of $(X', \text{weak} *)$.

105 In $\sigma(\varphi, \omega)$, Prob. 8-2-4, show that all bounded sets are finite-dimensional and all vector subspaces are closed ⟦Probs. 4-4-15 and 4-5-107⟧.

106 Show that no proper subspace of φ can be total over ω ⟦Prob. 8-2-105 and Theorem 8-1-10⟧.

107 The Fréchet topology of ω is minimal among lc separated topologies ⟦Probs. 8-2-106 and 8-2-4⟧. See Prob. 8-2-301.

108 In Prob. 8-2-107, "minimal" cannot be replaced by "minimum." ⟦See Example 5-2-9.⟧

109 In the dual pair (l, c, b), Example 8-2-7, l is not $\sigma(l, c)$ sequentially complete. ⟦⟦$\{\delta^n\}$ is Cauchy but not convergent since $\lim [\delta^n, \delta^k] = 0$ for each k so if $\delta^n \to x$, $x = 0$. But $\lim [\delta^n, 1] = 1$.⟧

110 In the dual pair (l, c, b_1), Example 8-2-7, l is $\sigma(l, c)$ sequentially complete ⟦Theorem 3-3-13⟧.

111 In Example 2-3-4 and Prob. 2-3-3 it is shown that c_0' and c' are both l. Show that the resulting weak $*$ topologies are not comparable. (Compare Example 8-2-7.)

112 Let φ have the largest lc topology. Show that φ satisfies Prob. 4-5-105(c) but is not locally bounded ⟦Prob. 8-2-105⟧.

113 Define p, q on l by $p(x) = \sum |x_{2n}|$, $q(x) = \sum |x_{2n-1}|$. Show that $w \vee p$ and $w \vee q$ [where $w = \sigma(l, l^\infty)$] are noncomparable compatible topologies for (l, l^∞).

---

201 Give an example of a property P of vector topologies such that for some vector space, $\vee \{T : T \text{ has } P\}$ does not have property P.

---

301 The Fréchet topology of ω is minimal among Hausdorff topologies. ⟦See [77], Proposition 4.1.⟧

8-3 POLARS

There are two types of polar in common use, the (absolute) polar and the real part polar. The latter will play an important though brief role and will not be seen again after this section.

1. Definition Let (X, Y) be a dual pair and $A \subset X$. The polar of A, written A°, is $\{y \in Y : |[x, y]| \leq 1 \text{ for all } x \in A\}$. The real part polar of A, written A^r is $\{y \in Y : R[x, y] \leq 1 \text{ for all } x \in A\}$.

2. Remark The following very simple properties may be read through. They all hold also for A^r instead of A°:

(a) $A^\circ = \cap \{x^\circ : x \in A\}$, where $x^\circ = \{x\}^\circ$.

(b) $B \supset A$ implies $B^\circ \subset A^\circ$. A special case is $x \in B$ implies $B^\circ \subset x^\circ$. ⟦By (a).⟧

(c) $A^{\circ\circ} \supset A$. Here $A^{\circ\circ} = (A^\circ)^\circ$ is the polar of a subset of Y and is defined in the obvious (dual) way.

(d) $A^{\circ\circ\circ} = A^\circ$. ⟦$\supset$: By (c). $\subset$: In (b), take $B = A^{\circ\circ}$.⟧

(e) $(tA)^\circ = (1/t)A^\circ$ if $t \neq 0$. ⟦$\supset$: Let $y \in (1/t)A^\circ$, $z \in tA$. Then $|[y, z]| = |[ty, (1/t)z]| \leq 1$. $\subset$: $(tA)^\circ = (1/t)[t(tA)^\circ] \subset (1/t)[(1/t)(tA)]^\circ = (1/t)A^\circ$.⟧

(f) For all A, A° is absolutely convex; A^r is convex and contains 0.

(g) If A is balanced, $A^\circ = A^r$. ⟦It is always true that $A^\circ \subset A^r$. Now let $y \in A^r$, $x \in A$. Let $[x, y] = re^{i\theta}$, $u = e^{-i\theta}x$. Then $[u, y] = r = R[u, y] \leq 1$ since $u \in A$. Thus $|[x, y]| = r \leq 1$ and so $y \in A^\circ$.⟧

(h) *For a collection C of subsets of X,* $[\cup\{A : A \in C\}]° = \cap\{A° : A \in C\}.$ [If $B \in C$, $B \subset \cup A$ so $B° \supset (\cup A)°$. Since this is true for all B, $\cap A° \supset (\cup A)°$. Conversely, let $y \in \cap A°$, $x \in \cup A$. Say $x \in B$. Then $y \in B°$ so $|[x, y]| \leq 1$ and hence $y \in (\cup A)°.$]

(i) *For a collection C of subsets of X,* $[\cap\{A°° : A \in C\}]° = [\cup\{A° : A \in C\}]°°.$ [Using (h), $(\cup A°)°° = (\cap A°°)°.$] Clearly, (i) is false without the last two polars since $\cup A°$ need not be convex.

The above properties are purely linear. The next are topological.

3. Lemma Let (X, Y) be a dual pair, $A \subset X$. Then $A°$, A^r are closed in any compatible topology for Y.

PROOF It is sufficient to prove this for $A = \{x\}$ by Remark 8-3-2(a). Now $x° = \{y : |\hat{x}(y)| \leq 1\}$ and $\hat{x}$ is continuous. A similar proof works for A^r.

The *absolutely convex closure* of a set A is the intersection of the absolutely convex closed sets which include A. It is the smallest absolutely convex closed set which includes A. A similar definition is given for the *convex closure* of A.

4. Lemma Let (X, Y) be a dual pair, $A \subset X$, T any compatible topology for X, and H the absolutely convex closure of A in (X, T). Then $H° = A°$.

PROOF Since $H \supset A$ we have $H° \subset A°$. Conversely, let $y \in A°$. By Remark 8-3-2 (b) and (c), $A \subset A°° \subset y°$. Since $y°$ is absolutely convex and closed, $H \subset y°$. Thus $y \in y°° \subset H°$.

5. Theorem Let (X, Y) be a dual pair, $A \subset X$. Then A^{rr} is the convex closure C of $A \cup \{0\}$ where the closure is taken in any compatible topology T for X.

PROOF First $A^{rr} \supset C$ by Lemma 8-3-3 and Remark 8-3-2 (f). Next, let $x \notin C$. By the separation Theorem 7-3-4 there exist $f \in (X, T)'$ and a real number r such that $Rf(x) > r > \sup\{Rf(z) : z \in C\}$. [We also used Theorem 1-5-2. Note that if $g = Rf$ is continuous, so is f, since $f(x) = g(x) - ig(ix)$.] Since T is compatible, $f = \hat{y}$ for some $y \in Y$. Let $y' = (1/r)y$. [$r > 0$ since $0 \in C$.] Then $y' \in A^r$. [For $a \in A$, $R[a, y'] = (1/r)Rf(a) < (1/r)r = 1$.] But $R[x, y'] = (1/r)f(x) > 1$. Thus $x \notin A^{rr}$.

The first duality invariant now makes its long-awaited appearance!

6. Corollary Let (X, Y) be a dual pair, $A \subset X$. The convex closure of A is the same in all compatible topologies. In particular, all compatible topologies for X have the same closed convex sets, the same closed vector subspaces, and the same fundamental subsets.

PROOF Suppose A is a convex set and T, T' are compatible topologies. By a translation we may assume $0 \in A$. [A translation preserves convexity, T closure, T' closure.] The T' closure of A is A^{rr} and this is also the T closure. So they are equal. The rest follows since vector subspaces are convex and fundamental is defined in terms of the closure of the span.

7. Remark Given a dual pair (X, Y), the phrases: closed convex subset of X (or Y), closed vector subspace of X, fundamental subset of X, linear closure of a subset of X, and barrel (Definition 3-3-1) may be used without referring to any topology. It will be understood that they mean in any compatible topology. The same applies to other duality invariants which will appear. The phrase "dense subset of X" cannot be used in this way [Prob. 8-1-6]; nor can "compact" [Prob. 8-1-105].

8. Theorem : Bipolar Theorem Let (X, Y) be a dual pair, $A \subset X$. Then $A^{\circ\circ}$ is the absolutely convex closure H of A. (See Remark 8-3-7.)

PROOF The proof is $A^{\circ\circ} = H^{\circ\circ} = H^{\circ r} = H^{rr} = H$ where the first three equalities are justified respectively by Lemma 8-3-4 and Remark 8-3-2 (g), and the last by Theorem 8-3-5.

9. Theorem Let (X, Y) be a dual pair, $S \subset Y$. Then S is fundamental in Y if and only if S is total over X. If S is a vector subspace it is dense in Y if and only if total over X.

PROOF Recall, as in Remark 8-3-7, that fundamental and dense refer to any compatible topology for Y. To say that S is total over X means that $[x, y] = 0$ for all $y \in S$ implies that $x = 0$. Equivalent statements are that $\hat{S}$ is total over X and (X, S) is a dual pair. The second statement of the theorem is a special case of the first. $\Rightarrow$: $\hat{S}$ is weak $*$ fundamental in $\hat{Y}$ which, in turn, is weak $*$ fundamental in $X^{\#}$, hence total over X by two applications of Theorem 8-1-10. $\Leftarrow$: By Theorem 8-1-10, $\hat{S}$ is weak $*$ fundamental in $X^{\#}$, a fortiori in $\hat{Y}$. Thus S is $\sigma(Y, X)$ fundamental in Y.

10. Remark Every result can be written in two ways: in the language of dual pairs or the (nondual) language of lc spaces. Examples are Theorems 8-3-9 and 8-1-10; and also Probs. 8-3-3 and 8-3-4. Such statements are equivalent, as pointed out in Remark 8-2-4, at least when we are dealing with separated spaces.

Note. We call special attention to Prob. 8-3-108 in the following list. This was one of the basic motivating results for the whole theory.

11. Remark (This is very important.) Let (X, Y) be a dual pair, $y \in Y$, and T any vector topology for X such that $y \in (X, T)'$. [Really, $\hat{y} \in (X, T)'$.] Then $y^{\circ} \in \mathcal{N}(T)$. [$y^{\circ} = \{x : |\hat{y}(x)| \leq 1\}$.]

PROBLEMS

In this list (X, Y) is a dual pair or (X, T) is a lcs space.

1 Show that $A^\circ \cup B^\circ \subset (A \cap B)^\circ$. $\llbracket A \supset A \cap B$. Use Remark 8-3-2 (b).$\rrbracket$

2 Let $A \subset X$, $B \subset Y$. Show that $A \subset B^\circ$ if and only if $B \subset A^\circ$. More generally, show that B° absorbs A if and only if A° absorbs B.

3 Show that $[Y, \sigma(Y, X)]$ is complete if and only if $Y = X^*$ $\llbracket$Theorem 8-3-9 and Prob. 8-1-5$\rrbracket$.

4 Let X be a lcs space. Show that X is weakly complete if and only if $X = X'^*$ $\llbracket$Prob. 8-3-3$\rrbracket$.

5 Let X be a normed space and D its unit disc. Show that D° is the unit disc of X' and $D^{\circ\circ} = D$.

6 Let X be a normed space, $U = \{f \in X' : \|f\| \leq \varepsilon\}$, $A = \{x \in X : \|x\| \leq 1/\varepsilon\}$. Show that $U = A^\circ$.

---

101 Let $X = Y = R^2$, $[x, y] = x_1 y_1 + x_2 y_2$. Find A° and $A^{\circ\circ}$ if A is
  (a) a point
  (b) a line segment
  (c) a square centered at 0
  (d) a circle centered at 0

102 If two vector topologies for a vector space X have the same closed convex sets, they must have the same dual. (This is the converse of Corollary 8-3-6.)

103 Although density is not a duality invariant, the property of being separable is $\llbracket$Corollary 8-3-6$\rrbracket$.

104 For $A \subset X$, let $A^\perp = \{y : [x, y] = 0$ for all $x \in A\}$. If A is a vector subspace, show that $A^\perp = A^r = A^\circ$.

105 Let X be a lcs space, $g \in X^\#\backslash X'$, $S = g^\perp$. In the dual pair (X, X') show that $S^\circ = S^\perp = \{0\}$.

106 Let X be a nonreflexive normed space, $g \in X''\backslash X$, $S = g^\perp \subset X'$. Let $Y = X'$ with the norm topology T. In the dual pair (X, Y) show that $S^{\circ\circ}$ is not equal to the absolutely convex T closure of S. What does this say (with Theorem 8-3-8) about T? $\llbracket$Prob. 8-3-105$\rrbracket$.

107 Let C be a collection of subsets of X. Show that $(\cap\{A : A \in C\})^\circ \supset (\cup\{A^\circ : A \in C\})^{\circ\circ}$ and give an example in the dual pair of Prob. 8-3-101 in which they are not equal. However, if each A is absolutely convex and closed they are equal $\llbracket$Remark 8-3-2 (i)$\rrbracket$.

108 Let (X, T) be a lc metric space, $A \subset X$ and $x \in$ weak closure of A. Then $x \in$ convex T closure of A (so that x is the T limit of a sequence of points of A of a very special form). $\llbracket$See Corollary 8-3-6.$\rrbracket$

109 Let (X, Y) be a dual pair and fix $J \subset \mathcal{K}$. For $A \subset X$ let $A^{\cdot} = \{y \in Y : [x, y] \in J$ for all $x \in A\}$. Show that $A \subset B$ implies $A^{\cdot} \supset B^{\cdot}$, that $A \subset A^{\cdot\cdot}$, and that $A^{\cdot\cdot\cdot} = A^{\cdot}$. Discuss what choice of J leads to $A^\circ, A^r, A^\perp$.

110 Let (X, Y) be a dual pair and $A \subset X$. Show that $A^{\perp\perp}$ is the linear closure of A $\llbracket$Corollary 7-2-12$\rrbracket$.

111 Extend Corollary 8-3-6 as follows. Let a vector space X have two topologies T, T_1 which are compatible with each other. Show that T, T_1 have the same closed convex sets. $\llbracket$If they are separated, this is the same as Corollary 8-3-6. In general, just read through the proof of Theorem 8-3-5.$\rrbracket$ This extends Prob. 8-1-2 which deals with whether $\{0\}$ is closed.

112 Extend Prob. 8-3-111 as follows. Replace "compatible" by $(X, T_1)' \supset (X, T)'$ and assume T is locally convex. Show that each convex set that is T closed is T_1 closed.

113 A real function f is called *lower semicontinuous* if $x \to a$ implies $\liminf f(x) \geq f(a)$. Show that this holds if and only if $\{x : f(x) \leq t\}$ is closed for all t. Deduce that the norm on a normed space is weakly lower semicontinuous. (This is a generalization of the classical *Fatou's lemma*.)

114 A sequence $x_n \to x$ weakly, if and only if x belongs to the convex closure of every subsequence of $\{x_n\}$.

115 Suppose that $(X', \text{weak } *)$ is separable. Show that $|X| \leq \mathfrak{c}$ $\llbracket$Prob. 5-4-111 and Theorem 8-3-9$\rrbracket$. Compare Prob. 9-5-101.

116 Two vector topologies are called compatible if they have the same closed vector subspaces. Show that vector topologies with the same dual need not be compatible $\llbracket$Prob. 2-3-203$\rrbracket$.

117 If two topologies are compatible in the sense of Prob. 8-3-116, show that they have the same dual and that if one is separated so is the other.

118 Let (X, Y) be a dual pair and S a subspace of X. Show that $(X/S, S^\perp)$ is a dual pair with $[x + S, y] = [x, y]$ except possibly for the first separation condition (a) of Definition 8-2-1. By Prob. 8-3-110, this holds if and only if S is closed.

119 Let X be a Banach space such that X'' is weakly sequentially complete. Show that X is weakly sequentially complete ⟦Corollary 8-3-6 and Prob. 8-1-101⟧.

---

201 In Prob. 8-3-118, $\sigma(X/S, S^\perp)$ is the quotient topology of $(X, w)/S$ where $w = \sigma(X, Y)$. ⟦See [82], 16.11.⟧

202 Suppose $\liminf \| x_n + x \| \geq \| x \|$ for all x (in a Banach space). Must $x_n \to 0$ weakly? (Compare Prob. 8-3-113.)

8-4 BOUNDEDNESS

Recall the criterion for weak boundedness given in Prob. 8-2-5.

1. Theorem Boundedness is a duality invariant. For a dual pair (X, Y), all compatible topologies for X have the same bounded sets.

PROOF Let T_1, T be compatible topologies for X and let B be T_1 bounded. Then B is $\sigma(X, Y)$ bounded. To prove that B is T bounded it is sufficient to prove that each T continuous seminorm p is bounded on B ⟦Theorem 7-2-6⟧. Let $Z = (X, p)$, a seminormed space; then Z' is a Banach space ⟦Example 3-1-8⟧. For $b \in B$, let $\breve{b}(f) = f(b)$ for all $f \in Z'$. the natural embedding. So $\breve{b} \in Z''$. We are going to show that $\breve{B} = \{\breve{b}(f) : b \in B\}$ is weak $*$ bounded, that is, $\sigma(Z'', Z')$ bounded. By Prob. 4-4-12, it is sufficient to show that $\{\breve{b}(f) : b \in B\}$ is bounded for each $f \in Z'$. Now, each such f is p continuous, hence T continuous. ⟦$x \to 0$ in T implies $p(x) \to 0$ since p is T continuous. Thus $f(x) \to 0$.⟧ Since T is compatible with the dual pair, it follows that $f = \hat{y}$ for some $y \in Y$. Since B is $\sigma(X, Y)$ bounded, $\{[b, y] : b \in B\}$ is bounded for each $y \in Y$. But $[b, y] = \hat{y}(b) = f(b) = \breve{b}(f)$, so the proof that $\breve{B}$ is weak $*$ bounded is complete. It follows that $\breve{B}$ is norm bounded in the Banach space Z''. ⟦See the uniform boundedness Theorem 3-3-6; note that weak $* =$ pointwise, as in Remark 8-1-6.⟧ Since the natural embedding is an isometry ⟦Theorem 3-2-2⟧, B is norm bounded in Z. This is the same as saying that p is bounded on B.

2. Example *Let X be a normed space. Then weakly bounded sets are norm bounded*, by Theorem 8-4-1. (A more general form was given in Theorem 3-3-11.) This looks like the uniform boundedness principle which says that weak $*$ bounded sets are norm bounded, and, indeed, is referred to by that name. However, there are major differences. First, the result of this example does not require X to be a Banach space, whereas the earlier result does ⟦Prob. 3-3-5⟧. Second, the present result deals only with compatible topologies, whereas the earlier result yields boundedness in the norm topology of X' which is not compatible if X is not reflexive. The extra assumption (that X is a

Banach space) yields an extra result (boundedness in a noncompatible topology)—and therein lies the power of this result.

3. Remark The two forms of uniform boundedness, Example 8-4-2, show a lack of symmetry. Certainly, any result for the weak topology $\sigma(X, X')$ will be true for the weak $*$ topology $\sigma(X', X)$. The difference lies in the norm topology which is compatible on X, but not on X' if X is not reflexive. For example, weak $*$ bounded certainly implies $\tau(X', X)$ bounded just as weak bounded implies $\tau(X, X')$ bounded.

4. Example Two lc topologies are compatible if and only if they have the same closed convex sets. In contrast, the converse of Theorem 8-4-1 fails. *Noncompatible topologies may have the same bounded sets.* For example, if X is a nonreflexive normed space the weak $*$ and norm topologies have the same bounded sets.

Add to the list (Remark 8-3-7) of duality invariants which can be mentioned without reference to a specific topology: bounded and bornivore (Definition 4-4-6), but not totally bounded, (sequentially, boundedly), complete, or BTB [Probs. 8-4-101, 8-4-102, and 8-2-6].

5. Definition A *bornological space* is a lcs space in which every absolutely convex bornivore is a neighborhood of 0.

6. Example *A lc metric space is bornological* [Example 4-4-7].

But a bornological space need not be metrizable. Consider, for example, the largest lc topology [Probs. 8-4-4 and 7-1-8].

7. Remark Definition 8-4-5 is the first in a sequence of properties which do not belong to pure duality. Certainly a set is a convex bornivore in one compatible topology if and only if in all; however, the definition mentions a neighborhood of 0 and this depends on more than just the dual pair. Theorem 8-4-9 points up this comment.

8. Remark Definition 8-4-5 says that (X, T) is bornological provided $\mathcal{N}$ is large enough to include all the convex bornivores. Thus the next result comes as no surprise.

9. Theorem A bornological space (X, T) is relatively strong.

PROOF (The converse is false, see Prob. 8-4-6.) Let T_1 be a compatible (with T) topology for X and $U \in \mathcal{N}(T_1)$, absolutely convex. Then U is a T bornivore since the T bounded sets are T_1 bounded by Theorem 8-4-1, hence absorbed by U. So $U \in \mathcal{N}(T)$, whence $T \supset T_1$.

10. Example *A lc metric space is relatively strong*, by Example 8-4-6 and Theorem 8-4-9. This is an identification of the Mackey topology: if (X, T) is a lc metric space, $\tau(X, X') = T$. The earliest form of the Mackey–Arens theorem to appear in the literature was this one: namely, for a lc metric space, a lc topology is compatible if and only if it lies between the weak and metric topologies.

11. Remark: Dual properties It is important to relate properties of a set with those of its polar. For example, if $A \subset X$, (X, Y) a dual pair, $A = \{0\}$ *if and only if* $A^\circ = Y$. [If $A^\circ = Y$, $A \subset A^{\circ\circ} = Y^\circ = \{0\}$.] It is not to be expected that $A = X$ if and only if $A^\circ = \{0\}$; for example, A might be a dense vector subspace. We do get $A^{\circ\circ} = X$ if and only if $A^\circ = \{0\}$. In general, a set is "small" if and only if its polar is "large," and a set A is "essentially large" (i.e. $A^{\circ\circ}$ is large) if and only if A° is "small." The next result is an example of this, as is Prob. 8-4-2.

12. Theorem Let (X, Y) be a dual pair, $A \subset X$. Then A is bounded if and only if A° is absorbing.

PROOF $\Rightarrow$: Let $y \in Y$. Then $|[x, y]| \leq M$ for all $x \in A$, where M is some positive number [Prob. 8-2-5]. Then $y/M \in A^\circ$. $\Leftarrow$: Fix $y \in Y$. Choose $M > 0$ so that $y \in MA^\circ$. Fix $x \in A$, $|[x, y]| = M|[x, y/M]| \leq M$; so A is bounded, by Prob. 8-2-5.

PROBLEMS

1 Show that the weak topology of an infinite-dimensional normed space is not bornological. Hence it is not metrizable. (Normed may *not* be replaced by lc Fréchet, by Prob. 8-2-4.)

2 Show that A° is bounded if and only if $A^{\circ\circ}$ is absorbing.

3 If A is bounded, show that $A^{\circ\circ}$ is also bounded.

4 Show that the largest lc topology for a vector space is bornological [Prob. 7-1-5].

5 Let X be bornological and Y a lc space. Show that every linear map from X to Y which preserves bounded sets is continuous [Theorem 4-4-9]. Hence every sequentially continuous linear map from X to Y is continuous [Theorem 4-4-3].

6 Let X be a nonreflexive normed space, D the unit disc in X', and let $Y = [X', \tau(X', X)]$. Show that D is a bornivore and a barrel in Y but not a neighborhood of 0. [D is closed since it is a polar; $D \in \mathcal{N}$ would imply $\tau(X', X) = $ norm.]

---

101 Show that the unit disc in an infinite-dimensional normed space is weakly totally bounded but not norm totally bounded [Prob. 8-2-6 and Theorem 6-4-2].

102 In the duality (c_0, l), $[x, y] = \sum x_i y_i$; $\|\cdot\|_\infty$ and $\sigma(c_0, l)$ are compatible. Show that the first is complete and the second not even sequentially complete [Prob. 8-1-108].

103 Give an example with $X = Y = R$, $[x, y] = xy$, of $A \subset X$ which is not absorbing but such that A° is bounded.

104 Let (X, T) be a lc space and U an absolutely convex bornivore. Let T_1 be the smallest extension of T such that $U \in \mathcal{N}(T_1)$ [Prob. 7-2-118]. Show that T, T_1 have the same bounded sets.

105 Let X be a lc separated space. Show that X is bornological if and only if its topology is the largest lc topology with the same class of bounded sets [Prob. 8-4-104.] Thus any topology with the same class of bounded sets must be compatible. Compare Example 8-4-4.

106 Let X be a lcs space. Show that X is bornological if and only if every linear $f : X \to Y$ (Y any lc space) which is bounded on bounded sets must be continuous. [Consider $T \vee \sigma f$ and Prob. 8-4-105.]

107 Let X be l with its weak topology. Show that X does not have the last-mentioned property of Problem 8-4-5 [Prob. 8-1-8].

108 Let T, T_1 be lc topologies for a vector space X such that $(X, T_1)' \supset (X, T)'$. Show that each T bounded set is T_1 bounded. The spaces are not assumed to be separated.

109 Let X be a normed space, $Y = (X', \text{weak } *)$. Show that $Y' = X$, $Y^b = X''$.

110 Let X be a lc metric space with each of two topologies T, T_1 such that $(X, T)' = (X, T_1)'$. Show that $T = T_1$ [Example 8-4-10]. (This modifies Prob. 5-5-108 where it was necessary to assume completeness. It also gives a converse to Prob. 2-3-4.)

111 A lc metric space has the stronger property : (*) every bornivore is a neighborhood of 0. (Compare Definition 8-4-5 and Example 8-4-6.) Show that in general a bornological space does not have property (*). [See Probs. 8-4-4, 1-5-102, 4-4-107, and 7-1-7. A different proof follows from Prob. 7-2-113. Such a product is bornological but (*) is hereditary. This also shows that a closed subspace of a bornological space need not be bornological. Compare Prob. 8-4-302.]

112 The sup of a collection of relatively strong (even normed) topologies need not be relatively strong [Prob. 4-5-111].

113 Let (X, T) be a lcs space. Let T^b be the largest lc topology with the same class of bounded sets (Prob. 4-4-106). Show that T^b is bornological and is equal to $\tau(X, X^b)$.

114 A set U is called a *sequential neighborhood of* x if, for each sequence $x_n \to x$, x_n must belong to U eventually. Show that every sequential neighborhood of 0 is a bornivore. [If U does not absorb B, let $x_n \in B \backslash nU$ and apply Theorem 4-4-1.] Problem 8-4-126 shows that the converse is false.

115 Let X be a vector space with its largest lc topology. Show that each bornivore is a sequential neighborhood of 0 [Prob. 7-1-7].

116 A TVS is called N *sequential* if every sequential neighborhood of $0 \in \mathcal{N}$. Show that X is N sequential if and only if $x \in \bar{A}$, $A \subset X$, implies that x is the limit of a sequence of points in A.

117 A set is called *sequentially open* if its complement is sequentially closed. Show that G is sequentially open if and only if G is a sequential neighborhood of each of its points.

118 A TVS is called *sequential* if every sequentially closed set is closed, or, equivalently, if every sequentially open set is open. Show that every N-sequential TVS is sequential. Compare Probs. 8-4-122 and 8-4-201.

119 Let X be an infinite-dimensional Banach space. Show that $X^{\#}$ with $\sigma(X^{\#}, X)$ is not sequential [Prob. 8-1-12].

120 Let $S = \{n\delta^n\} \subset (c_0, \text{weak})$. Show that $0 \in \bar{S}$ but S is sequentially closed.

121 In $C[0, 1]$ with the pointwise topology, let $S = \{f : \|f\|_\infty \leq 2, \int_0^1 f = 1\}$. Show that $0 \in \bar{S}$ but S is sequentially closed [Lebesgue's bounded convergence theorem].

122 Show that a sequential TVS is N sequential if and only if for each sequential neighborhood U of 0 there exists a sequential neighborhood V of 0 with $V + V \subset U$. [$\Leftarrow$: Let $W = \{x : U$ is a sequential neighborhood of $x\}$. Then $x + V \subset W$ (since if $y_n \to x + v$, $y_n - x - v \in V$ eventually, so $y_n \in x + U$ eventually ; hence $x + v \in W$). Also, W is sequentially open (since if $x \in W$ and $x_n \to x$ then $x_n - x \in V$ eventually, so $x_n \in x + V \subset W$ eventually). So W is open. Since $0 \in W \subset U$, we have $U \in \mathcal{N}$.]

123 Show that φ, with its largest lc topology is not N sequential [Probs. 8-4-111 and 8-4-115].

124 Let (X, T) be a lcs space and $T^+ = \vee \{T' : T'$ is a lc topology with the same convergent sequences as $T\}$. Show that T^+ is the largest lc topology with the same convergent sequences as T [Theorem 1-6-8].

125 Let (X, T) be a lcs space. Show that if X is either sequential or bornological, then $T^+ = T$. [A T^+ closed set is T sequentially closed ; see Prob. 8-4-105 and Theorem 4-4-3.]

126 Let X be a normed space with its weak topology w. Show the equivalence of
  (a) $\{x: \|x\| < 1\}$ is a sequential neighborhood of 0.
  (b) $C = \{x: \|x\| = 1\}$ is sequentially closed.
  (c) 0 is not a sequential limit point of C.
  (d) Weak and norm convergence of sequences coincide.
  (e) w^+ is relatively strong.

127 A TVS is called C *sequential* if every absolutely convex sequential neighborhood of $0 \in \mathcal{N}$. Show that every bornological space is C sequential [Prob. 8-4-114].

128 Let X be a C-sequential lcs space. Show that every sequentially continuous linear map from X to a lcs space is continuous. [$f^{-1}[U]$ is a sequential neighborhood of 0.] See Prob. 8-4-301.

129 Let $X = Y = \{f \in R^R: \text{the support of } f \text{ is finite}\}$, $[x, y] = \sum x_i y_i$, $T = \sigma(X, Y)$. Show that (X, T) is not metrizable but is N sequential and that every bornivore $\in \mathcal{N}(X, T)$.

---

201 Let $\mathcal{B}$ be the set of absolutely convex sequential neighborhoods of 0 in a lcs space (X, T). Show by Theorem 4-3-5 that $\mathcal{B}$ generates a lc topology and that it is T^+ (Prob. 8-4-124). [See [146], Proposition 1-1.] Deduce that (X, T) is C sequential if and only if $T = T^+$. Thus N sequential $\Rightarrow$ sequential $\Rightarrow C$ sequential [Probs. 8-4-118 and 8-4-125]. (Neither converse holds; see Probs. 12-3-113, 13-3-301, and 12-3-302.)

202 Let (X, p) be a normed space with dual (X', p'). Let q_1 be a norm on X', $A = \{f: q_1(f) \leq 1\}$. Finally, let q be the gauge of A°, a norm on X. Show that if q_1 is stronger than p', then q is equivalent to p. [Use Theorem 8-4-12.] (This is unexpected! In particular, Prob. 5-2-302 cannot be done in this way.)

203 Let X be c_0 with its weak topology. Show that each point in X is a G_δ. Compare Prob. 4-5-201.

---

301 The converse of Prob. 8-4-128 is true. [See [135], theorem 2. This article also characterizes spaces on which every bounded linear map is sequentially continuous.]

302 Under a natural set–theoretic assumption, every lcs space is a closed subspace of some bornological space. [See [160].]

303 (Nachbin–Shirota theorem). Let H be a $T_{3\frac{1}{2}}$ space. Then $C(H)$ is bornological if and only if H is real-compact. [See [108], [111], and [133]. For real-compact, see [22] and [156].]

304 Two comparable compatible (Prob. 8-3-116) topologies must have the same bounded sets. If the larger is Fréchet and the smaller is metrizable, they must be equal. [See [77], sec. 5.]

8-5 POLAR TOPOLOGIES

Let (X, Y) be a dual pair. Topologies for X can be generated by taking polars of subsets of Y as neighborhoods of 0. Since these neighborhoods of 0 must be absorbing, the subsets of Y must be bounded [Theorem 8-4-12]. One can easily imagine assumptions which imply the standard sufficient conditions for the neighborhood filter. Such assumptions are now given and checked in Theorem 8-5-3.

1. Definition Let (X, Y) be a dual pair. A collection $\mathcal{A}$ of subsets of Y is called *polar* if it is not empty, all its members are nonempty bounded sets, and $\mathcal{A}$ is directed under containment and doubling.

The last condition means that for $A \in \mathcal{A}$, there exists $B \in \mathcal{A}$ with $B \supset 2A$.

2. Example Two examples of polar families are the collections of all the (a) finite sets and (b) bounded sets in Y.

3. Theorem Let (X, Y) be a dual pair and $\mathscr{A}$ a polar family in Y. Let $\mathscr{B} = \{A^\circ : A \in \mathscr{A}\}$. Then $\mathscr{B}$ is an additive filterbase of absolutely convex absorbing sets.

PROOF Absorbing is by Theorem 8-4-12; absolutely convex since we are dealing with polars. Now if A_1° and A_2° are members of $\mathscr{B}$ let $A_3 \in \mathscr{A}$ with $A_3 \supset A_1 \cup A_2$. Then $A_3^\circ \subset (A_1 \cup A_2)^\circ = A_1^\circ \cap A_2^\circ$ so $\mathscr{B}$ is a filterbase. Finally, if $A^\circ \in \mathscr{B}$, let $A_1 \in \mathscr{A}$ with $A_1 \supset 2A$. Then $A_1^\circ \subset (2A)^\circ = \frac{1}{2}A^\circ$ and so $\mathscr{B}$ is additive by the criterion of Lemma 7-1-3.

It follows by Theorem 7-1-4 that $\mathscr{B}$ generates a lc topology for X. It is written $T_\mathscr{A}$ and called the *polar topology* associated with $\mathscr{A}$.

4. Example *Let $\mathscr{A}$ be the finite subsets of Y. Then $T_\mathscr{A} = \sigma(X, Y)$.* Let U be a $T_\mathscr{A}$ neighborhood of 0, and we may assume $U = A^\circ$ with $A \in \mathscr{A}$, that is, A is a finite set. Then $U = \cap\{y^\circ : y \in A\}$. Each y° is a $\sigma(X, Y)$ neighborhood of 0 and hence U is. Conversely, if $U \in \mathscr{N}[\sigma(X, Y)]$ we may assume $U = \{x : |[x, y]| \leq \varepsilon\}$ since finite intersections of such sets form a local base for $\mathscr{N}$. Then $U = \{x : |[x, y/\varepsilon]| \leq 1\} = y^\circ \in \mathscr{N}(T_\mathscr{A})$.

For each dual pair (X, Y) there is obviously a largest polar family in Y, namely, the collection $\mathscr{A}$ of bounded sets. The resulting topology $T_\mathscr{A}$ is called the *strong topology* for X and written $\beta(X, Y)$. It is the largest polar topology. As usual, if X is a lc space, $\beta(X, X')$, the strong topology for X, is $T_\mathscr{A}$, where $\mathscr{A}$ is the collection of weak $*$ bounded sets in X'. The *strong $*$ topology* for X' is $\beta(X', X)$.

5. Remark Let (X, Y) be a dual pair and $\mathscr{B}$ the set of barrels in X. Then $\mathscr{B}$ is a base for $\mathscr{N}[\beta(X, Y)]$. [For each bounded $A \subset Y$, A° is a barrel in X — and conversely, by Theorem 8-4-12.] On the other hand, the set of barrels which are $\tau(X, Y)$ neighborhoods of 0 forms a base for $\mathscr{N}[\tau(X, Y)]$. [This is true of any lc topology.] Thus the difference between β and τ is measured exactly by the set of τ barrels which are not τ neighborhoods of 0.

6. Example *If X is a normed space, $\beta(X', X)$, the strong $*$ topology, is the norm topology for X'.* Let $U \in \mathscr{N}(\text{norm})$. We may assume $U = \{f \in X' : \|f\| \leq \varepsilon\}$. Then $U = A^\circ$ for $A = \{x : \|x\| \leq 1/\varepsilon\}$, a bounded set. So $U \in \mathscr{N}(\beta)$. Conversely, if $U \in \mathscr{N}(\beta)$, say $U = A^\circ$ where A is bounded. There exists M such that $A \subset B = \{x : \|x\| \leq M\}$. Then $U = A^\circ \supset B^\circ = \{f : \|f\| \leq 1/M\}$. So $U \in \mathscr{N}(\text{norm})$.

However, in Example 8-5-6, $\beta(X, X')$ need not be the norm topology for X [Example 9-2-7].

Convergence in a polar topology can be described as uniform convergence (Theorem 8-5-7). Let $A \subset Y$ and $x = x_\delta$, a net in X. We say that $x \to 0$ *uniformly on A* if for every $\varepsilon > 0$ there exists δ_0 such that $|[x_\delta, y]| < \varepsilon$ for $\delta > \delta_0$ and all $y \in A$. (This is the same as saying that the net $\hat{x}_\delta$ of functions $\to 0$ uniformly on A.)

7. Theorem Let (X, Y) be a dual pair and $T_{\mathscr{A}}$ a polar topology on X. For a net x in X, $x \to 0$ in $T_{\mathscr{A}}$ if and only if $x \to 0$ uniformly on each $A \in \mathscr{A}$.

PROOF $\Rightarrow$: Let $A \in \mathscr{A}$, $\varepsilon > 0$. Now $x \in \varepsilon A^\circ$ eventually since $\varepsilon A^\circ \in \mathscr{N}$. Thus, eventually, $|[x, y]| \le \varepsilon$ for all $y \in A$. $\Leftarrow$: Let $U \in \mathscr{N}$, say $U = A^\circ$. Then, eventually, $|[x, y]| \le 1$ for all $y \in A$, that is, $x \in A^\circ$ eventually. Hence $x \to 0$ in $T_{\mathscr{A}}$.

8. Theorem Let (X, Y) be a dual pair. A vector topology T for X is a polar topology if and only if it has a local base $\mathscr{B}$ of neighborhoods of 0 which are barrels (in the duality).

PROOF Of course, $\mathscr{N}(T)$ always has a base of T barrels, by Theorem 7-1-2. The barrels referred to here are closed in compatible topologies; see Prob. 8-5-3. $\Rightarrow$: If $T = T_{\mathscr{A}}$, each A°, for $A \in \mathscr{A}$, is a barrel. $\Leftarrow$: Let $\mathscr{A} = \{U^\circ : U \in \mathscr{B}\}$. This is a polar family. [Its members are bounded by Prob. 8-4-2. Also $U^\circ \cup V^\circ \subset (U \cap V)^\circ \in \mathscr{A}$, and $2U^\circ = (\frac{1}{2}U)^\circ \in \mathscr{A}$.] Now let $U \in \mathscr{N}(T_{\mathscr{A}})$, say $U = A^\circ$. Then $A = V^\circ$ for some $V \in \mathscr{B}$ so $U = V^{\circ\circ} \supset V$, that is, $U \in \mathscr{N}(T)$. Conversely, let $U \in \mathscr{B}$. Then $U^\circ \in \mathscr{A}$ so $U = U^{\circ\circ} \in \mathscr{N}(T_{\mathscr{A}})$.

In order to compare polar topologies by comparing the corresponding polar families (Theorem 8-5-11) it will be convenient to make the family as large as possible. There is no loss of generality, by Theorem 8-5-10.

9. Definition A polar family $\mathscr{A}$ is called saturated if $A \in \mathscr{A}$, $B \subset A^{\circ\circ}$ imply $B \in \mathscr{A}$.

10. Theorem Let (X, Y) be a dual pair and $\mathscr{A}$ a polar family in Y. Let $\mathscr{B} = \{B \subset Y : B \subset A^{\circ\circ}$ for some $A \in \mathscr{A}\}$. Then $\mathscr{B}$ is a saturated polar family and $T_{\mathscr{B}} = T_{\mathscr{A}}$. ($\mathscr{B}$ is called the *saturated hull* of $\mathscr{A}$.)

PROOF If $B_1 \subset B^{\circ\circ}$ for some $B \in \mathscr{B}$, then $B \subset A^{\circ\circ}$ for some $A \in \mathscr{A}$, so $B_1 \subset A^{\circ\circ\circ\circ} = A^{\circ\circ}$ and hence $B_1 \in \mathscr{B}$. If $B_1, B_2 \in \mathscr{B}$, $B_1 \subset A_1^{\circ\circ}$, $B_2 \subset A_2^{\circ\circ}$, $A_1 \cup A_2 \subset A$, then $B_1 \cup B_2 \subset A^{\circ\circ}$. Thus $\mathscr{B}$ is actually closed under finite unions. Similarly, $B \in \mathscr{B}$ implies $2B \in \mathscr{B}$. Clearly, $T_{\mathscr{B}} \supset T_{\mathscr{A}}$ since $\mathscr{B} \supset \mathscr{A}$. Finally, if $U \in \mathscr{N}(T_{\mathscr{B}})$, say $U = B^\circ$. Then $B \subset A^{\circ\circ}$ so $U \supset A^{\circ\circ\circ} = A^\circ \in \mathscr{N}(T_{\mathscr{A}})$.

11. Theorem Let X, Y be a dual pair and $\mathscr{A}, \mathscr{B}$ polar families in Y with $\mathscr{A}$ saturated. Then $T_{\mathscr{A}} \supset T_{\mathscr{B}}$ if and only if $\mathscr{A} \supset \mathscr{B}$.

PROOF ⇒: Let $B \in \mathscr{B}$. Then $B^\circ \in \mathscr{N}(T_\mathscr{B})$ and hence $B^\circ \in \mathscr{N}(T_\mathscr{A})$. Thus $B^\circ \supset A^\circ$ for some $A \in \mathscr{A}$. This implies $B \subset B^{\circ\circ} \subset A^{\circ\circ}$ so $B \in \mathscr{A}$.

12. Definition A polar topology for X is called admissible for the dual pair (X, Y) if it is larger than $\sigma(X, Y)$.

Not strictly! Thus $\sigma(X, Y)$ is admissible by Example 8-5-4. There are a largest, β, and a smallest, σ, admissible topology.

13. Theorem Every compatible topology T is admissible. The converse is false.

PROOF First $T \supset \sigma(X, Y)$ since the latter is the smallest compatible topology. Also T is a polar topology by Theorem 8-5-8 and Theorem 7-1-2. If X is a nonreflexive normed space, $\beta(X', X)$ is not compatible for the dual pair (X', X) by Example 8-5-6.

14. Corollary Let (X, T) be a lcs space. Then T is admissible for the dual pair (X, X').

PROOF Indeed, it is compatible.

Note that T is the topology of uniform convergence on the polars (in X') of the neighborhoods of 0 in X.

15. Theorem Let $\mathscr{A}$ be a polar family in Y. Then $T_\mathscr{A}$ is admissible for the dual pair (X, Y) if and only if $\cup\{A^{\circ\circ} : A \in \mathscr{A}\} = Y$.

PROOF Let $\mathscr{A}_1$ be the saturated hull of $\mathscr{A}$. By Theorems 8-5-10 and 8-5-11, and Example 8-5-4, $T_\mathscr{A} \supset \sigma(X, Y)$ if and only if every finite set in Y belongs to $\mathscr{A}_1$. This is equivalent to $Y = \cup\{A : A \in \mathscr{A}_1\} = \cup\{A^{\circ\circ} : A \in \mathscr{A}\}$.

16. Definition An admissible family is a polar family which satisfies the condition of Theorem 8-5-15.

Thus $\mathscr{A}$ is admissible if and only if $T_\mathscr{A}$ is.

17. Remark Most of the examples which arise in practice satisfy $\cup\{A : A \in \mathscr{A}\} = Y$. This is much easier to check and remember.

The very natural problem of characterizing those polar families corresponding to compatible topologies (Mackey–Arens theorem) is more conveniently treated in the next chapter.

18. Definition A lcs space (X, T) is called *absolutely strong* if $T = \beta(X, X')$.

19. Example *The largest lc topology* T *for a vector space* X *is absolutely strong.* T *is compatible with* (X, X') *hence* $T \subset \beta(X, X')$. *The converse is trivial. Recalling that* $X' = X^{\#}$ [Prob. 7-1-7] *we have* $T = \tau(X, X^{\#}) = \beta(X, X^{\#})$. *(Of course, "absolutely strong" implies "relatively strong.") (But* $T \neq \sigma(X, X^{\#})$, *Prob. 8-5-101.)*

20. Example Consider the dual pair (ω, φ), $[x, y] = \sum x_i y_i$. Then $\omega = \varphi^{\#}$ in the sense that the natural embedding of ω in $\varphi^{\#}$ is onto. Thus, by Example 8-5-19, $\tau(\varphi, \omega) = \beta(\varphi, \omega) = $ largest lc topology. Moreover, this topology is bornological by Prob. 8-4-4.

PROBLEMS

In this list, (X, Y) is a dual pair.

1 Show that the collections of finite and of bounded sets in Y are admissible (see Remark 8-5-17) and are, respectively, the smallest and the largest admissible families.

2 Let T be a vector topology for Y which is larger than $\sigma(Y, X)$. Show that the collections of compact and of totally bounded sets in (Y, T) are admissible.

3 Where would the proof of Theorem 8-5-8 break down if it was assumed only that the members of $\mathcal{B}$ are T barrels?

4 Let T be a polar topology for X and T_1 a lcs topology for X. Show that T_1 is a polar topology if and only if it is F linked to T. [$\Rightarrow$: Theorem 8-5-8. $\Leftarrow$: Let $U \in \mathcal{N}(T_1)$ be T closed, $U \supset V \in \mathcal{N}(T_1)$ with V absolutely convex. Apply Theorem 8-5-8 to $\bar{V}$.]

5 Let T be an admissible topology for X. Show that every T complete set is complete in any larger admissible topology. Hence if (X, T) is complete, sequentially or boundedly complete, so is (X, T_1) where T_1 is any larger admissible topology [Prob. 8-5-4 and Theorems 6-1-13 and 6-1-16]. Larger cannot be omitted, by Prob. 8-4-102.

6 Let X be a normed space and $\mathcal{A} = \{nD : n = 1, 2, \ldots\}$ where D is the unit disc in X'. Show that $T_{\mathcal{A}}$ is the norm topology.

7 For a vector space X, show that $\sigma(X^{\#}, X) = \beta(X^{\#}, X)$ [the last sentence of Prob. 7-1-7]. Thus $X^{\#}$ has only one topology admissible for this dual pair; in particular, ω has only one topology admissible for (ω, φ).

8 In the dual pair (l, φ), $[x, y] = \sum x_i y_i$, show that $\beta(l, \varphi)$ is given by $\|\cdot\|_1$. [Note that $(\varphi, \|\cdot\|_\infty)' = l$ by Example 2-3-4 since φ is a dense subspace of c_0.]

---

101 If X is an infinite-dimensional vector space, the largest lc topology is not equal to $\sigma(X, X^{\#})$ [Probs. 7-1-7, 7-1-110, and 2-2-8].

102 Let (X, T) be a lc metric space which is not complete. Show that there is no sequentially complete topology which is compatible with T.

103 Let $\mathcal{A}, \mathcal{B}$ be polar families in Y. Show that $T_{\mathcal{A}} \subset T_{\mathcal{B}}$ if and only if for every $A \in \mathcal{A}$, there exists $B \in \mathcal{B}$ with $B^{\circ\circ} \supset A$ (that is, B "essentially" includes A).

104 Let $\mathcal{A}$ be a polar family. Show that $T_{\mathcal{A}}$ is separated if and only if $\cup \{A : A \in \mathcal{A}\}$ is fundamental in Y.

105 Give an example of a separated nonadmissible polar topology.

106 Let $\mathcal{A}$ be a polar family. For each $A \in \mathcal{A}$ let p be the gauge of A° and P the set of all such p. Show that $T_{\mathcal{A}} = \sigma P$.

107 Let $\mathcal{A}$ be an admissible family in Y and $T = T_{\mathcal{A}}$. Now consider the dual pair $(X, X^{\#})$. Show that $(X, T)' = \cup \{A^{\circ\circ} : A \in \mathcal{A}\}$. (Note that each $A \subset X^{\#}$.) [For $\supset$, use Prob. 4-5-2.]

108 Let (X, Y) be a dual pair, $f \in X^*$. Let $w = \sigma(X, Y)$, $T = w \vee \sigma f = \sigma(Y \cup \{f\})$. Show that T is admissible if and only if $f \in Y$. ⟦⇒: $f^\circ \supset U \in \mathcal{N}(T)$ with U a w barrel, by Remark 8-3-11 and Theorem 8-5-8. So $U \supset V \cap W$ with open $V \in \mathcal{N}(w)$, $W \in \mathcal{N}(\sigma f)$. If $f \notin Y$, $W \supset f^\perp$ is w dense, by Theorem 4-5-10 and Prob. 4-2-5, so $U = \bar{U} \supset V$. Thus f is bounded on V.⟧

109 Let (X, Y) be a dual pair such that X has an admissible topology T_1 which is not compatible (Theorem 8-5-13). Let $f \in (X, T_1)' \backslash Y$. Show that T, Prob. 8-5-108, is a nonadmissible topology which lies between two admissible topologies. Compare Prob. 8-2-1. (A much neater example is given in Example 10-2-8.)

110 Let T be a lcs topology for X. Let $\mathcal{A} = \{U^\circ : U \in \mathcal{N}(T)\}$. Show that T is a polar topology if and only if $T = T_\mathcal{A}$. ($T_\mathcal{A}$ is called the *polar topology associated with T*.)

111 Let K be the family of norm compact sets in c_0. Show that $P_n \to 0$ in (c_0', T_K) where $P_n(x) = x_n$. Equivalently, $\delta^n \to 0$ in (l, T_K) considering the dual pair (c_0, l), $[x, y] = \sum x_i y_i$ ⟦Prob. 3-1-102⟧. However, if "norm" is replaced by "weakly" in the first sentence, the result becomes false. ⟦Let $A = \{\delta^n\} \cup \{0\} \subset c_0$.⟧

112 Let $\mathcal{A}$ be a saturated polar family. Show that $\mathcal{A}$ is closed under finite unions and doubling.

113 Let $\mathcal{A}, \mathcal{B}$ be saturated polar families. Show that $T_{\mathcal{A} \cup \mathcal{B}} = T_\mathcal{A} \vee T_\mathcal{B}$, $T_{\mathcal{A} \cap \mathcal{B}} = T_\mathcal{A} \wedge T_\mathcal{B}$. ⟦Problem 8-5-112 will help in showing that $\mathcal{A} \cap \mathcal{B}$ is polar.⟧

114 Let A be a $\beta(X, Y)$ bounded set. Show that $A^{\circ\circ}$ is $\beta(X, Y)$ bounded.

115 In the dual pair (l, c, b), Example 8-2-7, show that $\beta(c, l) = \|\cdot\|_\infty$. ⟦Let $B = \{\delta^n\} \subset l$. Then $B^\circ \subset \{x : \|x\|_\infty \le 1\}$. Conversely, if $A^\circ \in \mathcal{N}[\beta(l, c)]$, A is $\sigma(l, c_0)$ bounded so, by uniform boundedness, $\|x\|_1 \le M$ for $y \in A$; thus $A^\circ \supset \{x : \|x\|_\infty \le 1/M\}$.⟧

116 In Prob. 8-5-115 show that $[c, \beta(c, l)]' = l + [\lim]$ ⟦Prob. 2-3-3⟧.

117 In the dual pair (l, c, b_1), Example 8-2-7, show that $[c, \beta(c, l)]' = l$.

118 The bipolar Theorem 8-3-8 fails for admissible topologies ⟦Prob. 8-3-106⟧. Reconcile this with the fact that $A^{\circ\circ}$ is T closed if T is admissible.

201 Let (X, Y) be a dual pair and Z a subspace of X^* strictly larger than Y. Show that $\sigma(X, Z)$ is not admissible for (X, Y).

202 The *solid hull* of a set $S \subset \omega$ is $\{y : |y_n| \le |x_n| \text{ for all } n; x \in S\}$. Let X be a sequence space, $X \supset \varphi$; let $\mathcal{A}$ be the family of subsets of the solid hulls of $\{y\}$ for all $y \in X^\alpha$ with $y_n \ge 0$ for all n. Show that $T_\mathcal{A} = \sigma\{p_y : y \in X^\alpha\}$, where $p_y(x) = \sum |x_i y_i|$. This is in the dual pair (X, X^α), $[x, y] = \sum x_i y_i$. This topology is written $|\sigma|(X, X^\alpha)$ and is called the *normal topology*. ⟦See [88], §30, for this and the following problems.⟧

203 Show that $|\sigma|(X, X^\alpha)$ is always compatible with (X, X^α).

204 Show that $|\sigma|(\omega, \omega^\alpha) = \sigma(\omega, \varphi)$; $|\sigma|(\varphi, \omega) = \beta(\varphi, \omega)$, $|\sigma|(l, l^\alpha) = \|\cdot\|_1$; $|\sigma|[l^p, (l^p)^\alpha]$ is strictly smaller than the Mackey topology for $1 < p \le \infty$.

301 A sequence space X is called *perfect* if $X^{\alpha\alpha} = X$. These are equivalent (assuming $X \supset \varphi$):
  (a) X is perfect.
  (b) $\sigma(X, X^\alpha)$ is sequentially complete.
  (c) $|\sigma|(X, X^\alpha)$ is complete.
⟦See [88], 30.5.⟧

302 The topology T^+, Probs. 8-4-124 and 8-4-201, is admissible for (X, X'). ⟦See [146], Proposition 1-3.⟧

8-6 SOME COMPLETE SPACES

The main result of this section is the strong ∗ completeness of the dual of a bornological space, Corollary 8-6-6. However, just as classical analysts pride themselves on pushing down the bounds of an inequality, so functional analysts try to push down topologies which can be proved to be (boundedly, sequentially) complete. For if (X, T_1) is complete, so is (X, T_2) where T_2 is any larger admissible topology [Prob. 8-5-5]. In the case at hand we can get down as far as T_{c_0} (Definition 8-6-1 and Corollary 8-6-6). Other writers have pushed it to a best possible result (Prob. 8-6-301).

A *null sequence* is a sequence which converges to 0. Any finite set is considered to be a null sequence terminating in zeros.

1. Definition Let (X, T) be a lcs space and $\mathscr{A}$ the family of null sequences in X. Then T_{c_0} is the (admissible) topology for X' of uniform convergence on the members of $\mathscr{A}$, that is, $T_{c_0} = T_{\mathscr{A}}$.

Of course, T_{c_0} depends on T, and it is intended that this be reflected in the notation. Thus, for example, $\sigma(X, Y)_{c_0}$ is the topology for Y of uniform convergence on the $\sigma(X, Y)$ null sequences in X. It must be checked that $\mathscr{A}$ is admissible. For example, the union of two null sequences can be arranged to be a null sequence; also, $T_{c_0} \supset \sigma(X', X)$ since each singleton is a null sequence.

2. Lemma Let (X, T) be a lcs space and $f = f_\delta$ a T_{c_0} Cauchy net in X'. Then $f \to g \in X^\#$ pointwise, and g is sequentially continuous on (X, T).

PROOF Since f is T_{c_0} Cauchy, it is weak $*$ Cauchy and the first part follows by completeness of $X^\#$, Prob. 8-1-5. Now let $x_n \to 0$ in (X, T), $\varepsilon > 0$. Let $N = \{x_n\}$. Then $\frac{1}{2}\varepsilon N^\circ \in \mathscr{N}(T_{c_0})$ so there exists δ such that $f_\alpha - f_\delta \in \frac{1}{2}\varepsilon N^\circ$ for $\alpha \geq \delta$. Choose m so that $|f_\delta(x_n)| < \varepsilon/2$ for $n > m$. Then $n > m$ implies that $|g(x_n)| \leq |g(x_n) - f_\delta(x_n)| + |f_\delta(x_n)| < \varepsilon$. [$|g(x_n) - f_\delta(x_n)| = \lim |f_\alpha(x_n) - f_\delta(x_n)| \leq \varepsilon/2$.]

It is natural to consider spaces for which such functions are continuous. These are named after S. Mazur.

3. Definition A Mazur space is a lcs space on which every sequentially continuous linear functional is continuous, that is, $X^s = X'$ where X^s is the set of sequentially continuous linear functionals.

A fairly easy example of a non-Mazur space is given in Prob. 9-3-117. Spaces on which every sequentially continuous linear map is continuous are characterized in Prob. 8-4-301. These are less important.

4. Example *A bornological space must be a Mazur space.* [See Prob. 8-4-5. More precise information is contained in Probs. 8-4-125, 8-4-128, and 8-4-201.]

5. Theorem Let (X, T) be a Mazur space; then (X', T_{c_0}) is complete.

PROOF The Cauchy net of Lemma 8-6-2 converges weak * to $g \in X'$. Hence it converges in T_{c_0}. ⟦T_{c_0} is F linked to $\sigma(X', X)$ by Prob. 8-5-4, and the result follows by Lemma 6-1-11.⟧

6. Corollary If X is a Mazur space, $(X', \text{strong} *)$ is complete. If (X, T) is bornological, (X', T_{c_0}) and $(X', \text{strong} *)$ are complete.

PROOF By Prob. 8-5-5 and Example 8-6-4.

Corollary 8-6-6 is a generalization of the fact (Example 3-1-8) that the dual of a normed space is complete. This is because the norm topology on X is bornological, by Example 8-4-6, and the norm topology on X' is the strong * topology, by Example 8-5-6.

7. Definition A dual pair (X, Y) is called bornological if X has a compatible bornological topology; equivalently, if $\tau(X, Y)$ is bornological.

For example, if X is a nonreflexive Banach space, (X, X') is bornological, but (X', X) is not ⟦Prob. 8-4-6⟧.

8. Remark The style of Definition 8-6-7 is used for other properties; e.g., a *metrizable dual pair* (X, Y) is one such that X has a compatible metrizable topology; equivalently, such that $\tau(X, Y)$ is metrizable. In these two cases the property implies relatively strong. In contrast, complete does not, yet it is still true that X has a complete topology compatible with (X, Y) if and only if $\tau(X, Y)$ is complete ⟦Prob. 8-5-5⟧, so again a dual pair (X, Y) is complete, that is, X has a compatible complete topology, if and only if $\tau(X, Y)$ is complete.

9. Remark The question "Does relatively strong imply complete?" is equivalent to "Is every dual pair complete?". (The answer is "no," of course.) Similarly, relatively strong does not imply bornological exactly because there exist nonbornological dual pairs ⟦Prob. 9-3-103⟧.

Consider now this part of Corollary 8-6-6: (X, T) bornological implies X' strong * complete. The conclusion, about $\beta(X', X)$, is pure duality, so the converse could not possibly be true. ⟦Replace T by a smaller compatible topology.⟧ As expressed in Corollary 8-6-10, the converse could be true, but is not ⟦Prob. 9-3-101⟧.

10. Corollary If (X, Y) is a bornological dual pair, $\beta(Y, X)$ is complete.

There are cases where $\beta(X, Y)$ is not complete ⟦Prob. 9-3-2⟧.

PROBLEMS

In this list (X, T) is a lcs space.

1 Suppose that X is a Mazur space. Show that it remains a Mazur space with any smaller compatible topology. Hence show an example of a Mazur space which is not bornological. Compare Prob. 9-5-108.

2 Show that $X' \subset X^s \subset X^b$ [Theorem 4-4-3].

---

101 Show that a dual pair (X, Y) is Mazur (Remark 8-6-8) if and only if $[X, \sigma(X, Y)]$ is Mazur [Prob. 8-6-1], and sequentially (or boundedly) complete if and only if $[X, \tau(X, Y)]$ is the same.

102 If $T \supset T_1$, show that $T_{c_0} \subset (T_1)_{c_0}$.

103 Show that g, Lemma 8-6-2, has the property that $g(x) \to 0$ if x is a bounded null net.

104 Let K be the family of norm compact sets in c_0. Show that (l, T_K) is a Mazur space [Probs. 8-6-2, 8-5-111, and 8-4-109]. Hence, by Prob. 8-6-1, so is $[l, \sigma(l, c_0)]$. This is a special case of Prob. 12-2-3.

105 Let X be a Mazur space and K the family of compact sets in X. Show that (X', T_K) is complete [Theorem 8-6-5 and Prob. 8-5-5].

106 Show that both (l, φ) and (φ, l) are metrizable dual pairs and neither is complete. [See Prob. 4-1-5, Example 2-3-4, l is dense in ω, and $\tau(\varphi, l) = \| \cdot \|_\infty$.]

107 Give an example in which both (X, Y) and (Y, X) are normed dual pairs.

108 Show that $\sigma(l, \varphi) = \tau(l, \varphi) = \sigma(\omega, \varphi)|_l$.

109 A sequence $\{x^n\}$ in a TVS is called *local-null* if there exists a sequence $\{t_n\}$ with $t_n > 0$, $t_n \to \infty$, $t_n x^n \to 0$. Show that each local-null sequence is null and the converse is true in a linear metric space.

110 Let X be a Banach space, $Y = (X', \text{weak} *)$. Show that $\{f^n\} \subset Y$ is local-null if and only if $\| f^n \| \to 0$. Hence show an example of a null sequence which is not local-null.

111 Show that a bounded linear functional maps local-null sequences into null sequences.

112 Suppose that each null sequence in X is local-null. Show that $X^s = X^b$ [Prob. 8-6-111].

113 Show that a sequence x is local-null if and only if it lies in an absolutely convex bounded closed set B such that $p(x) \to 0$, p the gauge of B. [$\Rightarrow$: $B = $ absolutely convex closure of $\{t_n x^n\}$; Lemma 4-5-1. $\Leftarrow$: $x^n \in \varepsilon_n^2 B$ implies x^n / ε_n^2 is bounded; use Theorem 4-4-1.]

114 Show that a set which absorbs all local-null sequences must be a bornivore.

115 Let X be called *semibornological* if $X' = X^b$. Show that this is a duality invariant and implies Mazur [Prob. 8-6-2].

116 Show that X is bornological if and only if it is semibornological and relatively strong [Prob. 8-4-113].

117 Suppose that X has a topology T compatible with (X, Y) such that (X, T) is Mazur and null sequences are local-null. Show that (X, Y) is bornological [Prob. 8-6-116].

118 Show that a semibornological space need not be C sequential. [See Probs. 8-4-126 or 9-1-106. The opposite implication is also false. See [54].]

119 Let $w = \sigma(c_0, l)$ and let K be the collection of compact sets in (c_0, w). Show that the following four topologies on l all have the same convergent sequences: $\| \cdot \|_1$, $\sigma(l, l^\infty)$, w_{c_0} (Definition 8-6-1), and T_K. The last part of Prob. 8-5-111 is a special case. [See Prob. 8-1-8 for the first two; also $w_{c_0} \subset T_K \subset \beta(l, c_0) = \| \cdot \|_1$. Finally, if $a^n \to 0$ in w_{c_0} and $\| a^n \|_1 \geq 2$, we have $\lim a_k^n = 0$ for each n since $w_{c_0} \supset \sigma(l, c_0)$. Let $k_1 = n_1 = 1$, $\sum_{k=k_i+1}^{k_{i+1}} |a_k^{n_i}| > 1$. Let $x_k^i = 0$ for $k \leq k_i$ and $k > k_{i+1}$ and $x_k^i = \text{sgn } a_k^{n_i}$ otherwise; so each $x^i \in c_0$ and $x^i \to 0$ weakly. Let $A = \{x^i\}$, then $A^\circ \in \mathcal{N}(w_{c_0})$ and $a^n \in A^\circ$ eventually. But $|[a^{n_i}, x^i]| > 1$.]

---

201 Show that the topology given in Prob. 8-6-104 is the largest making l a Mazur space.

202 Let (X, T) be a Mazur space. Let $T_m = \vee \{T' : T'$ is compatible with T and (X, T') is a Mazur space$\}$. Is (X, T_m) a Mazur space?

203 Show that R^R has null sequences which are not local-null. [See [146], p. 344.]

204 Let K be the family of norm compact sets in c_0. Show that in l, T_K and $\sigma(l, c_0)$ have the same convergent sequences.

---

301 Completeness of the dual of a bornological space (Corollary 8-6-6) can be pushed down to a topology smaller than T_{c_0}, namely, T_N where N is the set of local-null sequences (Prob. 8-6-109). This is the greatest extension since a relatively strong space X is bornological if and only if (X', T_N) is complete. [See [88], 28.5 (1), (4).]

NINE

EQUICONTINUITY

9-1 EQUICONTINUOUS SETS

The concept of boundedness in normed spaces was extended to boundedness in general TVS by abstracting the property of being absorbed by a neighborhood of 0. There is another equivalent property which may be used to extend this concept. That it is an extension is shown in Example 9-1-6.

1. Definition Let X, Y be TVSs, S a set of linear maps from X to Y. Then S is called equicontinuous if for each $V \in \mathcal{N}(Y)$ there exists $U \in \mathcal{N}(X)$ such that $f(U) \subset V$ for each $f \in S$.

2. Remark It is obviously equivalent to assume that $\cap \{f^{-1}[V] : f \in S\} \in \mathcal{N}(X)$ for each $V \in \mathcal{N}(Y)$. Moreover, each member of an equicontinuous family is clearly continuous.

Equicontinuity arises mostly for sets of functionals. Thus it will be convenient to express it in the language of duality.

3. Definition Let (X, Y) be a dual pair and T a vector topology for X. A set $S \subset Y$ is called T equicontinuous if S is an equicontinuous set of functions on (X, T).

The next result shows a pair of dual properties: S is "small" if and only if S° is "large." The other formulation is Prob. 9-1-102.

4. Theorem Let (X, Y) be a dual pair, T a vector topology for X and $S \subset Y$. Then S is T equicontinuous if and only if $S° \in \mathcal{N}(T)$.

PROOF $\Rightarrow$: $S° = \cap\{\hat{y}^{-1}[D] : y \in S\}$ where D is the unit disc in $\mathcal{K}$. So $S° \in \mathcal{N}(T)$ by Remark 9-1-2. $\Leftarrow$: Let $V \in \mathcal{N}(\mathcal{K})$. We may assume $V = \{z : |z| \le \varepsilon\}$. Then $\cap\{\hat{y}^{-1}[V] : y \in S\} = \varepsilon S° \in \mathcal{N}(T)$ so S is equicontinuous by Remark 9-1-2.

5. Corollary With X, Y, T as in Theorem 9-1-4, let $U \in \mathcal{N}(T)$. Then $U°$ is T equicontinuous.

PROOF For $U°° \supset U \in \mathcal{N}(T)$.

6. Example Let X be a seminormed space, $S \subset X'$. Then S is equicontinuous if and only if it is norm bounded.

PROOF $\Rightarrow$: Let D be the unit disc in $\mathcal{K}$, $D_\varepsilon = \{x : \|x\| \le \varepsilon\} \subset X$. Then $\cap\{f^{-1}[D] : f \in S\} \supset D_\varepsilon$ for some $\varepsilon > 0$. Thus $\|f\| \le 1/\varepsilon$ for all $f \varepsilon S$. $\Leftarrow$: Let $\varepsilon > 0$, $\|f\| \le M$ for $f \in S$. Then $\|x\| < \varepsilon/M$ implies $|f(x)| < \varepsilon$ for all $f \in S$.

It now turns out that if $T = T_{\mathscr{A}}$ is a polar topology, the T equicontinuous sets are precisely the members of $\mathscr{A}$, at least if $\mathscr{A}$ is saturated. Otherwise the collection of T equicontinuous sets is the saturated hull of $\mathscr{A}$. (See Prob. 9-1-5.)

7. Theorem Let (X, Y) be a dual pair, $S \subset Y$, $\mathscr{A}$ a saturated polar family in Y. Then S is $T_{\mathscr{A}}$ equicontinuous if and only if $S \in \mathscr{A}$.

PROOF $\Leftarrow$: $S° \in \mathcal{N}(T_{\mathscr{A}})$. The result follows by Theorem 9-1-4. $\Rightarrow$: By Theorem 9-1-4, $S° \in \mathcal{N}(T_{\mathscr{A}})$, hence $S° \supset A°$ for some $A \in \mathscr{A}$. Thus $S \subset S°° \subset A°°$ and so $S \in \mathscr{A}$.

Corollary 8-5-14 and the succeeding note say that *if (X, T) is a lcs space, T is the topology of uniform convergence on the T equicontinuous sets in X'.*

Equicontinuity is a measure of smallness. In many respects equicontinuous sets behave as if they were finite dimensional. Evidence of this is given below, in Prob. 9-1-104, and in Corollary 12-1-7. Like boundedness, equicontinuity is inherited by subsets. It is important to expose the relationship between equicontinuity and boundedness. (In Example 9-1-6 they are the same.) A trivial result is that if (X, Y) is a dual pair and $S \subset Y$ is T equicontinuous (T any vector topology for X), then S is bounded (in the dual pair, i.e., in any compatible topology for Y). $[\![S° \in \mathcal{N}(T)$; hence $S°$ is absorbing.$]\!]$ This is the most we can say in this generality since T may be chosen so that T equicontinuous = bounded. $[\![$Take $T = \beta(X, Y)$ and apply Theorem 9-1-7; or see Example 9-1-6.$]\!]$ Indeed, the family of T equicontinuous sets can be any saturated polar family. The situation is different for compatible topologies. Much stronger forms of boundedness are obtained—specifically, strong boundedness (Theorem 9-1-8) and weak compactness (Theorem 9-1-10). The converses, which do not hold in general, are related to important properties of lc spaces (Theorems 9-4-2 and 10-1-11).

8. Theorem Let (X, Y) be a dual pair, $S \subset Y$, and T a compatible topology for X. Then each T equicontinuous set S in Y is $\beta(Y, X)$ bounded, hence bounded in all polar topologies.

See Prob. 9-1-6. PROOF Let $U \in \mathcal{N}[\beta(Y, X)]$; say $U = B^\circ$ where B is a bounded set in X. Now $S^\circ \in \mathcal{N}(T)$ and B is T bounded. Thus S° absorbs B. It follows that B° absorbs S.

The second consequence of equicontinuity is quite a different story. By Theorem 9-1-8, such S is weakly bounded, that is, $\sigma(Y, X)$ bounded. Thus it is weakly totally bounded. ⟦This is the same thing, by Prob. 8-2-6.⟧ What is surprising is that it is weakly relatively compact! Thus the crucial property is completeness, since compact equals totally bounded and complete ⟦Theorem 6-5-7⟧.

9. Definition Let (X, T) be a topological space and $S \subset X$. Then S is called relatively compact if the closure of S is compact.

Thus a set in a TVS is relatively compact if and only if it is included in some compact set ⟦Theorem 4-2-8⟧.

It is not trivial to see that compatible cannot be omitted in Theorem 9-1-10. This is shown in Example 10-2-5.

10. Theorem: Alaoglu–Bourbaki theorem Let (X, Y) be a dual pair and T a compatible topology for X. Then each T equicontinuous set S in Y is $\sigma(Y, X)$ relatively compact. For each $U \in \mathcal{N}(X, T)$, U° is $\sigma(Y, X)$ compact.

PROOF Only the second statement needs proof. ⟦Then $S \subset S^{\circ\circ}$ and $S^\circ \in \mathcal{N}(X, T)$, by Theorem 9-1-4.⟧ As pointed out in the preceding discussion, it is sufficient to prove U° complete. Consider the dual pair $(X, X^\#)$ and write $U^\bullet$ for the polar in this duality. Now $U^\bullet$ is weak $*$ closed, hence weak $*$ complete ⟦Prob. 8-1-5⟧. Also, $U^\bullet \subset Y$. ⟦Let $f \in U^\bullet$. It is bounded on U, hence is T continuous by Theorem 4-4-9. Thus $f \in Y$ since T is compatible.⟧ So $U^\bullet = U^\circ$. Since $\sigma(X^\#, X)|_Y = \sigma(Y, X)$, U° is $\sigma(Y, X)$ complete.

This important result deserves to be stated in its nondual form.

11. Corollary Let X be a lcs space and $U \in \mathcal{N}$. Then U° is weak $*$ compact; also, each equicontinuous set in X' is weak $*$ relatively compact.

12. Theorem: Banach–Alaoglu theorem Let X be a normed space. The unit disc in X' is weak $*$ compact.

PROOF It is the polar of the unit disc in X.

13. Remark An important type of observation is suggested by Theorem 9-1-8. Let (X, T) be a lcs space. Let $\mathscr{E}, \mathscr{S}$ be, respectively, the equicontinuous and the strong * bounded sets in X'. [Strong * stands for $\beta(X', X)$.] Then $\mathscr{E} \subset \mathscr{S}$ by Theorem 9-1-8, and so $T = T_{\mathscr{E}} \subset T_{\mathscr{S}}$. This shows that $T_{\mathscr{S}}$ is larger than every compatible topology, so $T_{\mathscr{S}} \supset \tau(X, X')$. Similarly, Theorem 9-1-10 shows that $T_k \supset \tau(X, X')$ where k is the family of weak * compact sets.

PROBLEMS

In this list (X, Y) is a dual pair unless otherwise defined.

1 Show that any finite set of continuous functions is equicontinuous.

2 If S is T equicontinuous, it is equicontinuous in any vector topology larger than T.

3 Let $f_n(x) = nx$. Show that $\{f_n\}$ is not equicontinuous on R.

4 Let $X = \varphi$ with $\| x \|_\infty$. Let $f_n(x) = \sum_{k=1}^n x_k$. Show that $\{f_n\}$ is not equicontinuous on X.

5 Let T be a vector topology on X, $S \subset Y$. Show that if S is T equicontinuous so is every subset of $S^{\circ\circ}$ [Theorem 9-1-4].

6 Let X be a lcs space, and let $S \subset X'$ have the property that S° is a bornivore. Show that S is strong * bounded. [*Note.* Strong * $= \beta(X', X)$.] [This is contained in the proof of Theorem 9-1-8.]

7 Let S be a dense subspace of a TVS X and E a set of continuous maps from X to a TVS Z such that $\{f \mid S : f \in E\}$ is equicontinuous on S. Show that E is equicontinuous on X. [Consider a closed $V \in \mathcal{N}(Z)$.]

---

101 Let $P_n(x) = x_n$. Show that $\{P_n\}$ is not equicontinuous on ω.

102 Let T be a vector topology for X, $S \subset X$. Show that $S^\circ (\subset Y)$ is T equicontinuous if and only if $S^{\circ\circ} \in \mathcal{N}(T)$.

103 Prove Corollary 9-1-11 and Theorem 9-1-12 without assuming X separated.

104 Let $S \subset Y$. Show that S is $\sigma(X, Y)$ equicontinuous if and only if S is included in a polyhedron (= convex hull of a finite set).

105 Show that equicontinuity is not a duality invariant.

106 Show that the weak topology of an infinite-dimensional normed space is never C sequential. [S° with S a (norm) null sequence; use Prob. 9-1-104.] Hence it is never sequential [Prob. 8-4-201].

107 Let (X, T_X), (Y, T_Y) be FK spaces which include φ with φ dense in Y and $T_Y|_\varphi \supset T_X|_\varphi$. Show that $X \supset Y$. [If $u^n \to y$, then $u^n \to$ some $x \in X$, and $x = y$ since $u^n \to x$ and y in ω.] This result may be regarded as a converse of Corollary 5-5-8.

108 Let X, Y be FK spaces which include φ with φ dense in Y and such that every subset of φ which is bounded in Y is also bounded in X. Show that $X \supset Y$ [Prob. 9-1-107, Theorem 4-4-9, and Example 4-4-7].

109 Let Y be a BK space in which φ is dense. Let $D = \{x \in \varphi : \| x \| \leq 1\}$. If X is a FK space containing D as a bounded subset, show that $X \supset Y$ [Prob. 9-1-108].

110 Let X be an FK space which includes φ. Show that $X \supset l$ if and only if the absolutely convex hull of $\{\delta^n\}$ is bounded in X. If X is lc the condition is if and only if $\{\delta^n\}$ is bounded in X [Prob. 9-1-109].

111 Let $bv = \{x : \sum |x_n - x_{n+1}| < \infty\}$ with $\| x \| = |\lim x| + \sum |x_n - x_{n+1}|$. Show that a lc FK space X which includes φ also includes $bv \cap c_0$ if and only if $\{\sigma^n\}$ is bounded in X where $\sigma^n = \sum_{k=1}^n \delta^k$. Extend as in Prob. 9-1-110 [Prob. 9-1-109].

112 An FK space X is called *semiconservative* if it includes φ and $\sum f(\delta^k)$ is convergent for each $f \in X'$. Show that if $X \supset c_0$ it must be semiconservative and if X is locally convex and semiconservative then $X \supset bv \cap c_0$ [Corollary 5-5-8 and Prob. 9-1-111].

113 For an FK space $X \supset \varphi$, define $X^f = \{\{f(\delta^n)\} : f \in X'\}$. Show that X is semiconservative if and only if $X^f \subset$ cs.

114 Let X, Y be FK spaces which include φ. Suppose that X is locally convex and that φ is dense in Y. Show that $X \supset Y$ if and only if $X^f \subset Y^f$. [$\Rightarrow$: Corollary 5-5-8. $\Leftarrow$: In Prob. 9-1-108, weakly bounded is sufficient.]

115 Let X be a lc FK space which includes φ. Show that $X \supset c_0$ if and only if $\sum |f(\delta^k)| < \infty$ for all $f \in X'$ [Prob. 9-1-114]. Compare Prob. 9-1-112. The ideas of Probs. 9-1-107 to 9-1-115 are continued in [136].

116 Let X be a weakly sequentially complete lc FK space with $X \supset c_0$. Show that $X \supset l^\infty$ [Prob. 9-1-115].

---

201 Can "lc" be omitted in Prob. 9-1-115?

9-2 THE MACKEY–ARENS THEOREM

Let (X, Y) be a dual pair. The problem of characterizing compatible families $\mathscr{A}$ of sets in Y remains to be solved. These are defined by the requirement that $T_{\mathscr{A}}$ is compatible with the dual pair. The largest such family is the collection of T equicontinuous sets for all compatible topologies T [i.e., the collection of $\tau(X, Y)$ equicontinuous sets], but these are unrecognizable at present. We know that they are $\beta(Y, X)$ bounded and $\sigma(Y, X)$ relatively compact [Theorems 9-1-8 and 9-1-10]. However, the family of such sets is too large [Prob. 9-3-121]. Another easily derived condition turns out to be exactly the right one.

1. Lemma Let (X, Y) be a dual pair and T a compatible topology for X. Then each T equicontinuous set S in Y has the property that $S^{\circ\circ}$ is $\sigma(Y, X)$ compact.

PROOF $S^\circ \in \mathscr{N}(T)$ by Theorem 9-1-4, and the Alaoglu–Bourbaki Theorem 9-1-10 yields the result.

2. Lemma Let (X, Y) be a dual pair and $\mathscr{A}$ the family of absolutely convex $\sigma(Y, X)$ compact sets in Y. Then $\mathscr{A}$ is a compatible family, that is, $T_{\mathscr{A}}$ is compatible with the dual pair.

PROOF First, $\mathscr{A}$ is a polar family, Definition 8-5-1. The only nontrivial part is the proof that $\mathscr{A}$ is directed under containment. This is by Lemma 6-5-8. Next, $\mathscr{A}$ is admissible since $\cup \mathscr{A} = Y$. It remains to show that $T = T_{\mathscr{A}}$ is compatible. Certainly $(X, T)' \supset Y$ since T is admissible. Now let $f \in (X, T)'$. Let $U = \{x \in X : |f(x)| \le 1\}$. Then $U \in \mathscr{N}(T)$ so $U \supset A^\circ$ where $A \in \mathscr{A}$. Now consider the dual pair $(X, X^\#)$. Considering $A \subset X^\#$, A° has the same meaning as before, so $U \supset A^\circ$. Thus $U^\bullet \subset A^{\circ\bullet}$, where the solid circles denote polars in $X^\#$ of sets in X. Now $A^{\circ\bullet}$ is the $\sigma(X^\#, X)$ closure of A by the bipolar theorem. But this is A. [A is $\sigma(Y, X)$ compact, hence $\sigma(X^\#, X)$ compact, since $\sigma(X^\#, X)|_Y = \sigma(Y, X)$ and $A \subset Y$. Since $\sigma(X^\#, X)$ is separated this implies that A is closed.] We thus have $U^\bullet \subset A$. Since $f \in U^\bullet$, it follows that $f \in Y$.

3. Theorem: Mackey–Arens theorem The largest topology for X compatible with the dual pair (X, Y) is the topology $\tau(X, Y)$ of uniform convergence on the family $\mathscr{A}$ of absolutely convex $\sigma(Y, X)$ compact sets in Y. A lc topology for X is compatible if and only if it lies between $\sigma(X, Y)$ and $\tau(X, Y)$.

PROOF The truth of this theorem justifies our statement that the topology $T_{\mathscr{A}}$ is the same as the one we called τ earlier (Theorem 8-2-14). First, $T_{\mathscr{A}}$ is compatible by Lemma 9-2-2. Further, if T is any compatible topology, let $U \in \mathscr{N}(T)$ be a barrel. Then $U^\circ \in \mathscr{A}$ by Lemma 9-2-1, so $U = U^{\circ\circ} \in \mathscr{N}(T_{\mathscr{A}})$ and so $T_{\mathscr{A}} \supset T$.

Local convexity cannot be omitted; see Probs. 9-2-110 and 9-3-123.

4. Corollary Let (X, Y) be a dual pair. The largest compatible family for Y is the saturated hull of the collection of absolutely convex $\sigma(Y, X)$ compact sets. An admissible family $\mathscr{A}$ is compatible if and only if for each $A \in \mathscr{A}$, $A^{\circ\circ}$ is $\sigma(Y, X)$ compact.

5. Corollary Let X be a lc space. Then each absolutely convex weak $*$ compact set in X' is strong $*$ bounded.

PROOF Such a set is $\tau(X, X')$ is equicontinuous. The result follows by Theorem 9-1-8.

6. Example *"Absolutely convex" cannot be omitted in Corollary 9-2-5.* Let $X = (\varphi, \| \cdot \|_\infty)$, $P_n(x) = x_n$, $A = \{nP_n\} \cup \{0\} \subset X'$. Since $nP_n(x) = nx_n = 0$ eventually for each x, we have $nP_n \to 0$, weak $*$, and so A is weak $*$ compact. However, as pointed out in Prob. 8-5-8, $X' = l$ and $\beta(X', X)$ is given by $\| \cdot \|_1$. Since $\| nP_n \|_1 = n$, A is not strong $*$ bounded.

7. Example *"Absolutely convex" cannot be omitted in Lemma 9-2-2, Theorem 9-2-3, and Corollary 9-2-4.* Let k be the family of $\sigma(Y, X)$ compact sets in Y. It is easy to see that k is admissible and that $T_k \supset \tau(X, Y)$. It is possible that T_k is strictly larger: take X, A as in Example 9-2-6 and $Y = X'$. Then A is not equicontinuous in any compatible topology for X since it is not $\beta(Y, X)$ bounded [Theorem 9-1-8]. Thus T_k is not compatible. (Other proofs are given in Probs. 9-2-108 and 9-2-109. In Prob. 9-3-121 an example is given which is also strongly bounded!) Since $\beta(X, X') \supset T_k$ it follows that $\beta(X, X')$ is not the norm topology for X.

What is happening in Examples 9-2-6 and 9-2-7 is that A is weak $*$ compact but $A^{\circ\circ}$ is not, by Corollary 9-2-5. An important class of spaces is considered which avoids this pathology.

8. Definition A lc space X is said to have the convex compactness property if, for each compact set $K \subset X$, the absolutely convex closure K_0, of K, is also compact or, equivalently, is complete.

The equivalence is because K_0 is always totally bounded [Theorem 7-1-5] and hence compact if and only if complete [Theorem 6-5-7].

9. Example $[l, \sigma(l, \varphi)]$ *does not have the convex compactness property*, by Example 9-2-7, as pointed out above.

10. Example *A boundedly complete lc space has the convex compactness property*, for K_0 (Definition 9-2-8) is complete. The converse is false [Remark 9-2-16, or Prob. 10-2-1 with Theorem 14-2-4].

11. Theorem If a lc space (X, T) has the convex compactness property it does so in any larger admissible topology T_1.

PROOF Let K be T_1 compact, K_0, K_1 the absolutely convex T, T_1 closures of K respectively. Then K_0 is T complete, hence T_1 complete [Prob. 8-5-5]. Since K_1 is a T_1 closed subset, it is T_1 complete.

As usual a dual pair (X, Y) is said to have the *convex compactness property* if (X, T) has the property for some compatible T or, equivalently, by Theorem 9-2-11, if $[X, \tau(X, Y)]$ has the property.

12. Theorem Let X be a lc space and K the family of compact sets in X. Then X has the convex compactness property if and only if T_K is compatible with (X', X).

PROOF $\Rightarrow$: Each $A \in K$ has the property that $A^{\circ\circ}$ is compact, hence weakly compact. By Corollary 9-2-4, K is a compatible family. $\Leftarrow$: Let A be compact. Then $A^\circ \in \mathcal{N}(T_K)$ so $A^{\circ\circ}$ is weakly compact by the Alaoglu–Bourbaki Theorem 9-1-10. It follows that $A^{\circ\circ}$ is a complete subset of X by Prob. 8-5-5, so $A^{\circ\circ}$ is compact.

13. Example If (X, T) is a lc space with the convex compactness property, T_{c_0} (*Definition 8-6-1*) *is compatible with* (X', X). $T_{c_0} \subset T_K$ (Theorem 9-2-12) since $N \cup \{0\}$ is compact whenever N is a null sequence. The result follows by Theorem 9-2-12.

A useful completeness result (Example 9-2-15) follows from these ideas. A more general form isolates the ideas.

14. Theorem Let (X, T) be a Mazur space with the convex compactness property. Then (X', X) is complete.

PROOF The topology T_{c_0} is complete [Theorem 8-6-5] and compatible [Example 9-2-13]. Thus X' has a complete topology compatible with the dual pair (X', X); this is what the conclusion says [Remark 8-6-8].

15. Example *Let X be a lc Fréchet space. Then (X', X) is complete.* This follows from Theorem 9-2-14 and Example 9-2-10.

Example 9-2-15 is false for a normed space [Prob. 8-6-106]; hence "convex compactness" cannot be omitted in Theorem 9-2-14. Also, "Mazur" cannot be omitted in Theorem 9-2-14 [Prob. 10-2-106].

16. Remark The convex compactness property and sequential completeness are independent properties [Probs. 9-2-301 and 14-2-102]. They are both implied by bounded completeness, and often have the same consequences. Illustrations of this are Theorems 10-4-8 and 10-4-11; Remark 10-4-14; Example 9-2-13 and Prob. 11-1-116; Probs. 9-4-108 and 11-1-114; and Theorem 9-2-14 and Prob. 11-1-117.

PROBLEMS

In this list (X, Y) is a dual pair or (X, T) is a lcs space.

1 Show that the following conditions are equivalent for an absolutely convex closed set $U \subset X$:

    (a) $U \in \mathcal{N}(T)$ for some compatible topology T.

    (b) $U \in \mathcal{N}[\tau(X, Y)]$.

    (c) U° is $\sigma(Y, X)$ compact.

[Mackey–Arens and Alaoglu–Bourbaki Theorems.]

2 Show that the saturated hull of K (Examples 9-2-6 and 9-2-7) contains closed sets which are not $\sigma(l, \varphi)$ compact.

3 Explain why absolute convexity does not enter into the definition of such topologies as $\beta(X, Y)$, whereas it is crucial in the definition of $\tau(X, Y)$.

4 Show that $[X, \sigma(X, Y)]$ has the convex compactness property if and only if $[\sigma(X, Y)]_K = \tau(Y, X)$ where K is as in Theorem 9-2-12.

---

101 Let (X, T) be a lc space, not necessarily separated. Show that a lc topology T_1 for X is compatible with T if and only if weak $\subset T_1 \subset$ Mackey, that is, $\sigma(X, X') \subset T_1 \subset \tau(X, X')$.

102 Let $\mathcal{A}$ be a saturated admissible family in Y. Show that $\mathcal{A}$ is compatible if and only if every member of $\mathcal{A}$ is $\sigma(Y, X)$ relatively compact. (Examples 9-2-6 and 9-2-7 show that saturated cannot be omitted.)

103 In Prob. 9-2-102, omit "saturated" and replace the equivalent condition by "for each $A \in \mathcal{A}$, $A^{\circ\circ}$ is $\sigma(Y, X)$ compact."

104 Let X be the set of real continuous functions on $[-1, 1]$, each of which vanishes on a neighborhood of 0, with $\|\cdot\|_\infty$. Let $u_n(f) = nf(1/n)$. Show that $\{u_n\} \cup \{0\}$ is a weak $*$ compact set in X' which is not strong $*$ bounded [like Example 9-2-6].

105 Let $S \subset X$ be an absolutely convex closed set such that S° is $\sigma(Y, X)$ compact. Let $f \in X^*$ be bounded on S. Show that $f \in Y$.

106 In Prob. 9-2-105, "closed" cannot be omitted [$S = f^\perp$].

107 In Prob. 9-2-105, "absolutely convex" cannot be omitted.

108 In Examples 9-2-6 and 9-2-7, let $U = \{x : |x_n| \leq 1/n \text{ for all } n\}$. Show that $U \in \mathcal{N}(T_k) \backslash \mathcal{N}[\tau(\varphi, l)]$.

109 In Examples 9-2-6 and 9-2-7, show that $(\varphi, T_k)' = \varphi^{\#} = \omega$.

110 Let T_0 be a separated vector topology for X such that $(X, T_0)' = \{0\}$ (Prob. 2-3-119). Let $T = \tau(X, Y) \vee T_0$. Show that $(X, T)' = Y$ but T does not lie between $\sigma(X, Y)$ and $\tau(X, Y)$.

111 An N-complete space must have the convex compactness property.

112 Show that a TVS has the convex compactness property if and only if the convex closure of each compact set is compact.

113 Let X be a nonreflexive Banach space. Show that X is $\tau(X'', X')$ dense in X''; hence $\tau(X'', X')|_X \neq \tau(X, X')$. [The latter is complete.] It follows that a dense subspace of a relatively strong space need not be relatively strong.

114 Let S be a subspace of Y which is total over X and T a compatible topology for Y. Show that $T|_S$ is compatible with (S, X), i.e., it is smaller than $\tau(S, X)$. [Each $\sigma(X, Y)$ compact set is $\sigma(X, S)$ compact.]

115 Show that $\beta(l, \varphi) = \|\cdot\|_1$ has the convex compactness property but the dual pair (l, φ) does not. Example 9-2-9 gives a weaker result. [$\tau(l, \varphi)$ is Mazur by Prob. 8-6-106, but $\tau(\varphi, l) = \|\cdot\|_\infty$ is not complete.]

116 If X has the convex compactness property, show that its compact sets are strongly bounded [Theorems 9-2-12 and 9-1-8].

---

201 Let (X, Y) be a dual pair which is Mazur and has the convex compactness property. Must (Y, X) be complete?

---

301 A sequentially complete lcs space need not have the convex compactness property. [See [112], Example 2.1.]

9-3 BARRELLED SPACES

1. Definition A barrelled space is a lcs space in which every barrel is a neighborhood of 0.

Recall that a barrel is an absolutely convex absorbing closed set. Other names used for this important concept are tunneled, tonnelé, tonnelierte, and kegly.

An example of a nonbarrelled space was given in Prob. 3-3-4.

2. Example *A lcs space which is of second category is barrelled. In particular, every lc Fréchet space is barrelled.* This was given for normed spaces in Lemma 3-3-2, where it was shown that a barrel A must have nonempty interior. Then $A \in \mathcal{N}$ by Prob. 4-2-8.

3. Example : *A first-category barrelled space* The example is the largest lc topology on an infinite-dimensional vector space. The relevant citations are Probs. 7-1-5 and 7-1-10.

The initial appearance of barrelled spaces was in response to the question of finding an internal characterization of spaces for which the uniform boundedness principle holds. The uniform boundedness Theorem 3-3-6 gives a sufficient condition for a pointwise bounded family of linear maps to be equicontinuous. (Example 9-1-6 points out that "uniformly bounded" equals "equicontinuous.") The problem is solved by the equivalence of (a) and (b) in the following theorem, which is the most general form of the uniform boundedness principle for lcs spaces.

4. Theorem Let X be a lcs space. The following three conditions are equivalent:

(a) For any lc space Y, every pointwise bounded family S of continuous linear maps from X to Y is equicontinuous.
(b) X is barrelled.
(c) Every weak * bounded subset of X' is equicontinuous.

PROOF The condition S is pointwise bounded means that $S(x) = \{f(x): f \in S\}$ is a bounded set in Y for each $x \in X$. (b) $\Rightarrow$ (a). Let $V \in \mathcal{N}(Y)$ be a barrel. Then $A = \cap \{f^{-1}[V]: f \in S\}$ is a barrel in X. $[\![A$ is closed since each f is continuous; A is absorbing for if $x \in X, S(x)$ is bounded, hence absorbed by V. Thus x is absorbed by $A.]\!]$ So $A \in \mathcal{N}(X)$ and S is equicontinuous. (a) $\Rightarrow$ (c). Take $Y = \mathcal{K}$ in (a). The weak * topology is the pointwise topology $[\![$ Remark 8-1-6 $]\!]$. (c) $\Rightarrow$ (b). Let A be a barrel in X. Then $A°$ is weak * bounded since A is absorbing $[\![$ Theorem 8-4-12 $]\!]$ hence equicontinuous. Thus $A = A°° \in \mathcal{N}(X)$ by Theorem 9-1-4.

The Banach–Steinhaus closure theorem appeared along with uniform boundedness in Sec. 3-3, and will do so again (Theorem 9-3-7). First, we give a general closure theorem for equicontinuous sets.

5. Lemma Let X be a lcs space and $S \subset X'$ an equicontinuous set. Then the weak * closure of S in X' is equicontinuous.

PROOF $\bar{S} \subset S°°$ by the bipolar theorem, and $S°°° = S° \in \mathcal{N}$.

This is a special case of Lemma 9-3-6. We gave it because it is so easy.

6. Lemma Let X, Y be lcs spaces, S an equicontinuous set of linear maps from X to Y, and $g: X \to Y$ a linear map such that some net in S converges to g pointwise on X. Then g is continuous; moreover, $S \cup \{g\}$ is equicontinuous.

PROOF Let $V \in \mathcal{N}(Y)$ be closed. Let $U = \cap \{f^{-1}[V]: f \in S\}$ so that $U \in \mathcal{N}(X)$. The result will follow when it is proved that $g^{-1}[V] \supset U$. To this end, let $x \in U$. Then $f(x) \in V$ for each $f \in S$ and so, since V is closed, $g(x) \in V$.

7. Theorem: Banach–Steinhaus closure theorem Let X be a barrelled space, Y a lcs space, $\{f_n\}$ a sequence of continuous linear maps and $g: X \to Y$ a linear map such that $f_n \to g$ pointwise on X. Then g is continuous.

PROOF For each $x \in X$, $\{f_n(x)\}$ is convergent, hence bounded $[\![$Prob. 4-4-9$]\!]$. The result follows by Theorem 9-3-4 and Lemma 9-3-6.

Theorem 9-3-7 yields a sequential completeness result. This will be promoted to bounded completeness in Theorem 9-3-11.

8. Example *Let X be barrelled. Then $(X', weak *)$ is sequentially complete. Hence, by Prob. 8-5-5, X' is sequentially complete with every topology admissible for (X', X).* It is sufficient to prove that X' is sequentially closed in X^{*} since the latter is weak $*$ complete by Prob. 8-1-5. This is immediate from Theorem 9-3-7 and the fact that pointwise and weak $*$ convergence are the same $[\![$Remark 8-1-6$]\!]$.

9. Example : *A dense sequentially closed subspace* Let X be barrelled and $X' \neq X^{*}$ $[\![$Example 3-3-14$]\!]$. Then X' is sequentially closed in $(X^{*}, weak *)$, by Example 9-3-8, and dense by Theorem 8-1-10.

10. Theorem A lcs space (X, T) is absolutely strong, that is, $T = \beta(X, X')$, if and only if T is barrelled.

PROOF This is the content of Remark 8-5-5.

Thus $\tau(X, Y)$ is barrelled if and only if it is equal to $\beta(X, Y)$. (A natural conjecture is false: $\beta(X, Y)$ need not be barrelled in general, Theorem 15-4-6.)

Theorem 9-3-10 allows a brief deduction of the fact that (ω, φ) has only one admissible topology (on ω). (This is a special case of Prob. 8-5-7.) Since $\sigma(\omega, \varphi)$ is the Fréchet topology $[\![$Prob. 8-2-4$]\!]$ it is barrelled and so equal to $\beta(\omega, \varphi)$.

As usual a dual pair (X, Y) is called *barrelled* if X has a compatible barrelled topology. Equivalent conditions are that $\tau(X, Y)$ is barrelled, $\tau(X, Y) = \beta(X, Y)$, and every admissible topology is compatible.

Example 9-3-8 can now be improved.

11. Theorem If X is barrelled, $(X', weak *)$ is boundedly complete.

PROOF Let B be a bounded closed set. Then $B° \in \mathcal{N}[\beta(X, X')]$, hence $B°°$ is weak $*$ compact $[\![$Alaoglu–Bourbaki theorem; β is compatible by Theorem 9-3-10$]\!]$. Since B is a closed subset of $B°°$ it is compact.

12. Example : *A boundedly complete noncomplete space* This is the same as Example 9-3-9 except that now we know that X' is boundedly complete.

The converse of Theorem 9-3-11 could not be true. The conclusion is a duality result while the hypothesis depends on the topology of X; giving X a smaller compatible topology would not affect the conclusion. Theorem 9-3-13 states the result in a form in which both hypothesis and conclusion refer to duality properties. In this form the converse holds. The converse of a similar restatement of Example 9-3-8 is false [Prob. 9-3-108].

13. Theorem Let (X, Y) be a dual pair. Then Y is $\sigma(Y, X)$ boundedly complete if and only if (X, Y) is barrelled.

PROOF $\Leftarrow$: Theorem 9-3-11. $\Rightarrow$: Let U be a barrel in X. Then U° is bounded and closed, hence $\sigma(Y, X)$ compact. [Recall that bounded sets are totally bounded by Prob. 8-2-6; U° is compact by Theorem 6-5-7.] Since it is also absolutely convex $U = U^{\circ\circ} \in \mathcal{N}[\tau]$.

Theorem 9-3-13 may remind the reader of a superficially similar result: if (X, Y) is bornological, $\beta(Y, X)$ is complete (Corollary 8-6-10). The two results are actually of a dissimilar nature. For one thing, the converse of the latter result is not true [Prob. 9-3-101].

14. Example: *A sequentially complete space which is not boundedly complete* The space is l with its weak topology. The relevant citations are Prob. 8-1-11, Theorem 9-3-13, and Prob. 8-4-6.

PROBLEMS

1 Let X be a barrelled space. Show that (X', T) has the convex compactness property for all admissible T. [See Theorem 9-3-13. It also follows from Theorems 9-3-10 and 9-2-12.]

2 Give an example of a noncomplete normed space which is barrelled [Prob. 3-1-5]. Deduce that $\beta(X, Y)$ is not always complete.

3 Show that every linear map u onto a barrelled space (from a lc space) is almost open. [$u[V]$ is a barrel if V is.]

---

101 Let X be a nonreflexive Banach space. Show that $\beta(X, X')$ is complete but (X', X) is not bornological [Theorem 9-3-10 and Prob. 8-4-6].

102 The dual pair (φ, l) is not barrelled.

103 Let X be a nonreflexive Banach space. Then (X, X') is barrelled and bornological but (X', X) is neither [Prob. 8-4-6].

104 (Convergence lemma). Let X, Y be TVSs and f an equicontinuous net of linear maps from X to Y. Show that $\{x : f(x) \text{ is Cauchy}\}$ and $\{x : f(x) \to 0\}$ are closed vector subspaces of X. Compare Prob. 3-3-3.

105 A lcs space is barrelled if and only if every lower semicontinuous seminorm is continuous [Prob. 8-3-113].

106 Let X be an infinite-dimensional vector space with its largest locally convex topology. Show that X is barrelled but must have a closed absorbing set with empty interior [Prob. 5-2-301].

107 Let X be a barrelled space which has a Schauder basis. Show that X is a Mazur space [Theorem 9-3-7]. Hence X' is strong $*$ complete. Compare Prob. 10-3-301.

108 Let $X = [l^\infty, \tau(l^\infty, l)]$. Show that $\sigma(X', X)$ is sequentially complete [Prob. 8-1-11] but X is not barrelled [Prob. 8-4-6]. See Prob. 9-3-302.

109 Let X, Y be lcs spaces and $B(X, Y)$ the set of continuous linear maps from X to Y. Give B the relative topology of Y^X. Rewrite Lemma 9-3-6 as a statement about the closure of an equicontinuous set.

110 In Lemma 9-3-6, the conclusion that g is continuous can be reached if either one (but not both) of these assumptions is dropped: Y is separated, g is linear.

111 Let B be a barrel in a lcs space X which is not a neighborhood of 0. Show that there is a discontinuous linear functional on X which is bounded on B. [If not, $B^\bullet = B^\circ$ so $B \in \mathcal{N}$.]

112 Let (X, T) be a RS space. Prove that it is barrelled if and only if T is larger than every lc topology which is F linked to T. [$\Rightarrow$: Theorem 9-3-10 and Prob. 8-5-4. $\Leftarrow$: Theorem 8-5-8.]

113 Let a separated TVS (X, T) be called *ultrabarrelled* if T is larger than every vector topology which is F linked to T. Show that a second-category TVS, e.g., a Fréchet space, must be ultrabarrelled.

114 Let (X, T) be ultrabarrelled. Show that (X, T_c) is barrelled if it is separated [Probs. 9-3-112 and 7-1-103].

115 Show that on $l^{1/2}$, T_c is given by $\|\cdot\|_1$. [See Prob. 7-1-103; each is relatively strong.]

116 Show that $(l^{1/2}, T_c)$ is barrelled but not ultrabarrelled. [See Probs. 9-3-114 and 9-3-115; also, $\|\cdot\|_{1/2}$ is F linked since $\{x : \|x\|_{1/2} \leq 1\} = \cap\{x : \sum_{k=1}^n |x_k|^{1/2} \leq 1\}$. Another proof follows from Prob. 12-4-107.]

117 Let X be normed, barrelled, and not complete (e.g., Prob. 9-3-2). Show that $(X', \text{weak} *)$ is not a Mazur space; hence "compatible" cannot be omitted in Prob. 8-6-1 [Corollary 8-6-6].

118 Show that $(X, X^\#)$ is barrelled [Example 8-5-19].

119 Show that $\sigma(X^\#, X)$ is barrelled [Prob. 8-5-7].

120 Show that (l, φ) is not barrelled [Probs. 8-5-8 and 8-6-106].

121 Let X be normed, barrelled, and not complete (e.g., Prob. 9-3-2). Let $\{x_n\}$ be as in Prob. 5-1-104, $S = \{2^n x_n\} \cup \{0\}$. Show that S is weakly compact and strongly bounded [Theorem 9-3-10] but not equicontinuous in any compatible topology for X'. [$u_n = \sum_{k=1}^n x_k \in S^{\circ\circ}$; if $S^{\circ\circ}$ is weakly compact, the weak closure of $\{u_n\}$ is weakly complete so $\sum x_k$ converges by Lemma 6-1-11.]

122 Let $f \in X^\# \backslash X'$ be weak $*$ sequentially continuous (see, for example, Prob. 9-3-117). Show that $f^\perp$ is sequentially closed and not closed [Theorem 4-4-10].

123 Show that $\|\cdot\|_{1/2}$ does not lie between $\sigma(l^{1/2}, l^\infty)$ and $\tau(l^{1/2}, l^\infty) = \|\cdot\|_1$ even though $(l, \|\cdot\|_{1/2})' = l^\infty$ [Probs. 9-3-115, 9-3-116, and 7-1-103].

124 Suppose that X is a lcs space which has a dense barrelled subspace S. Show that X is barrelled. Hence show that any subspace which includes S is barrelled.

125 In the dual pair $(c, l, \sum x_i y_i)$, show that $\tau(c, l) \neq \|\cdot\|_\infty$ [even though $\tau(c, c') = \|\cdot\|_\infty$] [Example 9-3-8 and Prob. 8-2-109].

126 Let T be a vector topology for X such that if $T' \supset T$ and T' is F linked to T then $T' = T$. Show that (X, T) is ultrabarrelled. [Consider $T \vee T_1$.]

127 Let (X, T) be a TVS and T' a lc topology for X with $T' \supset T$. Let T'' be generated by the set of absolutely convex T' neighborhoods of 0 which are T closed. Show that $T'' \supset T$.

128 Let (X, T) be ultrabarrelled and not lc. Show that X has no lc topology $\supset T$ [Prob. 9-3-127].

129 Let (X, T) be a Fréchet space with $X' = \{0\}$, $X \neq \{0\}$. Show that X has no lcs topology which is comparable with T [Prob. 9-3-128].

130 Let (X, T) be ultrabarrelled and F a pointwise bounded set of continuous linear maps from X to a TVS Y. Show that F is equicontinuous. [For closed $V \in \mathcal{N}(Y)$, let $U = \cap\{f^{-1}[V] : f \in F\}$. The set of such U generates a topology which is F linked to T.]

131 State and prove an analogue of the Banach–Steinhaus closure Theorem 9-3-7 for X ultra-barrelled and Y a TVS [Prob. 9-3-130].

132 Let X be an FK space such that $\{P_n\}$ is equicontinuous, where $P_n(x) = x_n$. Show that $c_0 \cap X$ is closed in X [Prob. 9-3-104]. This gives another solution of Prob. 9-1-101.

133 Let X be an FK space $\supset \varphi$ such that $c_0 \cap X$ is not closed in X. Show that X must contain an unbounded sequence [Prob. 9-3-132 and Theorem 9-3-4].

134 Let A be a conservative matrix such that c_A contains a bounded divergent sequence. Show that it must contain an unbounded sequence. $[\![c_A \subset l^\infty$ implies A coregular by Prob. 8-1-112. Apply Probs. 9-3-133 and 7-2-123.$]\!]$

135 Suppose that a conservative matrix A has a left inverse B with $\|B\| < \infty$. Show that the topology of c_A is given by $\|\cdot\|_\infty$ on c. Hence c is closed in c_A and A is coregular. $[\![$In Example 5-5-15, $\|x\|_\infty \le \|B\| \, p_0(x); |P_{2n}(x)| \le \|x\|_\infty; |P_{2n-1}(x)| \le \|A\| \, \|x\|_\infty$. See also Prob. 8-1-112(b).$]\!]$

136 (J. Copping). With the assumptions of Prob. 9-3-135 show that c_A contains no bounded divergent sequence $[\![$Probs. 9-3-135 and 7-2-123$]\!]$. Compare Prob. 5-5-302.

137 Suppose that a conservative matrix A has an inverse B with $\|B\| < \infty$. Show that B is conservative. $[\![$For $x \in c$, $Bx \in c_A \cap l^\infty = c$ by Prob. 9-3-135 and the hint to Prob. 7-2-122.$]\!]$ If B is row-finite, $c_A = c$ $[\![x = B(Ax)]\!]$. "Row-finite" cannot be omitted $[\![$Prob. 5-5-116$]\!]$.

138. Let X, Y be lcs spaces and $\{u_n\}$ a sequence of continuous linear maps: $X \to Y$ which converges to 0 on a dense barrelled subspace of X. Show that $u_n \to 0$ on X $[\![$Theorem 9-3-4, Prob. 9-3-104$]\!]$.

139. Prove the result of Prob. 138 without local convexity and with ultrabarrelled instead of barrelled $[\![$Prob. 9-3-130$]\!]$.

140 Suppose that $\{\delta^n\}$ is a weak basis for a lc FK space X, that is, $f(x) = \sum x_k f(\delta^k)$ for all $f \in X'$. Show that $\{\delta^n\}$ is a basis. Compare Prob. 5-4-302. $[\![$Let $u_n(x) = x - \sum_{k=1}^n x_k \delta^k$. Apply Prob. 9-3-104.$]\!]$

---

301 (Nachbin–Shirota theorem). Let H be a $T_{3\frac{1}{2}}$ space. Then $C(H)$, with compact convergence, is barrelled if and only if for every noncompact closed set $F \subset H$ there exists $f \in C(H)$ which is not bounded on F. $[\![$See [108], [111], and [133].$]\!]$

302 If $(X',$ weak $*)$ is sequentially complete and $\beta(X, X')$ is separable, then (X, X') is barrelled. $[\![$See [75], Corollary, p. 407.$]\!]$

9-4 THE EQUIVALENCE PROGRAM

Consider the result: X is barrelled if and only if weak $*$ bounded sets in X' are equicontinuous (Theorem 9-3-4). This takes the form $P \equiv (Q \Rightarrow R)$, with P a property of X and Q, R properties of subsets of X' or of Y if (X, Y) is a dual pair. Theorems which take this form will be described as belonging to the *equivalence program*.

1. Remark Statements of the form $Q \Rightarrow R$ are, of course, inclusion statements. Let $\mathscr{A}(Q) = \{A \subset Y : A$ has property $Q\}$. Then $Q \Rightarrow R$ if and only if $\mathscr{A}(Q) \subset \mathscr{A}(R)$. If these are admissible families in Y with $\mathscr{A}(R)$ saturated, we have $Q \Rightarrow R$ if and only if $T_{\mathscr{A}(Q)} \subset T_{\mathscr{A}(R)}$ $[\![$Theorem 8-5-11$]\!]$. The most frequently occurring form of the equivalence under discussion is $P \equiv (Q \Rightarrow E)$ where E is "equicontinuous." The motivating example is of this form. Another way of stating it is that (X, T) has property P if and only if $T \supset T_{\mathscr{A}(Q)}$, since $T = T_E$. Consider again the above example which says that (X, T) is barrelled if and only if weak $*$ bounded sets in X' are equicontinuous. The present interpretation yields: (X, T) *is barrelled if and only if* $T \supset \beta(X, X')$. This is Theorem 9-3-10.

Let us examine in this light possible converses of results in which equicontinuity is a hypothesis. These are Theorems 9-1-8 and 9-1-10, in which the conclusions are compactness and strong boundedness. Example 9-2-6 shows that weak * compact sets need not be equicontinuous in any compatible topology, i.e. (Remark 9-4-1), $\tau(X, Y) \not\supset T_k$ where k is the set of $\sigma(Y, X)$ compact sets. However, the assumption of absolute convexity changes the picture.

2. Theorem: Equivalence program A lcs space (X, T) is relatively strong if and only if each absolutely convex weak * compact set is equicontinuous.

PROOF By Remark 9-4-1, the second condition is equivalent to $T \supset \tau(X, X')$.

In particular, each absolutely convex weak * compact set is equicontinuous in some compatible topology.

Sometimes the equivalence program takes the form of a definition, as in Prob. 9-4-103.

PROBLEMS

1 Express the result of Corollary 9-2-5 in the form $T_b \supset \tau(X, Y)$ ⟦Remark 9-4-1⟧.

101 Discuss conditions for equality in Prob. 9-4-1.

102 Show that $(X', \text{weak } *)$ has the convex compactness property if and only if weak * compact sets are $\tau(X, X')$ equicontinuous.

103 A lcs space is called ω *barrelled* if each weak * bounded sequence is equicontinuous, C *barrelled* if each weak * Cauchy sequence is equicontinuous, and *sequentially barrelled* if each weak * null sequence is equicontinuous. Show that each of barrelled, ω barrelled, C barrelled, and sequentially barrelled implies the next.

104 Show that if (X, T) has any of the properties of Prob. 9-4-103 it has it in any larger compatible topology.

105 A dual pair (X, Y) is called ω *barrelled* if X has a compatible ω-barrelled topology. Show that (X, Y) is ω barrelled if and only if $\tau(X, Y)$ is. The same is true for C and sequentially barrelled.

106 Show that (X, T) is sequentially barrelled if and only if $T \supset (\text{weak } *)c_0$, Definition 8-6-1 ⟦Remark 9-4-1⟧. Equivalently, (X, Y) is sequentially barrelled if and only if $[\sigma(Y, X)]_{c_0}$ is compatible with (X, Y).

107 Let X be C barrelled. Show that $(X', \text{weak } *)$ is sequentially complete ⟦as in Example 9-3-8⟧. Equivalently, if (X, Y) is C barrelled, $\sigma(Y, X)$ is sequentially complete.

108 Show that (X, Y) is sequentially barrelled if and only if, in $[Y, \sigma(Y, X)]$, the absolutely convex closure of each null sequence is compact. In particular, this holds if $\sigma(Y, X)$ has the convex compactness property. Compare Theorem 9-3-13.

109 Let X, Y be TVSs. A family of continuous linear maps from X to Y is called UN if it is uniformly bounded on some neighborhood of 0, E if it is equicontinuous, UB if it is uniformly bounded on all bounded sets, and P if it is pointwise bounded. Show that

    (a) UN $\Rightarrow E \Rightarrow$ UB $\Rightarrow P$.
    (b) No converse holds.
    (c) If X is locally bounded, UN $= E =$ UB.
    (d) If Y is locally bounded, UN $= E$.
    (e) If X is bornological, $E =$ UB.
    (f) If X is barrelled, $E =$ UB $= P$.

110 Two vector topologies (not necessarily lc) are called compatible with each other if they have the same closed vector subspaces. Show that two compatible topologies have the same dual but not conversely [Probs. 2-3-203, 7-1-103, and 9-3-116].

111 Let X be a barrelled metric space and H a Hamel basis. Show that all but a finite number of the coefficient functionals are noncontinuous. [If $P_n \in X'$, let $\alpha_n h^n \to 0$. Then $\{t_n \alpha_n\} = \{t_n P_n(\alpha_n h^n)\}$ is bounded for every $t \in \omega$ since $\{t_n P_n\}$ is weak $*$ null, hence equicontinuous.]

112 In Prob. 9-4-111, "metric" cannot be omitted [Prob. 7-1-7].

---

201 Let T_b, T_K be the topologies on X of uniform convergence on $\beta(Y, X)$ bounded and $\sigma(Y, X)$ compact sets respectively. Must $T_b \wedge T_K$ be compatible? (Compare Prob. 8-5-113.)

202 Must two compatible topologies (Prob. 9-4-110) have the same closed convex sets?

---

301 An ω-barrelled space need not be relatively strong and a relatively strong ω-barrelled space need not be barrelled. [See [96], p. 101.]

9-5 SEPARABLE SPACES

An important consequence of separability is the metrizability of equicontinuous sets, Theorem 9-5-3.

If (X, Y) is a dual pair and $F \subset Y$, let $\sigma(X, F)$ denote the relative topology of $\mathscr{K}^F$, that is, consider each $x \in X$ as a function on F, namely, $y \to [x, y]$ for $y \in F$; and a net $x \to 0$ if and only if $[x, y] \to 0$ for each $y \in F$. This is consistent with the meaning of $\sigma(X, F)$ if F is a vector subspace of Y, as pointed out in Prob. 8-1-4.

1. Lemma Let X be a lc space, K a weak $*$ compact set in X', and F a fundamental set in X. Then weak $* = \sigma(X', F)$ on K.

PROOF In symbols, $\sigma(X', X)|_K = \sigma(X', F)|_K$. Since the first-mentioned topology is compact and larger than the second, which is Hausdorff, the result follows from a topological fact (Prob. 1-6-14).

2. Lemma The result of Lemma 9-5-1 holds if K is assumed equicontinuous (instead of weak $*$ compact).

PROOF The Alaoglu–Bourbaki theorem asserts that K is included in a weak $*$ compact set. The result follows from Lemma 9-5-1.

3. Theorem Let X be a separable lc space, K a weak $*$ compact set in X', E an equicontinuous set in X', and $U \in \mathscr{N}(X)$. Then $(E, \text{weak } *)$ is metrizable and $(K, \text{weak } *)$ and $(U^\circ, \text{weak } *)$ are compact metric spaces.

PROOF By Lemma 9-5-2, $(E, \text{weak } *) = [E, \sigma(X', F)]$ for any fundamental set $F \subset X$. Choose F to be countable. Metrizability follows by Prob. 2-1-12 or 7-2-6. A special case is the result for U°. Lemma 9-5-1 allows the same argument to be applied to K.

4. Example *If X is a separable lc metric space, $(X', \text{weak} *)$ is separable.* Thus X' is separable in all topologies compatible with (X', X). We may express this as: *the dual pair (X', X) is separable.* For if $\{U_n\}$ is a local base for $\mathcal{N}(X)$, each $(U_n^\circ, \text{weak} *)$ is a compact metric space, hence is separable. Moreover, $X' = \cup U_n^\circ$. ⟦For $y \in X'$, $y^\circ \supset U_n$ for some n, so $y \in U_n^\circ$.⟧

5. Example (l, weak) *and* $(l, \|\cdot\|_1)$ *have the same compact sets.* Let K be weakly compact and y a sequence in K. Now let $X = [l^\infty, \sigma(l^\infty, l)]$. This is separable by Example 9-5-4; thus $(K, \text{weak}) = [K, \sigma(l, l^\infty)]$ is a compact metric space. ⟦See Theorem 9-5-3; weak $*$ in that result refers to $\sigma(X', X) = \sigma(l, l^\infty)$.⟧ Hence y has a convergent subsequence, i.e., weakly convergent. This subsequence converges in $(K, \|\cdot\|_1)$ by Prob. 8-1-8. So $(K, \|\cdot\|_1)$, being metrizable, is compact.

It follows that (l, weak) has the convex compactness property. This is extended to an arbitrary Banach space in Theorem 14-2-4. The argument of Example 9-5-5 is applied without separability in Prob. 14-1-106.

PROBLEMS

1 Let X be a separable normed space. Show that the unit disc in X' is a weak $*$ compact metric space ⟦Theorem 9-5-3⟧.

2 Show that l has the same compact sets in all topologies admissible for (l, l^∞) ⟦Example 9-5-5⟧.

3 Show that a dual pair (X, Y) is separable if and only if X is separable with all compatible topologies.

4 Let X be a normed space whose dual is separable. Show that X is separable. ⟦By Example 9-5-4, $(X'', \text{weak} *)$ is hereditarily separable and (X, weak) is a subspace; use Corollary 8-3-6.⟧

101 Let X be a separable normed space. Show that (X'', X') is a separable dual pair. Deduce that the converse of Example 9-5-4 is false in that a nonseparable lc metric space may have a weak $*$ separable dual. ⟦X is dense in X''.⟧ Compare Prob. 8-3-115.

102 Let B be a separable normed space with a nonseparable dual and let $X = [B'', \sigma(B'', B')]$. Show that X is separable ⟦Prob. 9-5-101⟧, but $(X', \text{weak} *)$ is not. Thus "metric" cannot be omitted in Example 9-5-4.

103 Show that a space may be separable with one admissible topology and not with another.

104 Let X be a separable infinite-dimensional normed space. Show that X' contains a sequence $\{f_n\}$ with $\|f_n\| = 1$, $f_n \to 0$, weak $*$. ⟦The unit circumference is not weak $*$ closed by Theorem 4-1-12.⟧ See Prob. 9-5-301.

105 Let X be a separable subspace of l^∞. Show that there exists a sequence $\{k(n)\}$ such that $\lim x_{k(n)}$ exists for all $x \in X$. ⟦Apply Prob. 9-5-1 to $\{P_n\}$ where $P_n(x) = x_n$.⟧

106 Deduce from Prob. 9-5-105 that l^∞ is not separable.

107 Show that $\{\delta^n\}$ is a basis for $[l^\infty, \tau(l^\infty, l)]$. (In Prob. 9-5-101 it is pointed out that this space is separable.) ⟦See Prob. 3-1-103 and Example 9-5-5.⟧

108 Show that $[l^\infty, \tau(l^\infty, l)]$ is a Mazur space ⟦Prob. 9-5-107 and Example 3-3-7⟧. Thus a relatively strong Mazur space need not be bornological ⟦Prob. 8-4-6⟧.

109 Deduce from Probs. 5-1-301 and 5-1-302 that every separable Fréchet space can be given a smaller compact metrizable topology.

110 Show that the converse of Theorem 9-5-3 is false. Weak * metrizability of the equicontinuous sets does not imply separability [Prob. 9-1-104]. Compare Prob. 9-6-201.

111 Let X be a normed space such that X' is separable. Show that on bounded sets in X the weak topology is metrizable. Deduce that a bounded, weakly sequentially closed set in X must be weakly closed. [Consider $B \subset X''$; Theorem 9-5-3.]

112 In Prob. 9-5-111, neither "separable" in the first sentence nor "bounded" in the third may be omitted [Probs. 8-1-6, 8-1-8, and 8-4-120].

113 Let X be a Banach space which is the dual of a separable space, for example, $X = l^\infty$, and let K be a weakly compact subset. Show that K is separable in the norm topology of X. [The weak and weak * topologies agree on K by Prob. 1-6-14. By Theorem 9-5-3, K is compact metric, hence separable. Apply Prob. 8-3-103 to the span of K.]

301 Problem 9-5-104 is true without "separable." [See [71] and [110].]

9-6 APPLICATIONS

The only parts of this section which are used later in this book are Corollary 9-6-3 and Prob. 9-6-102, an application of which occurs in Sec. 14-7.

1. Theorem Let X be a normed space. There exists a compact Hausdorff space H such that X is equivalent with a subspace of $C(H)$.

PROOF Let H be the unit disc of X' with the weak * topology. It is compact by the Banach–Alaoglu theorem. For each $x \in X$, consider $\hat{x} | H$ where $\hat{x} \in X''$, the natural embedding. It is easy to check that the map $x \to \hat{x} | H$ is the required equivalence, for example, $\| \hat{x} | H \|_\infty = \sup \{ |\hat{x}(h)| : h \in H \} = \| \hat{x} \| = \| x \|$ by Theorem 3-2-2.

2. Theorem Let H be a $T_{3\frac{1}{2}}$ space ($=$completely regular Hausdorff space). There exists a Banach space X such that H is homeomorphic with a subset of $(X', \text{weak } *)$.

PROOF Namely, let $X = C^*(H)$. For each $h \in H$, define $g_h \in X'$ by $g_h(f) = f(h)$ for $f \in X$ [$|g_h(f)| = |f(h)| \leq \| f \|_\infty$.] The map $h \to g_h$ is one to one, since X separates the points of H, and continuous, since if $h \to h_0$, $g_h(f) = f(h) \to f(h_0) = g_{h_0}(f)$ for each f; that is, $g_h \to g_{h_0}$ weak *. Finally, this map is a homeomorphism since if $g_h \to g_{h_0}$, then for every $f \in X$, $f(h) = g_h(f) \to g_{h_0}(f) = f(h_0)$. This implies that $h \to h_0$. [If not, for a certain $U \in \mathcal{N}_{h_0}$, $h \notin U$ frequently and $f \in X$ may be chosen with $f(h_0) = 1$, $f = 0$ on $H \backslash U$.]

3. Corollary Every $T_{3\frac{1}{2}}$ space H has a compactification βH in which it is C^* embedded.

PROOF This means that βH is a compact Hausdorff space, H is dense in βH, and every real continuous bounded function on H can be extended to a continuous real function on βH. In Theorem 9-6-2 (take real functions only) the map $h \to g_h$ carries H into the unit disc D of X'. Thus H is homeomorphic into $(D, \text{weak} *)$, a compact Hausdorff space by the Banach–Alaoglu theorem. To show that H is C^* embedded in D, let f be a real continuous bounded function on H. Then $f \in X$. Let $F = \hat{f} \,|\, D$ where $\hat{f}$ is the natural embedding of f into X''. Then $F \in C(H)$ since each member of X is weak $*$ continuous on X' and $F \,|\, H = f$. [[For $h \in H$, $F(g_h) = g_h(f) = f(h)$.]] Of course, it is not to be expected that H is dense in D (e.g., let $H = R$), but the closure of H in D fulfills all the requirements.

The space βH, called the *Stone–Cech compactification* of H, is studied in topology books such as [156].

The representation of Theorem 9-6-1 is not valuable for studying Banach spaces since X is usually a very small subset of $C(H)$. Far-reaching studies have been made of the points of $C(H)$ which are the images of X, and of the possibility of choosing H to be a subset of the unit disc of X', e.g., the extreme points. (See Sec. 14-3 and [72].) We shall sketch the Gelfand representation of a Banach algebra which takes for H the subset of the unit disc consisting of the nonzero scalar homomorphisms. The interested reader will require the results of Prob. 5-5-201. Let X be a commutative complex Banach algebra with identity. The *spectrum of* X is the set H of nonzero scalar homomorphisms; it is a subset of the unit disc D of X'. [[If $f \in H$ and $|f(x)| > \|x\|$, let $a = 1 - x/f(x)$. Then a is invertible by Prob. 5-5-201, but $f(a) = 0$ so $f(x) = f(x)f(a)f(a^{-1}) = 0$.]] Moreover, it is a weak $*$ closed subset. [[If f is a net in $H, f \to g \in D$, then $g(xy) = \lim f(xy) = \lim f(x)f(y) = g(x)g(y)$; also $g(1) = \lim f(1) = 1$.]] The map $x \to \hat{x}$, where $\hat{x}$ is the restriction to H of the natural embedding of x into X'', is called the *Gelfand representation* of X. It has the good property that it is an equivalence onto when $X = C(T)$, the complex continuous functions on a compact Hausdorff space T. Banach algebra is a major field in mathematics; its further study falls outside our scope.

The next application shows a special property of the space c_0. It is unknown whether this is the only Banach space with this property.

4. Theorem: Sobczyk theorem Let X be a separable normed space, S a closed subspace, and $T: S \to c_0$ a continuous linear map. Then T has an extension $T_1: X \to c_0$ with $\|T_1\| \leq 2\|T\|$.

PROOF (W. A. Veech). Let $u_n \in S'$ be defined by $u_n(x) = (Tx)_n$ for each positive integer n. Extend u_n to $v_n \in X'$ with $\|v_n\| = \|u_n\|$. Let D be the unit disc in X' and give it the weak $*$ topology so that D is a compact metric space [[Prob. 9-5-1]]. Let $K = \{f \in D: f = 0 \text{ on } S\}$. Then K is a compact nonempty subset of D. We may assume that $\|T\| \leq 1$ so that $v_n \in D$. We prove that $d(v_n, K) \to 0$ where d is a metric for $(D, \text{weak} *)$. [[If this is false there is a subsequence $\{v_n'\}$ with $d(v_n', K) \geq \varepsilon > 0$ and a subsequence $\{v_n''\}$ of $\{v_n'\}$ with $v_n'' \to v \in D$. Say $v_n'' = v_{k(n)}$. Then for each $x \in S$, $v(x) = \lim v_{k(n)} = \lim u_{k(n)}(x) = \lim (Tx)_{k(n)} = 0$ since $Tx \in c_0$. Thus $v \in K$ which contradicts $d(v, K) \geq \varepsilon$.]]

So there exists $\{w_n\} \subset K$ with $d(v_n, w_n) \to 0$. Let $T_1(x) = \{v_n(x) - w_n(x)\}$ for $x \in X$. Then $T_1(x) \in c_0$ for each $x \in X$ and for $x \in S$, $(T_1 x)_n = v_n(x) - w_n(x) = u_n(x) = (Tx)_n$ so T_1 extends T. The inequality holds since, for $\| x \| \leq 1$ and each n, $|(T_1 x)_n| = |v_n(x) - w_n(x)| \leq \| v_n \| + \| w_n \| \leq 2$, and $\| T_1 x \|_\infty = \sup |(T_1 x)_n|$.

5. Corollary Let c_0 be a closed subspace of a separable normed space X. Then there is a projection P of X onto c_0 with $\| P \| \leq 2$.

PROOF In Theorem 9-6-4 take $S = c_0$, $T = $ identity map.

Corollary 9-6-5 becomes false if "separable" is omitted. Indeed, there may be no continuous projection [Example 14-4-9]. Also, as shown by Prob. 3-2-101, there is no projection: $c \to c_0$ with norm < 2 so the preceding two results are the best possible.

PROBLEMS

In this list H is a $T_{3\frac{1}{2}}$ space.

101 Let X be a separable normed space. Show that there exists a compact metric space H such that $X \subset C(H)$. (See Prob. 9-6-301.)

102 Show that $C^*(H) = C(\beta H)$. In particular, $l^\infty = C(\beta N)$, N the positive integers [Corollary 9-6-3].

103 Choose $t \in \beta N \setminus N$. Define $L \in (l^\infty)'$ by $L(x) = \bar{x}(t)$, where $\bar{x}$ is the extension of x to be a real continuous function on βN. (See Prob. 9-6-102.) Show that $L(x) = \lim x$ for $x \in c$, that $L(xy) = L(x)L(y)$, but that L is not translative, that is, $L(y) \neq L(x)$, possibly, if $y_n = x_{n+1}$. Hence L is not a Banach limit [Prob. 2-3-108].

104 Let H be normal and $b \in \beta H \setminus H$. Show that no sequence $\{h_n\} \subset H$ can converge to b. [[$\{h_{2n}\}$, $\{h_{2n+1}\}$ are disjoint closed sets in H. If f separates them it cannot be continuous at b.]]

105 If H is not compact, βH is not metrizable. (Do not assume that H is normal.) [See Prob. 9-6-104.]

106 If $X = C^*(H)$ is separable, H must be a compact metric space [Probs. 9-6-105 and 9-5-1].

107 Let $H = [0, 1]$, $X = C(H)$, and define $g \in X'$ by $g(f) = \int_0^1 f(t) dt$. Show that there exists no $h \in H$ such that $g(f) = f(h)$ for all $f \in X$.

108 Spot the flaw in this reasoning. Continuing Prob. 9-6-107, embed H in $(X', \text{weak} *)$ as in Theorem 9-6-2. Then g, H can be separated, i.e., there exists $u \in (X', \text{weak} *)' = X$ such that $u(g) > r > u(g_h)$ for all $h \in H$ [Prob. 7-3-105]. Considering $u \in X$, $g(u) > r > u(h)$ for all $h \in H$. Then $g(u) = \int_0^1 u(t) dt \leq r < g(u)$.

109 Let $h_1, h_2 \in H$, $X = C^*(H)$. Define $g \in X'$ by $g(f) = \frac{1}{2}[f(h_1) + f(h_2)]$. Show that $\| g \| = 1$ but $g \notin \beta H$ as defined in Corollary 9-6-3.

110 Let S be a subspace of c_0 which is linearly homeomorphic with it. Show that S is complemented. [Extend $T: S \to c_0$ to $T_1: c_0 \to c_0$; take $P = T^{-1} \circ T_1$.]

111 Let $T: c_0 \to c_0$ be a linear homeomorphism (into). Show that T has a left inverse. [Let P project c_0 onto $T[c_0]$, $U = T^{-1} \circ P$.]

112 Suppose that $|H| > c$, $X = C^*(H)$. Show that $(X', \text{weak} *)$ has no basis [Theorem 9-6-2 and Prob. 5-4-112].

113 Accepting the result of Prob. 9-6-206, show that the space $(X', \text{weak} *)$ in Prob. 9-6-112 can be separable. [$X = l$; see Prob. 9-6-102 and Example 9-5-4.]

114 Give an example of a nonseparable lcs space all of whose bounded sets are separable. See Prob. 9-6-302 ⟦Prob. 7-1-7⟧.

115 Let $X = l^\infty$, $L \in X^*$, and assume that for each $x \in X$ some subsequence of x converges to $L(x)$. Show that $L(xy) = L(x)L(y)$. ⟦L is a cluster point of $\{P_n\}$ in the weak $*$ topology, hence belongs to the spectrum of X.⟧

---

201 Let H be a compact metric space. Show that $C(H)$ is separable. (Compare Prob. 9-6-106.) Deduce that the converse of Theorem 9-5-3 is true if X is a normed space. (Compare Prob. 9-5-110.) ⟦See Theorem 9-6-1.⟧ Deduce from Prob. 9-6-106 that a continuous T_2 image of a compact metric space is metrizable.

202 Show that H is dense in the spectrum of $C^*(H)$.

203 Let X be a complex Banach algebra with identity, $f, g \in$ spectrum. Show that $\| f - g \| < 2$ if and only if $\| g \mid f^\perp \| < 1$. ⟦⇐: Prob. 2-3-114.⟧

204 Let T be compact. We saw that the topology of T is $(H, \text{weak } *)|_T$ $[H = \text{spectrum of } C(T)]$. Also $(H, \text{norm})|_T$ is of interest. Let X be a closed separating subalgebra of $C(T)$ with H its spectrum. Show that $\| f - g \| < 2$ is an equivalence relation between $f, g \in H$. (Compare Prob. 9-6-203.) The equivalence classes are called *Gleason parts*. If $X = C(T)$ the parts are all singletons.

205 Prove Theorem 9-6-4 with c, 3 instead of c_0, 2.

206 Show that the cardinality of βN is 2^c. ⟦See [156], Prob. 8-3-108.⟧

---

301 Let X be a separable normed space. Then X is equivalent with a subspace of $C[0, 1]$. ⟦See [88], 21.3(6).⟧

302 Assuming the continuum hypothesis, there exists a nonseparable lc Fréchet space all of whose bounded sets are separable. ⟦See [88], p. 404.⟧

THE STRONG TOPOLOGY

Some important extensions of Banach space ideas can be expressed in terms of the strong topology. For example, the fact that the natural embedding is an isometry can be expressed by the statement "if (X, T) is a Banach space, $T = \beta(X'', X')|_X$." As a second example, the statement "a Banach space X is reflexive" can be written "$\beta(X', X)$ is compatible with the dual pair (X', X)."

The two preceding properties turn out to be independent of each other and they will be studied separately.

10-1 THE NATURAL EMBEDDING

Let X be a lcs space. A direct generalization of the normed space situation is to define the *strong dual* of X as $[X', \beta(X', X)]$. When no contrary statement is made, X' will stand for the strong dual of X.

The strong dual of this space is called the *bidual* and written X''. Thus $X'' = [X', \beta(X', X)]'$ endowed with the topology $\beta(X'', X')$; this topology will be abbreviated as $b(X'')$. The natural embedding is (just as in Remark 8-2-4) the map $x \to \hat{x}$ where $\hat{x}(f) = f(x)$ for $f \in X'$. Since $\hat{x}$ is weak $*$ continuous, it is strong $*$ continuous, that is, $\hat{x} \in X''$. [The two topologies just mentioned are $\sigma(X', X)$ and $\beta(X', X)$.]

At this stage in the normed space theory it was noted that X is isometric into X'', that is, the norm on X'' induces the norm on X when one pretends that $X \subset X''$. This equality no longer holds. It is possible that $b(X'')|_X$ is not the initial topology of X; indeed, it is trivially possible since the topology on X can be replaced by another compatible topology without affecting either X'' or $b(X'')$.

1. Remark In the presence of two dual pairs one needs unambiguous designations. For $A \subset X''$ use $A^\circ \subset X'$. For $A \subset X$ we may consider $A \subset X''$ by the natural embedding, and so write $A^\circ \subset X'$ as before. However, for $A \subset X'$, write $A^\bullet \subset X''$, $A^\circ \subset X$. Thus $A^\circ = A^\bullet \cap X$, a fact which will be used repeatedly.

2. Definition Let X be a lcs space, with X'' and $b(X'')$ as above. Then $b(X)$ stands for $b(X'')|_X$.

3. Remark $b(X) = T_{\mathscr{A}}$ where $\mathscr{A}$ is the family of strong $*$ bounded $(= \beta(X', X)$ bounded) sets in X'. For $b(X'')$ is generated by $\{A^\bullet : A$ a bounded set in X' in the dual pair $(X', X'')\}$. Now this type of boundedness is precisely $\beta(X', X)$ boundedness since this topology is compatible with the dual pair (X', X''). Thus $b(X'')$ is generated by $\{A^\circ : A \in \mathscr{A}\}$; hence $b(X)$ is generated by $\{A^\bullet \cap X : A \in \mathscr{A}\} = \{A^\circ : A \in \mathscr{A}\}$, that is, $b(X) = T_{\mathscr{A}}$ as asserted. This result shows in particular that $b(X)$ is admissible for the dual pair (X, X').

This result can be transformed into another useful form, Lemma 10-1-5, by means of the following example of a dual pair of properties (as in Remark 8-4-11).

4. Lemma Let (X, Y) be a dual pair, $A \subset X$. Then A is strongly bounded if and only if A° is a bornivore.

PROOF $\Rightarrow$: Let B be a bounded set in Y. Then B° is a strong neighborhood of 0 in X, so it absorbs A. Hence A° absorbs B. $\Leftarrow$: By Prob. 9-1-6.

5. Lemma Let (X, T) be a lcs space. The set of bornivore barrels in (X, T) is a local base for $\mathscr{N}[b(X)]$.

PROOF Let U be a bornivore barrel in (X, T). Then U° is strong $*$ bounded by Lemma 10-1-4, so $U = U^{\circ\circ} \in \mathscr{N}[b(X)]$ by Remark 10-1-3. On the other hand, let $U \in \mathscr{N}[b(X)]$; we may assume that $U = B^\circ$ with strong B strong $*$ bounded by Remark 10-1-3. By Lemma 10-1-4, $U = B^\circ$ is a bornivore. It is also a barrel.

6. Theorem The natural embedding of a lcs space (X, T) is always (relatively) open, that is, $b(X) \supset T$.

PROOF $\mathscr{N}(T)$ has a local base consisting of barrels. These are also bornivores since they are neighborhoods of 0. Hence, by Lemma 10-1-5, they belong to $\mathscr{N}[b(X)]$.

In particular, $b(X) \supset \tau(X, X')$. This fact was noted in a different way in Prob. 9-4-1; compare Remark 10-1-3.

The problem of giving an internal characterization of spaces for which the natural embedding is continuous (hence a homeomorphism into) is solved by Theorem 10-1-8.

7. Definition A lcs space is called quasibarrelled if every bornivore barrel is a neighborhood of 0.

8. Theorem The natural embedding of a lcs space X is a homeomorphism if and only if X is quasibarrelled.

PROOF This is clear from Lemma 10-1-5.

9. Theorem Every quasibarrelled space (X, T) is relatively strong.

PROOF This can be seen in several ways.
(a) If T_1 is compatible, $\mathcal{N}(T_1)$ has a base of bornivore barrels. These are in $\mathcal{N}(T)$ so $T_1 \subset T$.
(b) It is obvious from Theorem 10-1-8 and the fact that $b(X) \supset \tau(X, X')$, which is a special case of Theorem 10-1-6. •

10. Example *Every bornological space and every barrelled space is quasi-barrelled. In particular, every lc metrizable space is quasibarrelled.*

We already knew that such spaces are relative strong.

11. Theorem: Equivalence program A lcs space is quasibarrelled if and only if all strong * bounded sets are equicontinuous.

PROOF By Theorems 10-1-6 and 10-1-8 (X, T) is quasibarrelled if and only if $b(X) \subset T$. By Remark 10-1-3 this holds if and only if $\mathcal{A} \subset E$ where $\mathcal{A}$ and E are the families of strong * bounded and T equicontinuous sets respectively. $[\![T = T_E$ and E is saturated. We used Theorem 8-5-11. $]\!]$

A *quasibarrelled dual pair* (X, Y) is one such that X has a compatible quasibarrelled topology or, equivalently, $\tau(X, Y)$ is quasibarrelled. Theorem 10-1-11 says that (X, Y) is quasibarrelled if and only if $\beta(Y, X)$ bounded sets are $\tau(X, Y)$ equicontinuous.

12. Theorem If X is quasibarrelled, $\{U^{\circ \bullet} : U \in \mathcal{N}(X)\}$ is a base for $\mathcal{N}[b(X'')]$.

PROOF If $U \in \mathcal{N}(X)$, U° is $\beta(X', X)$ bounded by Lemma 10-1-4. Thus $U^{\circ \bullet} \in \mathcal{N}[b(X'')]$. Also let $V \in \mathcal{N}[b(X'')]$, say $V = B^\bullet$ with B being $\beta(X', X)$ bounded. By Theorem 10-1-11, B is equicontinuous so $U = B^\circ \in \mathcal{N}(X)$. Then $U^{\circ \bullet} = B^{\circ \circ \bullet} \subset B^\bullet = V$ since $B^{\circ \circ} \supset B$.

PROBLEMS

In this list (X, T) is a lcs space or (X, Y) is a dual pair.

1 Show the equivalence of
  (a) X is quasibarrelled.
  (b) $b(X)$ is compatible with (X, X').
  (c) The natural embedding of X is continuous.

2 Show that X is quasibarrelled if and only if every closed absolutely convex bornivore is a neighborhood of 0.

3 Let B be a Banach space, $X = [B', \tau(B', B)]$. Show that $b(X'') = \beta(B', B)$. Hence show that B is reflexive if and only if X is quasibarrelled. (Compare Prob. 8-4-6.)

4 Deduce Theorem 10-1-9 by means of the equivalence program [Theorems 10-1-11 and 9-4-2 and Corollary 9-2-5].

5 If X is metrizable, show that X'' is metrizable [Theorem 10-1-12].

---

101 Show that a relatively strong space need not be quasibarrelled; equivalently, there exists a non-quasibarrelled dual pair [Prob. 10-1-3].

102 Show that a quasibarrelled space need not be barrelled; indeed, a bornological space need not be barrelled.

103 Show that $T \subset b(X) \subset \beta(X, X')$.

104 Let X be an infinite-dimensional nonbarrelled normed space and let T be the weak topology of X. Show that the three topologies of Prob. 10-1-103 are all different [Example 8-1-4].

105 Let $\bar{T}$ be the topology for X'' generated by $\{U^{\circ \bullet} : U \in \mathcal{N}(X, T)\}$. Show that $\bar{T}|_X = T$ and that $\bar{T} = b(X'')$ if and only if (X, T) is quasibarrelled. [Part of this is Theorem 10-1-12.]

106 Let X be metrizable. If the strong dual Y of X is of second category, X must be a normed space. [$Y = \cup U_n^\circ$ so some U_n° has interior. Thus Y is locally bounded, hence a Banach space. Finally, $X \subset Y'$.] Deduce that if the strong dual of X is metrizable (as well as X itself), then X is a normed space [Corollary 8-6-6]. This is false if X is assumed to be barrelled and bornological instead of metrizable. [Consider φ.]

107 In contrast with Prob. 10-1-106, both (X, Y) and (Y, X) may be metrizable and not normed. [Consider (φ, φ).]

108 Let X be metrizable. Show that $[X', \beta(X', X)]$ has a countable cobase for its bounded sets. [$\{U_n^\circ\}$; Lemma 10-1-4 and Example 4-4-7.] Problem 10-1-106 may be deduced from this and Prob. 4-5-105.

109 Show that X is quasibarrelled if and only if each bounded lower semicontinuous seminorm on X is continuous. Compare Prob. 9-3-105.

110 (Equivalence program). Let X be relatively strong. Show that X is quasibarrelled if and only if each strong * bounded set in X' is weak * relatively compact [Lemma 10-1-4 or Prob. 8-5-114].

111 Show that (X, Y) is quasibarrelled if and only if each $\beta(Y, X)$ bounded set is $\sigma(Y, X)$ relatively compact [Prob. 10-1-110].

112 Show that $b(X) = \beta(X, X')$ if and only if every barrel in X is a bornivore [Lemma 10-1-5 and Remark 8-5-5].

113 Show that the natural embedding is a bounded map [Lemma 10-1-5].

114 Give an example in which the natural embedding is not sequentially continuous [some Banach space with its weak topology].

115 Give an example of a nonquasibarrelled space such that the natural embedding is sequentially continuous [Prob. 8-1-8].

---

201 Let $X = [l^\infty, \tau(l^\infty, l)]$. Show that the natural embedding of X is not sequentially continuous. [Consider $\{\delta^n\}$; use Prob. 3-1-103 and Example 9-5-5.]

202 Suppose that X is relatively strong and the natural embedding is sequentially continuous. Must X be quasibarrelled? (See Probs. 10-1-114 and 10-1-115.)

203 Give conditions under which $b(X) = T^b$, Prob. 8-4-113. [They are both admissible and $b(X) \subset T^b$ by Prob. 10-1-113.] Compare Prob. 10-5-101.

10-2 SEMIREFLEXIVITY

Having solved the problem of characterizing those spaces for which the natural embedding is a homeomorphism (into), it is natural to investigate those for which it is onto.

1. Definition A lcs space X is called semireflexive if $X = X''$, that is, the natural embedding is onto X''. It is called reflexive if the natural embedding is a homeomorphism of X onto X''.

Thus reflexive means semireflexive and quasibarrelled.

2. Example *A lc metric space is reflexive if and only if it is semireflexive*, for it is quasibarrelled. This is why the word reflexive, rather than semireflexive, was used for normed spaces from the very beginning.

The result of Example 10-2-2 holds for any bornological or barrelled space, since these properties imply quasibarrelled.

In terms of the strong topology, *X is semireflexive if and only if $\beta(X', X)$ is compatible with (X', X), and quasibarrelled if and only if $b(X)$ is compatible with (X, X').*

3. Definition The dual pair (X, Y) is called semireflexive if (X, T) is semireflexive for some (hence every) compatible topology T.

Note that semireflexivity is a property of a dual pair (X, Y) in the sense that if (X, T) is semireflexive it remains so with any other compatible topology. In contrast, a quasibarrelled space must be relatively strong. Put another way, (X, Y) is quasibarrelled if and only if $\tau(X, Y)$ is quasibarrelled; (X, Y) is semireflexive if and only if for every compatible topology T, (X, T) is semireflexive.

4. Theorem The following conditions are equivalent for a dual pair (X, Y):
  (*a*) (X, Y) is semireflexive.
  (*b*) $\beta(Y, X)$ is compatible.
  (*c*) (Y, X) is barrelled.
  (*d*) $[X, \sigma(X, Y)]$ is boundedly complete.

PROOF $(a) = (b)$. This is clear from the preceding discussion. $(b) = (c)$. $\beta(Y, X) = \tau(Y, X)$ if and only if τ is barrelled by Theorem 9-3-10. $(c) = (d)$ by Theorem 9-3-13.

5. Example *"Compatible" cannot be omitted in the Alaoglu–Bourbaki Theorem 9-1-10.* Let X be nonsemireflexive, e.g., a nonreflexive normed space. By Theorem 10-2-4, there exists a bounded closed set A in $[X, \sigma(X, X')]$ which is not compact. Then A is certainly $\beta(X', X)$ equicontinuous but not $\sigma(X, X')$ relatively compact.

6. Example *A normed space X is reflexive if and only if its unit disc D is weakly compact.* $\Rightarrow$: D is bounded, hence weakly totally bounded [Prob. 8-2-6] and complete by Theorem 10-2-4. $\Leftarrow$: Every bounded closed set is included in a multiple of D, hence is weakly compact. So X is semireflexive by Theorem 10-2-4.

7. Remark One tries to improve results in which sequential completeness is a *conclusion* by replacing it by bounded completeness. In this way Theorem 9-3-11 improves Example 9-3-8. On the other hand, a theorem is improved when, in its *hypothesis*, bounded completeness is replaced by sequential completeness. This cannot be done in Theorem 10-2-4. A weakly sequentially complete space need not be semireflexive. The space l is an example [Prob. 8-1-11].

8. Example These ideas lead to an extremely simple example of *a non-admissible topology lying between two admissible ones.* (A direct construction was given in Prob. 8-5-109.) Let X be a nonreflexive Banach space, $w^* = \sigma(X', X)$, $w = \sigma(X', X'')$, and $n =$ norm topology on X'. Then w is not admissible for the dual pair (X', X). [w^* is boundedly complete since (c) implies (d) in Theorem 10-2-4; w is not boundedly complete since (d) implies (a) in Theorem 10-2-4 (X' is not reflexive by Theorem 3-2-9); and $w \supset w^*$ so w is not admissible by Prob. 8-5-5.] But $w^* \subset w \subset n$, and n is admissible for X', X; indeed, $n = \beta(X', X)$ [Example 8-5-6].

PROBLEMS

In this list X is a lcs space or (X, Y) is a dual pair.

1 Let X be a normed space. Prove the equivalence of

    (a) X is reflexive.

    (b) X with its weak topology is boundedly complete.

    (c) The unit disc of X is weakly complete.

2 Let X be a nonreflexive Banach space. Show that (X', X) is semireflexive and not quasibarrelled and that (X, X') is quasibarrelled and not semireflexive.

3 Let X be a normed space, and D, D_2 the unit discs in X, X''. Show that D_2 is the closure of D in the weak $*$ topology, $\sigma(X'', X')$. [$D_2 = D^{\circ\circ}$.] Deduce sufficiency in Example 10-2-6. (This was the first proof. The text shows the elegant generalization.)

4 Show that $\tau(X, Y) = \beta(X, Y)$ if and only if (Y, X) is semireflexive.

101 For any vector space X, (X^*, X) and (X, X^*) are both semireflexive [Probs. 9-3-118 and 9-3-119].

102 Show that X is semireflexive if and only if each bounded set is relatively weakly compact. Compare Prob. 10-1-110.

103 The continuous image of a semireflexive space need not be semireflexive. ⟦Consider $i: l^2 \to c_0$, or the largest lc topology.⟧

104 Let X be a normed space with separable dual. Show that X is weak $*$ ⟦that is, $\sigma(X'', X')$⟧ sequentially dense in X'' ⟦Prob. 10-2-3 and Theorem 9-5-3⟧. Deduce that if X is weakly sequentially complete and has a separable dual it must be reflexive.

105 Show that whether or not (X, Y) is semireflexive depends on the bilinear form chosen to put X, Y in duality ⟦Example 8-2-7 and Prob. 8-2-109⟧.

106 Let Y be barrelled and not complete (Prob. 3-1-5) and let $X = (Y', \text{weak } *)$. Show that X has the convex compactness property and that (X', X) is not complete.

107 Show that a weak $*$ closed set S in the dual of a normed space has a point of minimum norm. ⟦By an analogue of Prob. 8-3-113, the norm assumes its minimum on $S \cap D$ for some large disc D.⟧ Why does $\{[1 + 1/n]\delta^n\} \subset (l^\infty)'$ not contradict this?

108 Show that a closed convex set in a reflexive Banach space has a point of minimum norm ⟦Prob. 10-2-107⟧ and that reflexive cannot be omitted ⟦Prob. 2-3-104⟧—nor can convex.

109 Show that a product of semireflexive spaces is semireflexive ⟦Theorems 10-2-4 and 6-1-7 and Probs. 8-1-104 and 4-4-13⟧.

110 Let X be semireflexive, Z a lcs space, and $Q: X \to Z$ a quotient map such that each bounded set in Z is the image of a bounded set in X. Show that Z is semireflexive ⟦Prob. 8-3-201⟧. Compare Prob. 10-2-302. A condition which implies this "lifting property" is given in [29].

111 Show that a closed subspace of a semireflexive space is semireflexive ⟦Theorem 10-2-4(d)⟧.

112 If the strong dual of X is semireflexive show that $\tau(X'', X')|_X = \tau(X, X')$. Compare Prob. 9-2-113. ⟦Absolutely convex weak $*$ compact implies strong $*$ bounded which implies $\sigma(X', X'')$ relatively compact.⟧

113 If the strong dual of X is semireflexive show that (X, X') is quasibarrelled ⟦Probs. 10-2-4 and 10-2-112⟧.

114 If X is complete and its strong dual is semireflexive, show that X is semireflexive. ⟦X is dense and closed in X'' by Prob. 10-2-112.⟧

115 In Prob. 10-2-114, complete may be replaced by boundedly complete. ⟦If $f \in X''$ then $f^\circ \supset B^\circ$, B absolutely convex, bounded, and closed. So $f \in B^{\circ \bullet} = \tau(X'', X')$ closure of B. But B is $\tau(X, X')$ complete. See Prob. 10-2-112.⟧

116 Let (X, Y) and (Y, X) be metrizable. Neither need be normed ⟦Prob. 10-1-107⟧, but if (Y, X) is barrelled (e.g., if it is complete) show that (X, Y) must be a reflexive Banach pair ⟦Prob. 10-1-106⟧.

117 Let K be the family of compact sets in X. Show that $T_K = \beta(X', X)$ ⟦or $T_K \supset \tau(X', X)$⟧ if and only if every bounded set (respectively, every absolutely convex weakly compact set) in X is included in $K^{\circ\circ}$ for some compact K. What do these conditions become if X is boundedly complete?

118 If X is an infinite-dimensional reflexive Banach space, T_K is not relatively strong ⟦Prob. 10-2-117 and Example 6-4-14⟧.

119 Let H be an infinite compact T_2 space. Show that $C(H)$ is not weakly sequentially complete. ⟦Find a strictly increasing sequence of open sets and use it to construct open G_n such that $\bar{G}_n \subset G_{n+1}$ (strictly). Let $A = \cup G_n = \cup \bar{G}_n$ a noncompact open set. Let $f_n \in C(H)$, $0 \le f_n \le 1$, $f_n = 1$ on G_n, 0 on $H \backslash A$. (See [156], Theorem 4.2.11.) Then $f_n \to f \notin C(H)$ weakly by Lebesgue's bounded convergence theorem. (See Prob. 2-3-301 and [57], Theorem 12.24.)⟧

120 Let H be a compact T_2 space. These are equivalent for $X = C(H)$: X is reflexive, X is weakly sequentially complete, H is finite ⟦Probs. 10-2-1 and 10-2-119⟧.

121 Show that every infinite-dimensional Fréchet space has an infinite-dimensional absolutely convex compact, hence weakly compact, subset. ⟦Apply Example 9-2-10 to a null sequence; Example 3-3-14 shows how to get one.⟧

122 Show that Prob. 10-2-121 becomes false if "Fréchet" is replaced by "normed." Indeed, there may be no such weakly compact set $(\varphi, \|\cdot\|_\infty)$; $\sigma(l, \varphi) = \tau(l, \varphi)$ by Prob. 4-1-5.

123 Let X be an infinite-dimensional lc Fréchet space. Show that $\tau(X', X)$ is not smaller than $\sigma(X', X'')$ [Prob. 10-2-121].

124 The result of Prob. 10-2-123 is false for a vector space X with its largest lc topology. [Prob. 8-5-7.]

125 Let X be a non-reflexive lc Fréchet space. Show that $\tau(X', X)$ and $\sigma(X', X'')$ are not comparable. [Prob. 10-2-123.]

201 Let $P_n(x) = x_n$ for $x \in l^2$. Find a point of minimum norm in the weak $*$ closure of $\{(1 + 1/n)P_n\}$. (It exists by Prob. 10-2-107.)

301 Call X *quasireflexive of order n* if $\dim X''/X = n$. Then X is quasireflexive of order n if X' is. Further X is quasireflexive of order n if X is linearly homeomorphic with a Banach space Y such that $Y' = A \oplus B$, $\dim B = n$ and the unit disc of Y is $\sigma(Y, A)$ compact. [See [19].]

302 A quotient of a reflexive Fréchet space need not be reflexive. Compare Prob. 10-2-110. [See [88], 31.5.]

10-3 REFLEXIVITY

As usual a dual pair (X, Y) is called *reflexive* if there is some compatible topology T for X such that (X, T) is reflexive, i.e., quasibarrelled and semireflexive. Since a quasibarrelled space is relatively strong, (X, Y) *is reflexive if and only if* $[X, \tau(X, Y)]$ *is reflexive.*

1. Lemma Let (X, T) be reflexive. Then $T = b(X) = \beta(X, X')$.

PROOF First, $b(X'') = \beta(X'', X') = \beta(X, X')$ since $X'' = X$. Thus $b(X) = b(X'')|_X = \beta(X, X')$. Further, $T = b(X)$ since X is quasibarrelled.

2. Theorem The following are equivalent for a dual pair (X, Y):
    (a) (X, Y) is reflexive.
    (b) (Y, X) is reflexive.
    (c) (X, Y) and (Y, X) are both barrelled.
    (d) (X, Y) and (Y, X) are both semireflexive.

PROOF $(a) \Rightarrow (c)$: $\tau(X, Y)$ is reflexive, hence absolutely strong by Lemma 10-3-1. Thus it is barrelled [Theorem 9-3-10]. (Y, X) is barrelled since (X, Y) is semireflexive [Theorem 10-2-4]. $(c) \Rightarrow (a)$: (X, Y) is semireflexive since (Y, X) is barrelled [Theorem 10-2-4] and quasibarrelled since it is barrelled. Since (a) and (c) are equivalent and (c) is symmetric, the equivalence of (a) and (b) follows. Also, (a) and (b) together obviously imply (d). $(d) \Rightarrow (c)$ by Theorem 10-2-4.

3. Example Thus a reflexive space must be barrelled. The converse is false, as any nonreflexive Banach space will testify.

PROBLEMS

In this list (X, Y) is a dual pair or (X, T) is a lcs space.

101 It is false that (X, Y), (Y, X) both quasibarrelled implies (X, Y) reflexive [Prob. 8-6-106 or Prob. 10-1-107].

102 Show that the strong dual of a reflexive space is reflexive.

103 The converse of Prob. 10-3-102 fails; even a barrelled space may have a reflexive strong dual without being reflexive [Prob. 3-1-5].

104 Show that a relatively strong complete space (e.g., a lc Fréchet space) is reflexive if and only if its strong dual is reflexive. This generalizes Theorem 3-2-9 [Probs. 10-3-102, 10-2-113, and 10-2-114]. "Complete" may be replaced by "boundedly complete" [Prob. 10-2-115].

105 Let (X, Y) be boundedly complete. Show that (X, Y) is reflexive if and only if $[Y, \beta(Y, X)]$ is reflexive [Prob. 10-3-104].

106 Let X be boundedly complete. Show that (X, X') is reflexive if and only if (X', X'') is [Prob. 10-3-105].

107 Show that $(X, X^{\#})$ is reflexive [Prob. 10-2-101].

108 Show that a closed subspace of a reflexive metric space must be reflexive [Prob. 10-2-111].

301 There exists a separable reflexive space X which is not complete. [See [85].] Then $[X', \tau(X', X)]$ is an example of a barrelled space [it is reflexive] which is not bornological [Corollary 8-6-6]—indeed, not even a Mazur space [Theorem 8-6-5].

10-4 BOUNDEDNESS

Some examples of strongly bounded sets previously encountered are: any absolutely convex, weakly compact set; any equicontinuous set—more generally, the polar of a bornivore. The references are Corollary 9-2-5, Theorem 9-1-8, and Lemma 10-1-4. One form of the uniform boundedness principle says that weak $*$ bounded sets in the dual of a Banach space are strong $*$ bounded, i.e., norm bounded. The last implication was generalized to "weak $*$ bounded implies equicontinuous in the dual of a barrelled space" in Theorem 9-3-4. It follows that $w*$ bounded implies strong $*$ bounded in the dual of a barrelled space. Since strong $*$ boundedness is a strictly weaker property than equicontinuity [Theorem 10-1-11] it may be expected that the last implication will hold under weaker, or other, assumptions.

1. Remark A reason for being interested in this implication is the great importance of barrelled spaces. Now if weak $*$ bounded sets in X' are strong $*$ bounded and if X is quasibarrelled (equivalently, strong $*$ bounded sets are equicontinuous), it follows immediately that weak $*$ bounded sets are equicontinuous and so X is barrelled. (We made two uses of the equivalence program, Theorems 9-3-4 and 10-1-11.)

Here is a useful criterion for strong boundedness.

2. Lemma A set in a lc space X is strongly bounded if and only if it is absorbed by every barrel.

PROOF This is because the set of barrels is a base for $\mathcal{N}[\beta(X, X')]$ [Remark 8-5-5].

The class of spaces with the property under consideration is now isolated. The definition is an example of the equivalence program.

3. Definition A lcs space is called a Banach–Mackey space if all its bounded sets are strongly bounded. A Banach–Mackey pair is a dual pair (X, Y) such that all bounded sets in X are $\beta(X, Y)$ bounded.

Bounded sets in X are, of course, those which are bounded in some, hence all, compatible topologies. *A Banach–Mackey space or pair is one such that all admissible topologies have the same bounded sets.* The name derives from the landmark result, Theorem 10-4-8. Note that no mention is made of a topology for X. It all depends on the dual pair (X, X').

4. Example A barrelled space (hence a barrelled pair) furnishes a trivial example since it already has its strong topology.

This is extended to lots of nonbarrelled pairs by the next result.

5. Theorem If (X, Y) is a Banach–Mackey pair, so also is (Y, X).

PROOF Let A be a bounded set in Y and B a barrel in Y. Then B° is bounded in X [Theorem 8-4-12] hence is absorbed by A°. [B° is strongly bounded by hypothesis; $A^\circ \in \mathcal{N}[\beta(X, Y)]$.] Thus A is absorbed by $B^{\circ\circ} = B$. The result follows by Lemma 10-4-2.

6. Example A semireflexive space must be a Banach–Mackey space [by Theorem 10-4-5 and Theorem 10-2-4 $(a) \Rightarrow (b)$].

Next is given an internal characterization.

7. Theorem A lcs space is a Banach–Mackey space if and only if every barrel is a bornivore.

PROOF $\Rightarrow$: A barrel absorbs all strongly bounded sets [Lemma 10-4-2] hence all bounded sets. $\Leftarrow$: Every bounded set is absorbed by every bornivore, hence by every barrel, and so is strongly bounded by Lemma 10-4-2.

It is very easy to see that a weakly boundedly complete space X must be a Banach–Mackey space. $[\![(X', X)$ is barrelled by Theorem 10-2-4. Now see Example 10-4-4 and Theorem 10-4-5.$]\!]$ This is improved in two directions by Theorem 10-4-8. "Boundedly complete" is replaced by "sequentially complete" and "weakly" is dropped, i.e., the space may be sequentially complete in any compatible topology. A different generalization is given in Theorem 10-4-11.

8. Theorem: Banach–Mackey theorem Every sequentially complete lcs space (X, T) is a Banach–Mackey space. Every sequentially complete pair is a Banach–Mackey pair.

PROOF The second part is the same as the first. To prove the first we use Theorem 10-4-7. Let B be a barrel and A a bounded set. We may assume A to be absolutely convex and closed. $[\![$Consider $A^{\circ\circ}.]\!]$ Let Z be the span of A and p the gauge of A defined on Z. By Theorem 6-1-17, (Z, p) is a Banach space. Now $B \cap Z$ is a barrel in Z. $[\![$It is closed since, if $\{b^n\} \subset B$, $z \in Z$ and $p(b^n - z) \to 0$; then $b^n \to z$ in (X, T). (It was pointed out in the proof of Theorem 6-1-7 that $p \supset T | Z$.) Thus $z \in B.]\!]$ Since Z is a Banach space, $B \cap Z \in \mathcal{N}(Z)$ and so $B \cap Z$ absorbs A. $[\![A$ is bounded in Z since it is included in the unit disc.$]\!]$ Thus B absorbs A and so B is a bornivore.

9. Remark The historical importance of Theorem 10-4-8 lay partly in the deduction from it of the duality invariance of boundedness, Theorem 8-4-1. This was done as follows: if B is weakly bounded and $U = A^\circ$ is a Mackey neighborhood of 0, A absolutely convex weak $*$ compact, the argument of Theorem 10-4-8 shows that B° absorbs A; hence A° absorbs B and so B is bounded in the Mackey topology.

An independent sufficient condition (Theorem 10-4-11) was given by R. A. Raimi. That it is independent is shown by Probs. 9-2-301 and 14-2-102. It follows, of course, that the converses of Theorems 10-4-8 and 10-4-11 do not hold—a further result of this nature is given in Prob. 12-3-111.

10. Lemma Let X be a vector space, $A, B \subset X$. Then A absorbs B if and only if for each sequence $\{b^n\} \subset B$, A absorbs $\{b^n/n\}$.

PROOF $\Leftarrow$: If A does not absorb B, $n^2 A \not\supset B$ for $n = 1, 2, \ldots$. Let $b^n \in B \backslash n^2 A$. Then $b^n/n \notin nA$ so A does not absorb $\{b^n/n\}$.

11. Theorem Let X be a lcs space with the convex compactness property. Then X is a Banach–Mackey space.

PROOF It is sufficient by Theorem 10-4-5 to show that each bounded set A in X' is strong $*$ bounded. Let $B^\circ \in \mathcal{N}[\beta(X', X)]$ with B a bounded set in X. To show that A is strong $*$ bounded it is sufficient to show that A° absorbs B. By Lemma 10-4-10 we may consider $S = \{b^n/n\}$, $b^n \in B$. Now $b^n/n \to 0$ so S is strongly bounded. $[\![S \cup \{0\}$ is compact; hence $S^{\circ\circ}$ is compact and so it is weakly compact. Since it is also absolutely convex it is strongly bounded by Corollary 9-2-5.$]\!]$ Since $A^\circ \in \mathcal{N}[\beta(X, X')]$, A° absorbs S.

The uniform boundedness principle for Banach spaces now appears in this light. Equivalent statements (if X is a Banach space) of the conclusion are:

(a) Weak ∗ bounded sets are equicontinuous.

(b) Weak ∗ bounded sets are strong ∗ bounded.

We now see that for (a) X barrelled is sufficient, while for (b) X sequentially complete (or with the convex compactness property) is sufficient.

It will now be seen that a quasibarrelled space must be barrelled if it has one of a long list of properties.

12. Theorem A quasibarrelled space which is also a Banach–Mackey space is barrelled.

PROOF The proof was given in Remark 10-4-1. It also follows trivially from Theorem 10-4-7.

13. Remark Theorem 10-4-12 is a good result to remember. From it we can deduce immediately that a quasibarrelled space is barrelled if it is sequentially complete $[\![$Theorem 10-4-8$]\!]$, has the convex compactness property $[\![$Theorem 10-4-11$]\!]$, or is semireflexive $[\![$Example 10-4-6. Of course, this is trivial directly since the space is reflexive.$]\!]$ Other entries in this list may be found in Remark 10-4-14 and Prob. 10-4-107.

The first result of Remark 10-4-13 generalizes the fact that a lc Fréchet space is barrelled.

14. Remark There is another way to look at the result of Theorem 10-4-8. Some authors have been interested in the class of spaces X such that X' is weak ∗ sequentially complete. See, for example, Theorem 12-5-13. The barrelled spaces belong to this class and Theorems 10-4-5 and 10-4-8 show that spaces of this type are Banach–Mackey spaces. Thus we may add this condition to the list of Remark 10-4-13, i.e., a quasibarrelled space X is barrelled if $(X', \text{weak} ∗)$ is sequentially complete. As often happens, the same conclusion holds if $(X', \text{weak} ∗)$ has the convex compactness property $[\![$Theorem 10-4-12$]\!]$. We can also view these as sufficient conditions for bounded completeness, viz. if $\sigma(X, Y)$ is sequentially complete or has the convex compactness property and (Y, X) is quasibarrelled, then $\sigma(X, Y)$ is boundedly complete. $[\![(X, Y)$ is Banach–Mackey by Theorem 10-4-8 or Theorem 10-4-11, so (Y, X) is also. Hence (Y, X) is barrelled.$]\!]$

15. Corollary Let X be a lc metric space. Then X', the strong dual, is a Banach–Mackey space. Hence, if quasibarrelled, it is barrelled.

PROOF The strong $*$ topology on X' is complete by Theorem 8-6-5. The results follow by Theorems 10-4-8 and 10-4-12.

PROBLEMS

In this list (X, T) is a lcs space or (X, Y) is a dual pair.

1 Prove the converse of Theorem 10-4-12.

2 Give an example of a lcs space which is not a Banach–Mackey space.

3 Give an example of a Banach–Mackey space which is not barrelled.

4 Show that a sequentially complete bornological space must be barrelled.

5 If X is a Mazur space (e.g., metrizable or bornological), show that (X', X'') is a Banach–Mackey pair [Corollary 8-6-6].

101 Show that X is a Banach–Mackey space if and only if $b(X) = \beta(X, X')$ [Theorem 10-4-7 and Prob. 10-1-112].

102 Show that (l, φ) is not a Banach–Mackey pair [Example 9-2-6 or Theorem 10-4-12]; hence neither (l, φ) nor (φ, l) has the convex compactness property. This improves Prob. 9-2-115.

103 Show where the Baire category theorem enters into the proof, Remark 10-4-13, that a sequentially complete quasibarrelled space is barrelled.

104 Let (X, T) be sequentially complete and let T_1 be a larger vector topology for X which is F linked to T. Show that each T bounded set is T_1 bounded. Theorem 10-4-8 is a special case.

105 A C-barrelled space must be a Banach–Mackey space [Prob. 9-4-107]. See the next problem.

106 A sequentially barrelled space X must be a Banach–Mackey space. This improves Prob. 10-4-105. [Imitate the proof of Theorem 10-4-11 with $A \subset X$, $B \subset X'$. Now S is strong $*$ bounded for a different reason.]

107 Add to the list in Remark 10-4-13: ω barrelled, C barrelled, and sequentially barrelled [Prob. 106].

108 Let X be quasibarrelled. Prove the equivalence of the following:

   (*a*) X is barrelled.
   (*b*) X is sequentially barrelled.
   (*c*) $(X', \text{weak } *)$ is sequentially complete.
   (*d*) $(X', \text{weak } *)$ has the convex compactness property.
   (*e*) (X', w^*) is boundedly complete.
   (*f*) X is a Banach–Mackey space.

201 In a sequentially complete space every barrel is a bornivore; however, let $B = \{x : \sum |x_i| \leq 1\} \subset (l, \| \cdot \|_\infty)$. Show that B is a sequentially complete barrel but not a bornivore.

202 Let A be a closed, absolutely convex set in a lcs space X and B an absolutely convex bounded set which is of second category in itself. Suppose that A absorbs each point of B. Show that A absorbs B [the proof of Theorem 10-4-8].

203 Give examples to show that in Prob. 10-4-202 no one of these assumptions can be omitted: A is closed, A is absolutely convex, B is absolutely convex, B is bounded, and B is of second category. Further, if A, B are assumed convex (not absolutely) the result is false, but truth is restored by assuming that A absorbs each point of $B \cup -B$.

204 Prove the converse of Prob. 10-4-106, assuming that X is a relatively strong Mazur space. ⟦See [75], Theorem 3.2.⟧

205 Show that if X is relatively strong, Mazur, and a Banach–Mackey space, then $(X', \text{weak} *)$ is sequentially complete. ⟦See [75], Theorem 3.3.⟧

10-5 METRIC SPACE

The main result of this section (Corollary 10-5-5) is the fact that the bidual of a lc metric space is a Fréchet space. (This section may be omitted.)

1. Lemma Let $\{U_n\}$ be a base for $\mathcal{N}(X)$, where X is a lc metric space, and let $c_n > 0$ for all n. Then $I = \cap c_k U_k$ is bounded.

PROOF If $U \in \mathcal{N}(X)$, $U \supset U_n$ for some n, so $c_n U \supset c_n U_n \supset I$.

2. Lemma With the notation of Lemma 10-5-1, assume that each U_n is a barrel. Let $V_n = \cap\{c_k U_k : 1 \le k \le n\}$, $A_n = \cap\{c_k U_k : k > n\}$. Then for each n, $A_n^\circ + V_n^\circ \supset I^\circ$.

PROOF The Alaoglu–Bourbaki theorem implies that V_n° is weak $*$ compact and so $A_n^\circ + V_n^\circ$ is weak $*$ closed ⟦Lemma 6-5-11⟧. Hence $A_n^\circ + V_n^\circ = (A_n^\circ + V_n^\circ)^{\circ\circ} \supset (A_n^\circ \cup V_n^\circ)^{\circ\circ} = (A_n^{\circ\circ} \cap V_n^{\circ\circ})^\circ = (A_n \cap V_n)^\circ = I^\circ$, using Remark 8-3-2(i).

3. Lemma Let X be a lc metric space. Let $\{E_n\}$ be a sequence of $\beta(X', X)$ equicontinuous sets in X'' such that $E = \cup E_n$ is bounded. Then E is also $\beta(X', X)$ equicontinuous.

PROOF Boundedness is in the bidual topology $b(X'')$. Let $\{U_n\}$ be a base for $\mathcal{N}(X)$ consisting of barrels. For each n, $E_n^\circ \in \mathcal{N}[\beta(X', X)]$ so $E_n^\circ \supset B_n^\circ$ where each B_n is a bounded set in X. For each n, k, $B_n \subset a_{nk} U_k$, $a_{nk} > 0$. Since E is bounded we have, for each k, $E \subset b_k U_k^{\circ\bullet}$, $b_k > 0$. ⟦Recall that X is quasi-barrelled so $U_k^{\circ\bullet} \in \mathcal{N}[b(X'')]$ by Theorem 10-1-12.⟧ Let $c_k = \max\{b_k, a_{nk} : n \le k\}$ so that $c_k \ge b_k$ for all k and $c_k \ge a_{nk}$ if $k \ge n$. It is sufficient to show that $E^\circ \supset (2I)^\circ$, where I is given in Lemma 10-5-1. For this it is sufficient to show that $2E_n^\circ \supset A_n^\circ + V_n^\circ$ where A_n, V_n are given in Lemma 10-5-2. ⟦If this is proved, it will follow from Lemma 10-5-2 that $2E_n^\circ \supset I^\circ$ so $E_n \subset (2I)^{\circ\bullet}$ for each n; hence $E \subset 2I^{\circ\bullet}$. Thus $2E^\circ \supset I^{\circ\bullet\circ} = I^\circ$ (pretend $I \subset X''$) and so $E^\circ \supset (2I)^\circ$.⟧ This will be done by showing that $E_n^\circ \supset A_n^\circ$ and $E_n^\circ \supset V_n^\circ$. First, $B_n \subset \cap\{a_{nk} U_k : k > n\} \subset A_n$ so $A_n^\circ \subset B_n^\circ \subset E_n^\circ$. Finally, $E_n \subset E \subset c_k U_k^{\circ\bullet}$ so $E_n^\circ \supset (1/c_k)U_k^\circ = (c_k U_k)^\circ$. Let C be the absolutely convex hull (not closure!) of $\cup\{(c_k U_k)^\circ : 1 \le k \le n\}$. Since each $(c_k U_k)^\circ$ is weak $*$ compact ⟦Alaoglu–Bourbaki⟧ C is weak $*$ compact ⟦Prob. 6-5-3⟧ hence weak $*$ closed, and so $C = C^{\circ\circ} = [\cup\{(c_k U_k)^\circ : 1 \le k \le n\}]^{\circ\circ} = [\cap(c_k U_k)]^\circ = V_n^\circ$. ⟦See Remark

8-3-2(i); note that $U_k^{\circ\circ} = U_k$.⟧ Since E_n° is absolutely convex, it includes C and the proof is complete.

4. Theorem Let X be a lc metric space. Then $\left[X'', \sigma(X'', X')\right]$ is sequentially complete.

PROOF Let S be a Cauchy sequence. Now (X'', X') is a Banach–Mackey pair ⟦Prob. 10-4-5⟧ and so S is $\beta(X'', X') = b(X'')$ bounded. Applying Lemma 10-5-3 with each E_n a singleton, we conclude that S is $\beta(X', X)$ equicontinuous. This topology is compatible with (X', X'') and so S is $\sigma(X'', X')$ relatively compact, by the Alaoglu–Bourbaki theorem. Thus $\overline{S}$ is compact; hence complete, and so S is convergent.

5. Corollary The bidual of a lc metric space X is a Fréchet space.

PROOF Since $b(X'')$ is admissible for (X'', X') and larger than $\sigma(X'', X')$, it is also sequentially complete ⟦Prob. 8-5-5⟧. Since it is metrizable ⟦Prob. 10-1-5⟧ the result follows.

6. Corollary *A reflexive lc metric space must be complete.*

PROBLEMS

1 Let X be a lc metric space and $\{H_n\}$ a sequence of absolutely convex closed $\beta(X', X)$ neighborhoods of 0 such that $H = \cap H_n$ is a $\beta(X', X)$ bornivore. Show that $H \in \mathcal{N}\left[\beta(X', X)\right]$. This result and Lemma 10-5-3 both say (taking account of the equivalence program, Theorem 10-1-11) that X' is "almost quasibarrelled." ⟦Apply Lemmas 10-5-3 and 10-1-4 with $E_n = H_n^\circ$.⟧

101 Show that the proof of Lemma 10-5-3 may be greatly abbreviated if X is reflexive. ⟦$E \subset c_k U_k^{\circ\circ}$ becomes $E \subset c_k U_k$ and so $E^\circ \supset I^\circ$.⟧

102 A (DF) space is a lcs space X which has a countable cobase for its bounded sets and such that $E = \cup E_n \subset X'$ is equicontinuous if each E_n is equicontinuous and E is strong * bounded. Show that every normed space and the strong dual of any lc metric space are (DF) spaces ⟦Lemma 10-5-3 and Prob. 10-1-108⟧.

103 Show that every Banach–Mackey (DF) space is ω barrelled. ⟦Take each E_n to be a singleton in Prob. 10-5-102.⟧ Compare Prob. 10-4-106.

104 Let X be a lc metric space. Show that the strong dual of X is ω barrelled ⟦Probs. 10-5-103 and 10-4-5⟧.

ELEVEN

OPERATORS

11-1 DUAL OPERATORS

An *operator* is a linear map between vector spaces. Now let X_1, X_2 be vector spaces and $u: X_1 \to X_2$ an operator. We first consider $u^{\#}: X_2^{\#} \to X_1^{\#}$ defined by: $u^{\#}(g)$, for $g \in X_2^{\#}$, is that member of $X_1^{\#}$ whose value at $x_1 \in X_1$ is $u^{\#}(g)(x_1) = g[u(x_1)]$, that is, $u^{\#}(g) = g \circ u$. Now let $(X_1, Y_1), (X_2, Y_2)$ be dual pairs, $u: X_1 \to X_2$ an operator, and define $u': Y_2 \to X_1^{\#}$ to be $u^{\#} \mid Y_2$. Thus $u'(y_2)(x_1) = [u(x_1), y_2]$. The question arises as to when $u'[Y_2] \subset Y_1$, that is, $u': Y_2 \to Y_1$. When this is true we refer to u' as the operator *dual to u*. (It is also called the adjoint, conjugate, and transpose.) To abbreviate the answer to this question we shall say that u is σ *continuous* to mean that $u: [X_1, \sigma(X_1, Y_1)] \to [X_2, \sigma(X_2, Y_2)]$ is continuous.

1. Lemma u' maps Y_2 into Y_1 if and only if u is σ continuous.

PROOF $\Rightarrow$: Let x_1 be a net in X_1 with $x_1 \to 0$, $\sigma(X_1, Y_1)$. Let $y_2 \in Y_2$. Then $[u(x_1), y_2] = u'(y_2)(x_1) = [x_1, u'(y_2)] \to 0$ since $u'(y_2) \in Y_1$. $\Leftarrow$: Let $y_2 \in Y_2$. To prove $u'(y_2) \in Y_1$ it is sufficient to show that it is $\sigma(X_1, Y_1)$ continuous [Theorem 8-2-12]. Let x_1 be a net in X_1 with $x_1 \to 0$, $\sigma(X_1, Y_1)$. Then $u'(y_2)(x_1) = [u(x_1), y_2] \to 0$ since $u(x_1) \to 0$, $\sigma(X_2, Y_2)$ by hypothesis.

2. Remark Note the formula $[ux_1, y_2] = [x_1, u'y_2]$ which figured in the proof of Lemma 11-1-1.

3. Corollary Let $u: (X_1, T_1) \to (X_2, T_2)$ be continuous, where T_i is compatible for (X_i, Y_i), $i = 1, 2$. Then u is σ continuous and $u': Y_2 \to Y_1$.

PROOF For $y_2 \in Y_2 = (X_2, T_2)'$, $u'(y_2) = y_2 \circ u \in (X_1, T_1)' = Y_1$. [For $y_2 \circ u$ is the composition of two continuous maps $(X_1, T_1) \xrightarrow{u} (X_2, T_2) \xrightarrow{y_2} \mathscr{K}$.] Thus $u': Y_2 \to Y_1$ and the result follows by Lemma 11-1-1.

4. Example Let E, F be lcs spaces. Then if $u: E \to F$ is continuous, it is weakly continuous. (All maps are assumed to be linear here and throughout.)

5. Definition An admissible topology α is called a Hellinger–Toeplitz topology if whenever $u: X_1 \to X_2$ is continuous when X_1, X_2 have some topologies compatible for (X_i, Y_i), $i = 1, 2$; u is also $\alpha(X_1, Y_1) - \alpha(X_2, Y_2)$ continuous.

Thus σ is a Hellinger–Toeplitz topology [Corollary 11-1-3]. Note that α in Definition 11-1-5 has to be "categorically" defined, that is, $\alpha(X, Y)$ has a meaning for every dual pair (X, Y). For example, σ, τ, β are such topologies whereas T_{c_0} is not, since it depends on some topology T which has to be placed on X.

6. Theorem Let E, F be lcs spaces and $v: F' \to E'$ a linear map. Then there exists $u: E \to F$ with $v = u'$ if and only if v is weak $*$ continuous.

PROOF This is precisely Lemma 11-1-1. Let $X_1 = F', Y_1 = F, X_2 = E', Y_2 = E$. Then weak $*$ continuous $= \sigma(X_1, Y_1) - \sigma(X_2, Y_2)$ continuous $= \sigma$ continuous. So $v': E \to F$ if and only if v is weak $*$ continuous. If v is weak $*$ continuous, $v': E \to F$ and, with $u = v'$, we have $v = u'$. [For $g \in F', e \in E, v(g)(e) = g[v'(e)] = g[u(e)] = u'(g)(e)$. Alternatively, write $[v''(e), g] = [e, v'(g)] = [v(e), g]$ so that $v = v''$.] Conversely, if $v = u', v' = u$ and v is weak $*$ continuous as just pointed out.

The basic relationship, now to be developed, is that an operator is one to one if and only if its dual is something like onto, i.e., has large range. The symmetric statement (interchange u, u') is also true since $u = u''$. In Lemma 11-1-7, X, Y are lcs spaces, $u: X \to Y$ is continuous and linear, $Nu = u^\perp = \{x : u(x) = 0\}$, and $Ru = $ range of u. For $A \subset X$, $\overline{A}$ is the closure of A; for $A \subset X'$, $cl_{w*}A$ is the weak $*$ closure of A. For $A \subset X$, $A^\perp = \{f \in X': f = 0 \text{ on } A\}$; for $A \subset X'$, $A^\perp = \{x \in X: f(x) = 0 \text{ for all } f \in A\}$.

7. Lemma
(a) $Nu = (Ru')^\perp$
(b) $Nu' = (Ru)^\perp$
(c) $\overline{Ru} = (Nu')^\perp$
(d) $cl_{w*}Ru' = (Nu)^\perp$

PROOF
(a) Let $x \in Nu$, $f \in Ru'$. Then $f(x) = (u'g)(x) = g(ux) = 0$ so $x \in (Ru')^\perp$. Let $x \in (Ru')^\perp$. For any $g \in u'$, $g(ux) = (u'g)(x) = 0$ so $u(x) = 0$ by the Hahn–Banach theorem. Thus $x \in Nu$.
(b) Apply (a) to $u': (Y', \text{weak } *) \to (X', \text{weak } *)$.
(c) $(Nu')^\perp = (Ru)^{\perp\perp} = (Ru)^{\circ\circ} = \overline{Ru}$. [For a vector subspace $S, S^\perp = S^\circ$.]
(d) Apply (c) as in the proof of (b).

8. Corollary u is one to one if and only if the range of u' is total over X; u has dense range if and only if u' is one to one.

9. Example Let X be a lcs space with a Schauder basis. As pointed out in Remark 5-4-2 we may consider X as a K-sequence space with basis $\{\delta^n\}$. For $f \in X'$ we have $f(x) = f(\sum x_k \delta^k) = \sum x_k f(\delta^k) = \sum f(\delta^k) P_k(x)$. Thus $(X', \text{weak } *)$ has $\{P_n\}$ as a Schauder basis. [Each map $f \to f_k = f(\delta^k)$ is continuous since $\delta^k \in X$.] Identifying $(X', \text{weak } *)$ with a sequence space in the same way, that is, $f = \{f_k\}$, we have $[x, f] = \sum x_i f_i$ as the value of $f \in X'$ at $x \in X$.

10. Example Let $u: X \to Y$ be continuous where X, Y are K-sequence spaces with basis $\{\delta^n\}$ as in Example 11-1-9. Let $a_{nk} = (u\delta^k)_n = P_n(u\delta^k)$. Then $(ux)_n = \sum_k a_{nk} x_k$, that is, $u(x) = Ax$ (matrix multiplication with x a column vector). $[u(x) = \sum (ux)_n \delta^n = \sum_n P_n(ux)\delta^n = \sum_n P_n[u(\sum x_k \delta^k)]\delta^n = \sum_n \sum_k x_k a_{nk} \delta^n = \sum_n (Ax)_n \delta^n.]$ For $g \in Y'$, identify g with $\{g_n\}$ where $g_n = g(\delta^n)$; then $g[ux] = \sum (Ax)_n g(\delta^n) = \sum (Ax)_n g_n = [Ax, g]$.

11. Example With X, Y, u, A as in Example 11-1-10, $u' = A^T$, the transpose of A, that is, $[Ax, g] = [x, A^T g]$ for $x \in X$, $g \in Y'$. Compare Remark 11-1-2. $[[Ax, g] = g(ux) = g[u(\sum_k x_k \delta^k)] = \sum_k x_k g(u\delta^k) = \sum_k x_k g(\sum_n a_{nk} \delta^n) = \sum_k \sum_n x_k a_{nk} g_n = [x, A^T g].]$ What this equation says is $\sum_n \sum_k = \sum_k \sum_n x_k a_{nk} g_n$. The interchange of limits was effected thus: in Example 11-1-10, we wrote $u(x) = \sum (ux)_n \delta^n$ and in Example 11-1-11, $u(x) = u(\sum x_k \delta^k)$.

PROBLEMS

1 Show that u is one to one if and only if Ru' is weak $*$ dense [Corollary 11-1-8 and Theorem 8-3-9].

2 Let X be a vector subspace of Y and $i: X \to Y$ the inclusion map. Show that $i^*: Y^* \to X^*$ is given by $i^*(g) = g \mid X$.

3 If S is a subspace of a lcs space X, the weak topologies of S, X coincide on S. [Apply Corollary 11-1-3 to i and i^{-1}.]

101 Define $u(x) = \{(1/n)x_n\}$. Show that $u: c_0 \to c_0$ has dense range but is not onto.

102 Show that an operator from l^∞ to itself is $\sigma(l^\infty, l)$ continuous if and only if it is given by a matrix. $[\Leftarrow: A = (A^T)'. \Rightarrow: (l^\infty, \text{weak } *)$ has $\{\delta^n\}$ as a Schauder basis.]

103 Let A be the matrix with $a_{1k} = 1$, $a_{nk} = 0$ for $n > 1$. Show that $A: l \to l$ is $\| \cdot \|_1$ continuous but not $\sigma(l, c_0)$ continuous (a) directly, (b) by noting that $A^T[c_0] \not\subset c_0$.

104 Show that the converse of Corollary 11-1-3 is false. [Consider $i: X \to X$.]

105 Let $u: c_0 \to c_0$ be given by the matrix whose nth row is $\delta^n - \delta^{n+1}$. Show that u is one to one but $u'': l^\infty \to l^\infty$ is not. [Use Example 11-1-11.]

106 Give an example of $u: c_0 \to c_0$ which is one to one but Nu'' is infinite dimensional. Compare Prob. 11-1-105.

107 Let X, Y be normed spaces, $u \in B(X, Y)$, $h \in X''$, $u''h = y \in Y$. Show that $y \in \overline{Ru}$ [Lemma 11-1-7(c)].

108 Let $X = \{x : \sum |x_{2n-1} + x_{2n}| < \infty\}$, $u: X \to l$ be the matrix whose nth row is $\delta^{2n-1} + \delta^{2n}$. Show that u is not $\sigma(X, X^\alpha) - \sigma(l, l^\beta)$ continuous. See Prob. 3-3-106. $[u'1 \notin X^\alpha.]$

109 Let X be the dual of a separable normed space (for example, $X = l^\infty$) and Y a reflexive BX space (Sec. 5-5). Show that Y is a separable subspace of X. In particular, l^∞ includes no dense reflexive BK space. [See Prob. 9-5-113 and Example 10-2-6. The unit disc of Y is weakly compact in X, by Corollary 11-1-3.]

110 Let X be a sequentially complete lcs space and $x^n \to 0$ weakly. Define $v : l \to X$ by $v(t) = \sum t_n x^n$ [Prob. 6-1-115] and $u : X' \to c_0$ by $u(f) = \{f(x^n)\}$. Show that $v = u'$, considering the dual pairs (X', X) and (c_0, l).

111 With $X, \{x^n\}$ as in Prob. 11-1-110, show that $K = \{\sum t_n x^n : t \in l, \|t\|_1 \leq 1\}$ is weakly compact [Prob. 11-1-110; Theorem 11-1-6; $K = v[D]$, D the unit disc in l].

112 Say that a lcs space has SCC (the *sequential convex compactness property*) if the absolutely convex hull of each null sequence is compact. Show that if (X, T) has SCC, so does (X, T_1), where T_1 is any larger topology admissible for (X, X') [Prob. 8-5-5 and Theorem 6-5-7].

113 Let X be a sequentially complete lcs space. Show that (X, T_1) has SCC, where T_1 is any topology admissible for (X, X') [Probs. 11-1-111 and 11-1-112].

114 If the dual pair (Y, X) is sequentially complete, show that (X, Y) is sequentially barrelled. Equivalently, if X is relatively strong and $\tau(X', X)$ is sequentially complete, then X is sequentially barrelled [Probs. 11-1-113 and 9-4-108].

115 Show that (X, Y) is C barrelled if and only if $\sigma(Y, X)$ is sequentially complete. Equivalently, if X is relatively strong, it is C barrelled if and only if $(X', \text{weak} *)$ is sequentially complete. Compare Prob. 9-4-108. [See Probs. 11-1-114 and 9-4-107.]

116 Let (X, T) be a sequentially complete lcs space. Show that T_{c_0} (Definition 8-6-1) is compatible with (X', X) [like Example 9-2-13, using Prob. 11-1-113].

117 Let X be a sequentially complete Mazur space. Show that (X', X) is complete [Prob. 11-1-116 and Theorem 8-6-5].

118 Show that $\tau(l^\infty, l)$ is C barrelled but not ω barrelled. [See Probs. 11-1-115 and 8-1-11. Consider $\{\delta^n\}$.] Compare Prob. 9-3-302.

119 Let S be a vector subspace of X of finite codimension, $U \in \mathcal{N}(X)$. Show that $S + U \in \mathcal{N}[X, \sigma(X, X')]$. [$q : X \to X/S$ is weakly continuous and $S + U = q^{-1}[qU]$.]

120 Show that $T \wedge \sigma(X, X^\#) = \sigma(X, X')$ where $X = (X, T)$ [Prob. 11-1-119].

121 Let S be a subspace of a lcs space X such that S is $\sigma(X'', X')$ closed in X''. Show that S is semireflexive. [If B is bounded and weakly closed, it is bounded in X by Prob. 11-1-3. Then $B^\circ \in \mathcal{N}[\beta(X', X)]$ and so $B^{\circ\circ}$ (in X'') and $B^{\circ\circ} \cap S$ are $\sigma(X'', X')$ compact. The latter is thus $\sigma(X, X')$ compact, hence, by Prob. 11-1-3, weakly compact in S.]

122 Show that c_0 is not linearly homeomorphic with the dual of a Banach space. [Apply Prob. 10-2-104 with $c_0 = X'$; X is weakly sequentially complete by Prob. 11-1-3 since it is a closed subspace of l.]

123 Let X be a nonreflexive Banach space and $j : X' \to X'''$ the natural embedding. Show that j is not the dual of any map: $X'' \to X$ [Theorem 11-1-6].

124 Let X be a nonreflexive Banach space and $j : X \to X''$, $j_2 : X'' \to X^{iv}$ the natural embeddings. Show that $j_2 \neq j''$ [Prob. 11-1-123].

11-2 THE HELLINGER–TOEPLITZ THEOREM

In 1910, E. Hellinger and O. Toeplitz proved a theorem which says roughly that an operator on separable Hilbert space (that is, l^2) is continuous if it has a dual. This has been refined (Corollary 11-2-6) to a form which is independent of both metrizability and completeness. Recall Lemma 11-1-1 which says that u has a dual if and only if it is σ continuous, i.e., weakly continuous. This is a result of the required type but does not give the Hellinger–Toeplitz theorem since that requires norm continuity. In any case, it is clear from Lemma 11-1-1 that the hypothesis "u' exists" may be replaced by "u is weakly continuous" in any theorem of the Hellinger–Toeplitz type. By Corollary 11-1-3, this may be replaced by "u is continuous in some pair of compatible topologies for X_1, X_2." One more conceptual refinement: the statement "α is a Hellinger–Toeplitz topology" (Definition 11-1-5) is equivalent, by the preceding remarks, to "if u has a dual, u is α continuous." So the original 1910 theorem says precisely that the norm topology is a Hellinger–Toeplitz topology, and this is the form we give it here. (See Example 11-2-5 and Corollary 11-2-6.)

1. Lemma Let (X_i, Y_i) be dual pairs, $i = 1, 2$, and $u: X_1 \to X_2$ a σ-continuous operator. For $A \subset Y_2$ we have $u^{-1}[A^\circ] = (u'[A])^\circ$.

PROOF $\subset$: Let $x_1 \in u^{-1}[A^\circ]$, $y_1 \in u'[A]$, say $y_1 = u'(y_2)$ with $y_2 \in A$. Then $\big|[x_1, y_1]\big| = \big|[x_1, u'y_2]\big| = \big|[ux_1, y_2]\big| \leq 1$. $\supset$: Let $x_1 \in (u'[A])^\circ$, $y_2 \in A$. Then $\big|[ux_1, y_2]\big| = \big|[x_1, u'y_2]\big| \leq 1$, so $ux_1 \in A^\circ$.

Now let α be some admissible topology defined for all dual pairs. (See the remarks following Definition 11-1-5.) Write $\alpha(X, Y) = T_{\mathscr{A}}$ where $\mathscr{A}$ is some saturated admissible family of subsets of Y, again defined for all dual pairs; for example, $\beta(X, Y)$ has for $\mathscr{A}$ the family of $\sigma(Y, X)$ bounded sets in Y.

2. Theorem With α as just defined, α is a Hellinger–Toeplitz topology if and only if whenever (X_i, Y_i) are dual pairs, $i = 1, 2$, and $u: X_1 \to X_2$ is linear and continuous in some pair of compatible topologies for X_1, X_2 (equivalently, u is σ continuous); then $u': \mathscr{A}_2 \to \mathscr{A}_1$ where $\alpha(X_i, Y_i) = T_{\mathscr{A}_i}$, $i = 1, 2$. This means that $A \in \mathscr{A}_2$ implies $u'[A] \in \mathscr{A}_1$.

PROOF $\Rightarrow$: Let $A \in \mathscr{A}_2$. Then $A^\circ \in \mathscr{N}[\alpha(X_2, Y_2)]$ and so $u^{-1}[A^\circ] \in \mathscr{N}[\alpha(X_1, Y_1)]$. By Lemma 11-2-1, $(u'[A])^\circ \in \mathscr{N}[\alpha(X_1, Y_1)]$ and so, since $\mathscr{A}_1$ is saturated, $u'[A] \in \mathscr{A}_1$. $\Leftarrow$: Let $u: X_1 \to X_2$ be σ continuous, $U \in \mathscr{N}[\alpha(X_2, Y_2)]$, say $U = A^\circ$ with $A \in \mathscr{A}_2$. Then, by Lemma 11-2-1, $u^{-1}[U] = (u'[A])^\circ \in \mathscr{N}[\alpha(X_1, Y_1)]$ since $u'[A] \in \mathscr{A}_1$.

3. Example β is a Hellinger–Toeplitz topology, for $\beta(X, Y) = T_{\mathscr{A}}$ with $\mathscr{A}$ the bounded sets in Y. By Theorem 11-1-6, u' is $\sigma(Y_2, X_2) - \sigma(Y_1, X_1)$ continuous, so it carries $\mathscr{A}_2$ into $\mathscr{A}_1$ [Theorem 4-4-3].

4. Remark The second half of the proof of Theorem 11-2-2 does not use the hypothesis that $\mathscr{A}_1, \mathscr{A}_2$ are saturated.

5. Example τ is a Hellinger–Toeplitz topology, for $\tau(X, Y) = T_{\mathscr{A}}$ with $\mathscr{A}$ the family of absolutely convex $\sigma(Y, X)$ compact sets. The result follows exactly as in Example 11-2-3, taking account of Remark 11-2-4.

We saw in Sec. 11-1 that σ is a Hellinger–Toeplitz topology. This is also because u' preserves finite sets.

6. Corollary Let $X, (Y, T)$ be lcs spaces and $u: X \to Y$ weakly continuous and linear. Then u is $\tau(X, X') - \tau(Y, Y')$ and $\beta(X, X') - \beta(Y, Y')$ continuous; a fortiori u is $\tau(X, X') - T$ continuous.

PROOF This is the content of Examples 11-2-3 and 11-2-5. The last part is because $T \subset \tau(Y, Y')$.

7. Corollary Let X, Y be normed spaces and $u: X \to Y$ linear. Then if u is weakly continuous (equivalently, u has a dual $u': Y' \to X'$), u is norm continuous.

PROOF By Corollary 11-2-6, since the norm is $\tau(X, X')$ [Example 8-4-10].

8. Remark With the hypotheses of Corollary 11-2-7 it is easy to prove that u has (norm) closed graph. [Let $x_n \to 0$, $ux_n \to y$. Then for all $g \in Y'$, $g(y) = \lim g(ux_n) = \lim u'g(x_n) = 0$ so $y = 0$.] So continuity of u follows if X, Y are Banach spaces. This was the earlier proof given for this result. However, completeness really plays no role.

9. Example Let X be a real Hilbert space. Take $X_1 = X_2 = Y_1 = Y_2 = X$. Then we get the following result: *if $u: X \to X$ is weakly continuous it is norm continuous*. In this form the result is correct even if X is not complete. The original Hellinger–Toeplitz theorem follows: *if $u: X \to X$ is linear and $[ux, y] = [x, uy]$ then u is continuous*. [For $u' = u$, $X' = X$.] In this form we cannot drop completeness, but this is not because of any defect in Corollary 11-2-7, but merely because $X' \neq X$. As an example let $X = (\varphi, \|\cdot\|_2)$, $u(x) = u'(x) = \{nx_n\}$.

PROBLEMS

In this list X, Y are lcs spaces and $u: X \to Y$ is linear.

1 Let X be relatively strong. Show that if u is weakly continuous it is continuous. [$T_Y \subset \tau(Y, Y')$.]

101 Show that b is a Hellinger–Toeplitz topology [Remark 10-1-3].

102 If α is a Hellinger–Toeplitz topology show that α_K and α_{c_0} are also. (For example, $\alpha_K(X, Y) = [\alpha(Y, X)]_K$, where K is the family of compact sets; see Sec. 8-5 and Definition 8-6-1.)

11-3 BANACH SPACE

(The part following Lemma 11-3-6 may be omitted.) There are important refinements of Lemma 11-1-7. They will be given for operators on Banach spaces first.

More general results (which do not use the ones given here) are given in Sec. 12-4.

In this section X, Y are Banach spaces, $X \neq \{0\}$, and $u \in B(X, Y)$. We shall vary our earlier usage by using the word "*isomorphism*" to indicate a linear homeomorphism (into).

1. Theorem $\|u'\| = \|u\|$.

PROOF $\|u'\| = \sup\{\|u'(g)\| : \|g\| \leq 1\} = \sup\{\sup\{|(u'g)(x)| : \|x\| \leq 1\} :$ $\|g\| \leq 1\} = \sup\{|g(ux)| : \|g\| \leq 1, \|x\| \leq 1\} = \sup\{\|ux\| : \|x\| \leq 1\} = \|u\|$. ⟦The third step used $\|y\| = \sup\{|g(y)| : \|g\| \leq 1\}$, which is Theorem 3-2-2.⟧

2. Lemma If u is range closed, $Ru' = (Nu)^{\perp}$.

PROOF The hypothesis means that Ru is a closed subspace of Y. By Lemma 11-1-7, it is sufficient to prove that $(Nu)^{\perp} \subset Ru'$. Let $f \in (Nu)^{\perp}$. Define g on Ru by $g(ux) = f(x)$. ⟦If $u(x) = u(a)$, $x - a \in Nu$ so $f(x) = f(a)$. Thus g is well defined.⟧ Clearly, g is linear. Also g is continuous. ⟦Let $y \in Ru$, $\|y\| \leq 1$. Now $u : X \to Ru$ is open by the open mapping Theorem 5-2-4; so $u[D] \supset D_{\varepsilon} \cap Ru$ where D is the unit disc in X, $D_{\varepsilon} = \{y \in Y : \|y\| \leq \varepsilon\}$. Hence $D_1 \cap Ru \subset u[(1/\varepsilon)D]$; in particular, $y = u(x)$ with $\|x\| \leq 1/\varepsilon$. Then $|g(y)| = |f(x)| \leq (1/\varepsilon)\|f\|$; so $\|g\| \leq 1/\varepsilon \|f\|$.⟧ By the Hahn–Banach theorem we may extend g to be defined on Y. Finally, $u'(g) = f$ since $u'(g)(x) = g(ux) = f(x)$ for $x \in X$.

3. Theorem The following are equivalent:
(a) u is range closed.
(b) u' is (norm) range closed.
(c) u' is (weak ∗) range closed.

PROOF $(a) \Rightarrow (c)$: $Ru' = (Nu)^{\perp}$ by Lemma 11-3-2. The latter is weak ∗ closed since it is $\cap\{\hat{x}^{\perp} : x \in Nu\}$. $(c) \Rightarrow (b)$ since the norm topology is larger than weak ∗.

$(b) \Rightarrow (a)$:
Case (1) *The range of u is dense.* We have to prove that u is onto and it is sufficient to prove that u is almost open ⟦Lemma 5-2-3⟧. Now u' is one to one by Corollary 11-1-8, and range closed, hence a (norm) isomorphism. Let $\delta = |(u')^{-1}|^{-1}$. We shall prove that

$$z \in Y, \|z\| < \delta \text{ implies } z \in \overline{u[D]}, D \text{ the unit disc in} X \qquad (11\text{-}3\text{-}1)$$

and so u is almost open. Note that

$$g \in Y', \|g\| = 1 \text{ implies } \|u'(g)\| \geq \delta \qquad (11\text{-}3\text{-}2)$$

⟦$1 = \|g\| = \|(u')^{-1}u'(g)\| \leq \|(u')^{-1}\| \cdot \|u'g\| = (1/\delta)\|u'g\|$.⟧ To prove (11-3-1), suppose $z \notin \overline{u[D]}$. By the Hahn–Banach Theorem 7-3-5, there exists $g \in Y'$ with $\|g\| = 1$, $|g(z)| \geq |g(y)|$ for all $y \in u[D]$. For all $x \in D$ we have

$|u'(g)(x)| = |g(ux)| \leq |g(z)| \leq \|z\|$ and so $\|u'(g)\| \leq \|z\|$. From (11-3-2), $\|z\| \geq \delta$, which proves (11-3-1).

Case (2) Let $Z = \overline{Ru}$. Let $v: X \to Z$ be defined by $v(x) = u(x)$ for $x \in X$. Then $v': Z' \to X'$ and $Rv' = Ru'$. $[\subset:$ Let $f \in Rv'$, say $f = v'(h)$, $h \in Z'$. By the Hahn–Banach theorem, extend h to $g \in Y'$. Then $u'(g) = f$ since for $x \in X$, $u'(g)(x) = g(ux) = h(vx) = (v'h)(x)$. $\supset:$ Let $f \in Ru'$, say $f = u'(g)$, $g \in Y'$. Let $h = g \,|\, Z$. Then $v(h) = f$ by the same argument.$]$ Thus v' is (norm) range closed. Since the range of v is dense it follows by case (1) that v is onto, that is, $Ru = Rv = \overline{Ru}$. Thus u is range closed.

4. Theorem
(a) u is an isomorphism if and only if u' is onto.
(b) u is onto if and only if u' is a (norm) isomorphism.

PROOF (a) $\Rightarrow$: The range of u' is weak $*$ dense $[$Prob. 11-1-1$]$ and weak $*$ closed $[$Theorem 11-3-3$]$. $\Leftarrow$: u is range closed $[$Theorem 11-3-3$]$ and one to one $[$Prob. 11-1-1$]$. By the open mapping theorem, Corollary 5-2-6, u is a (norm) isomorphism.

   (b) $\Rightarrow$: u' is one to one $[$Corollary 11-1-8$]$ and range closed $[$Theorem 11-3-3$]$. By the open mapping theorem it is a (norm) isomorphism. $\Leftarrow$: u has dense range $[$Corollary 11-1-8$]$ and is range closed $[$Theorem 11-3-3$]$.

5. Remark
In Theorem 11-3-4 (b) "onto" may be replaced by "an open map onto" or "a quotient map." These are equivalent by the open mapping theorem.

Certain parts of Theorem 11-3-4 should be regarded as trivialities. For example, if $X \subset Y$ and $i: X \to Y$ is inclusion, then $i'g = g \,|\, X$ for $g \in Y'$. That i' is onto is merely the Hahn–Banach theorem. If $u: X \to Y$ is an isomorphism the fact that u' is onto is essentially the same thing since X may be regarded as a subspace of Y. This part of Theorem 11-3-4 is thus applicable in the generality of Sec. 11-1. In the same way, one can say that if u is range closed and $u: X \to Ru$ is an open map (this is automatic in the Banach space setting) then $Ru' = (Nu)^\perp$ as in Lemma 11-3-2. This requires consideration of the diagram $X \xrightarrow{q} X/Nu \xrightarrow{u} Ru \xrightarrow{i} Y$ and the dual diagram. We omit the details since no further use will be made of the result. A detailed discussion may be found in $[82]$, Sec. 21, and in $[91]$.

6. Lemma
These are equivalent:
(a) u is not an isomorphism.
(b) For every $\varepsilon > 0$ there exists $x \in X$ with $\|x\| \geq 1$, $\|ux\| < \varepsilon$.
(c) For every $\varepsilon > 0$ there exists $x \in X$ with $\|x\| = 1$, $\|ux\| < \varepsilon$.

PROOF $(a) \Rightarrow (c)$. If u is not one to one we can make $ux = 0$. If u is one to one $\|u^{-1}\| = \infty$ where $u^{-1}: Ru \to X$. Choose $y \in Ru$ with $\|y\| = 1$, $t > 1/\varepsilon$, where

$t = \|u^{-1}y\|$. Let $x = (1/t)u^{-1}(y)$. Then $\|x\| = 1$, $\|ux\| = 1/t < \varepsilon$. $(b) \Rightarrow (a)$. If u is an isomorphism let $\delta = \|u^{-1}\|^{-1}$. Then if $\|x\| \geq 1$, $1 \leq \|x\| = \|u^{-1}ux\| \leq (1/\delta)\|ux\|$ so $\|ux\| \geq \delta$ and (b) is false.

The rest of this section may be omitted. Let $A = B(X, X)$. We call $u \in A$ a *left divisor* of 0 if there exists $v \in A$, $v \neq 0$, $uv = 0$, and a *left topological divisor of* 0 if for every $\varepsilon > 0$ there exists $v \in A$ with $\|v\| = 1$, $\|uv\| < \varepsilon$. (Here $(uv)(x) = u[v(x)]$.) Similarly for *right* instead of left.

7. Definition For $a \in X$, $f \in X'$, $a \otimes f$ is the operator in A given by $(a \otimes f)(x) = f(x)a$ for $x \in X$.

The range of this operator is the span of a.

8. Theorem
(a) u is a left divisor of 0 if and only if u is not one to one.
(b) u is a left topological divisor of 0 if and only if u is not an isomorphism.

PROOF (a) $\Rightarrow$: If $uv = 0$ then $u[v(x)] = 0$ for all x and $v(x) \neq 0$ for some x. $\Leftarrow$: If $u(a) = 0$ choose $f \in X'$ with $f(a) \neq 0$ and set $v = a \otimes f$. Then $uv = 0$.
  (b) $\Rightarrow$: Given $\varepsilon > 0$, choose v with $\|v\| = 1$, $\|uv\| < \varepsilon/2$. Let $\|a\| = 1$, $\|v(a)\| > \frac{1}{2}$, $x = v(2a)$. Then $\|x\| \geq 1$ and $\|ux\| = 2\|u[v(a)]\| < \varepsilon$. The result follows by Lemma 11-3-6. $\Leftarrow$: By Lemma 11-3-6, there exists a with $\|a\| = 1$, $\|ua\| < \varepsilon$. Choose $f \in X'$ with $\|f\| = |f(a)| = 1$ and set $v = a \otimes f$. Then $\|v\| = 1$. [Let $\|x\| \leq 1$; then $\|vx\| = |f(x)|\|a\| \leq 1$ so $\|v\| \leq 1$. But $\|v\| \geq \|va\| = 1$.] Also $\|uv\| \leq \varepsilon$. [Let $\|x\| \leq 1$. Then $\|u(vx)\| = |f(x)|\|ua\| < \varepsilon$.]

9. Theorem
(a) u is a right divisor of 0 if and only if its range is not dense.
(b) u is a right topological divisor of 0 if and only if u is not onto.

PROOF (a) $\Rightarrow$: u' is a left divisor of 0 $[(vu)' = u'v']$ so the result follows by Theorem 11-3-8 and Corollary 11-1-8. $\Leftarrow$: Let $f \in X'$, $f \neq 0$, $f = 0$ on Ru. Choose $x \in X$ with $f(x) = 1$ and set $v = x \otimes f$. Then $vu = 0 \neq v$.
  (b) $\Rightarrow$: u' is a left topological divisor of 0 as in (a), so the result follows by Theorems 11-3-4 and 11-3-8. $\Leftarrow$: u' is not an isomorphism, by Theorem 11-3-4, so by Lemma 11-3-6 there exists $f \in X'$, $\|f\| = 1$, $\|u'f\| < \varepsilon/2$. Let $a \in X$ with $\|a\| = 1$, $|f(a)| > \frac{1}{2}$, and set $v = (2a) \otimes f$. Then $\|v\| \geq 1$ [[$\|v\| \geq \|va\| = |f(a)|\|2a\| > 1$] and $\|vu\| < \varepsilon$. [Let $\|x\| \leq 1$. Then $\|vux\| = |f(ux)|\|2a\| = |u'(f)(x)| \cdot 2 < \varepsilon$.] To satisfy the definition exactly we may replace v by $v/\|v\|$.

10. Corollary u is singular if and only if it is either a left or right topological divisor of 0.

PROBLEMS

In this list X, Y are Banach spaces unless specified otherwise, and $u \in B(X, Y)$.

101 Show that $u'' \mid X = u$.

102 Show that u'' is a norm isomorphism if and only if u is.

103 Show that u'' is onto if and only if u is. Deduce that every (separated) quotient (in the sense of Sec. 6-2) of a reflexive Banach space is reflexive [Corollary 6-2-15].

104 Show that if u'' is one to one, u must also be [Prob. 11-3-101]—but not conversely [Prob. 11-1-105].

105 Show that Ru'' is weak $*$ dense if and only if u has dense range.

106 Let $f \in X''\backslash X$. Show that $f^\perp$ cannot be the range of a dual map from Y' [Theorem 11-3-3].

107 Show that u' is (a) an isomorphism onto and (b) an isometry onto, if and only if u is [Prob. 11-3-101].

108 Let u be an isometry. Show that for each $f \in X'$, there exists $g \in Y'$ with $u'(g) = f$, $\|g\| = \|f\|$. [See the proof of Lemma 11-3-2.]

109 Deduce the Hahn–Banach Theorem 2-3-7 from Prob. 11-3-108 [Prob. 11-1-2].

110 Show that u'' is an isometry if and only if u is [Probs. 11-3-101 and 11-3-108].

111 Suppose that u' is an isometry and $\varepsilon > 0$. Show that for each $y \in Y$ there exists $x \in X$ with $u(x) = y$, $\|x\| \le \|y\| + \varepsilon$. Compare Prob. 11-3-108. [The induced map $\bar{u} : X/Nu \to Y$ is an isometry onto, by Theorem 11-3-4 and Prob. 11-3-107, so there exists x with $u(x) = y$, $\|x + Nu\| = \|y\|$.]

112 Show that Prob. 11-3-111 is false with $\varepsilon = 0$ even if Y is one dimensional. [Take $u = f$ in Prob. 2-3-104, $f'(a) = af$ for $a \in \mathcal{K}' = \mathcal{K}$.]

113 Show that $Ru' = (X, p)'$ where $p(x) = \|u(x)\|$, that is, $f \in Ru'$ if and only if $|f(x)| \le M\|u(x)\|$. Deduce that Ru' always has a larger complete norm. [Imitate the proof of Lemma 11-3-2.]

114 Consider Theorem 11-3-4(b) where X, Y are normed spaces. Show that "$\Rightarrow$" holds if Y is complete, but fails if X is complete but Y is not. Show that "$\Leftarrow$" holds if X is complete (indeed, the hypothesis then implies that Y is complete), but fails if Y is complete but X is not. [Extend u to γX, Remark 3-2-13, by means of Prob. 2-1-11. In Example 5-2-8(a), consider the identity map. For the third part, u is onto γY. For the fourth part, consider an inclusion map.]

115 Let X, Y be normed spaces. Show the equivalence of the following:

    (a) u is almost open.

    (b) $u : \gamma X \to \gamma Y$ is onto.

    (c) u' is a (norm) isomorphism.

116 Let S be a reflexive subspace of X'. Define $u : X \to S'$ by $u(x) = \hat{x} \mid S$. Show that $u' : S \to X'$ is the inclusion map. Deduce that S must be weak $*$ closed in X' [Theorem 11-3-3].

117 Suppose that X' contains a reflexive subspace which is total over X. Show that X is reflexive [Prob. 11-3-116 and Theorem 8-3-9].

118 Let S be a subspace of X. Show that S is reflexive if and only if S is $\sigma(X'', X')$ closed in X'' [Probs. 11-3-116 and 11-1-121].

119 Let X be separable. Show that both X and X' can be embedded isomorphically in l^∞. [For X' use Prob. 6-2-201 and Theorem 11-3-4. For X, let $\{u_n\}$ be weak $*$ dense in the unit disc of X' by Theorem 9-5-3. Map $x \to \{u_n(x)\}$. This is an isometry!]

120 Let X be a Banach space, S a reflexive subspace of X', and $\varepsilon > 0$. Show that for each $f \in X''$ there exists $x \in X$ with $\|x\| \le \|f\| + \varepsilon$, $f(s) = s(x)$ for all $s \in S$. [Let $g = f \mid S$. Then u' in Prob. 11-3-116 is an isometry; so by Prob. 11-3-111 there exists x with $u(x) = g$, $\|x\| \le \|g\| + \varepsilon$.]

121 Problem 11-3-120 is false if X is only a normed space. [Take $S = X'$.]

122 (Helly's theorem). Prove the result of Prob. 11-3-120 with X a seminormed space and S finite dimensional. [The same proof. Now X/Nu is finite dimensional!]

123 In Probs. 11-3-120 and 11-3-122, ε cannot be taken to be 0 [as in Prob. 11-3-112].

124 Show that u is an isometry if and only if $u'[D] = D$; u' is an isometry if and only if $u[D]$ is a dense subset of D. Here D stands for the unit disc in X, Y, X', Y'.

125 Let Y be a closed subspace of X, $q : X \to X/Y$ the quotient map. Let $S \subset X$ be a subset such that $q[S]$ is fundamental in X/Y. Show that $S \cup Y$ is fundamental in X. [If $f = 0$ on $S \cup Y$, $f = q'(g)$ by Lemma 11-3-2; $g = 0$ on $q[S]$.]

126 Let $\{f_n\} \subset X'$ be bounded and assume that $\lim f_n(s)$ exists for $s \in S \cup Y$, Prob. 11-3-125. Show that $\{f_n\}$ is weak $*$ convergent [Probs. 11-3-125 and 9-3-104]. What does this say if $Y = \{0\}$?

127 Let X, Y, Z be normed spaces, $u \in B(X, Z)$, $v \in B(Y, Z)$. Prove the equivalence of

(a) $\overline{u[D]} \supset v[D_\varepsilon]$ for some $\varepsilon > 0$, where D, D_ε are the discs of radii 1, ε in X, Y.

(b) There exists k such that $\| v'(f) \| \leq k \| u'(f) \|$ for all $f \in Z'$.

Taking $Y = Z$, $v = i$ yields part of Prob. 11-3-115.

128 Let $A = B(X, X)$, $w \in X''\backslash X$, $\Gamma = \{u \in A : w$ is an eigenvector of $u''\}$, that is, $u''(w) = \chi(u)w$. Show that

(a) Γ is a closed subalgebra of $B(X)$. [Consider the map $u \to u''(w)$.]

(b) $\chi(uv) = \chi(u)\chi(v)$.

(c) $\chi \in A'$ [Prob. 5-5-201(c)].

(d) $\Gamma_0 = \{u : \chi(u) = 0\}$ is a left ideal in A. (See [157].)

129 Take $X = c$, $w = \delta^1 \in l^\infty$ in Prob. 11-3-128. Show that u is a matrix $[(ux)_n = \sum b_{nk} x_k]$ if and only if $u \in \Gamma$, and if this is true $\chi(u) = \lim_k \sum_k b_{nk} - \sum_k \lim_n b_{nk}$.

130 Let $u = a \otimes f$ (Definition 11-3-7). Show that $u' = f \otimes \hat{a}$, $u'' = \hat{a} \otimes \hat{f}$, and the equivalence of: $u \in \Gamma$ (Prob. 11-3-128), $u \in \Gamma_0$, $w(f) = 0$ or $a = 0$.

---

201 Show the equivalence of: u'' is one to one, Ru' is norm dense, if x is a weakly Cauchy bounded net in X with $u(x) \to 0$ weakly then $x \to 0$ weakly.

202 Let X be a closed subspace of the Banach space Y, $i : X \to Y$ the inclusion map. Then $i'' : X'' \to Y''$ is an isometry so we may consider $X'' \subset Y''$. Prove that $X'' + Y$ is a closed subspace of Y''.

203 Write the details of the following proof that there exists a null sequence which is not the sequence of Fourier coefficients of any $f \in L$. If this is false we get an isomorphism from c_0 onto L by the open mapping theorem. By Prob. 11-3-107, $c_0' = L'$ but L' is not separable.

204 Let $u : l^\infty \to l^\infty$ be continuous and linear. Show that $Nu \neq c_0$. [See [163].]

205 Give an example of Banach spaces such that the conditions of Prob. 11-3-127 hold but $Ru \not\supset Rv$. [See [151].]

11-4 WEAKLY COMPACT OPERATORS ON BANACH SPACES

The part of this section through Theorem 11-4-4 is of great importance but is not used in the remainder of this book. The succeeding part has not yet found a permanent place in mathematics and may be omitted entirely.

In this section X, Y are Banach spaces and $u \in B(X, Y)$. We consider $X \subset X''$ and note that $u'' \,|\, X = u$.

1. Definition u is called weakly compact if $u''[X''] \subset Y$.

The reason for the name is as follows. (See also Prob. 11-4-107.)

2. Theorem u is weakly compact if and only if $\overline{u[D]}$ is weakly compact, where D is the unit disc in X.

PROOF $\Rightarrow$: $u[D] = u''[D] \subset u''[D_2]$ where D_2 is the unit disc in X''. Now D_2 is weak $*$ compact [Banach–Alaoglu theorem] and u'' is weak $*$ continuous [Theorem 11-1-6]; so $u''[D_2]$ is weak $*$ compact, that is, $\sigma(Y'', Y')$ compact. Since $u''[D_2] \subset Y$ it is weakly compact, that is, $\sigma(Y, Y')$ compact; thus $u[D]$ is included in a weakly compact set. $\Leftarrow$: First, $D_2 = D^{\circ\circ} =$ weak $*$ closure of D so, since u'' is weak $*$ continuous, $u''[D_2] \subset$ weak $*$ closure of $u[D] =$ weak $*$ closure of $u[D]$. Now $\overline{u[D]}$ ($\subset X$) is weakly compact, that is, $\sigma(Y, Y')$ compact, hence $\sigma(Y'', Y')$ compact, and so it is $\sigma(Y'', Y')$ closed in Y''. Thus $u''(D_2) \subset u[D]$ and so $u''[X''] \subset Y$.

3. Lemma u is weakly compact if and only if $u': [Y', \sigma(Y', Y)] \to [X', \sigma(X', X'')]$ is continuous.

PROOF This condition may be referred to as weak $*$–weak continuity. $\Rightarrow$: Let g be a net in Y' with $g \to 0$, weak $*$, and $h \in X''$. Then $h(u'g) = (u''h)(g) \to 0$ since $u''h \in Y$. $\Leftarrow$: Let $h \in X''$. It is sufficient to show that $u''(h)$ is $\sigma(Y', Y)$ continuous [Theorem 8-1-7]. Let g be a net in Y' with $g \to 0$ weak $*$. Then $(u''h)(g) = h(u'g) \to 0$ since $u'g \to 0$, $\sigma(X', X'')$.

4. Theorem u is weakly compact if and only if u' is.

PROOF $\Rightarrow$: Let $G \in Y'''$, $g = G \mid Y \in Y'$. Then $u'''(G) = u'(g)$. [For $h \in X''$, $u'''(G)(h) = G(u''h)$ and $h(u'g) = (u''h)(g)$. These are equal because $u''(h) \in Y$.] $\Leftarrow$: As in Theorem 11-4-2, $D_2 =$ weak $*$ closure of D so $u''(D_2) \subset$ weak closure of $u[D] = \sigma(X'', X''')$ closure of $u[D]$ by Lemma 11-4-3. Now $u[D] \subset Y$ which is a norm closed, hence $\sigma(Y'', Y''')$ closed set in Y''; hence $u''[D_2] \subset Y$. Thus $u''[X''] \subset Y$.

Further information on weakly compact operators is given in [37], p. 549.
An exactly opposite property to that given in Definition 11-4-1 is: $h \in X'' \backslash X$ implies $u''(h) \notin X$. Operators $u \in B(X, Y)$ with this property are called *tauberian*. The reason for the name is that a matrix map $A: c_0 \to c_0$ is tauberian if and only if $Ax \in c_0$, $x \in l^\infty$ implies $x \in c_0$.

5. Theorem Consider
(a) u is tauberian.
(b) $Nu'' \subset X$ (equivalently, $Nu'' = Nu$).
(c) Nu is reflexive.
Then $(a) \Rightarrow (b) \Rightarrow (c)$, and if u is range closed the three conditions are equivalent.

PROOF $(a) \Rightarrow (b)$. Trivial. $(b) \Rightarrow (c)$. $Nu = Nu''$ is $\sigma(X'', X')$ closed; the result follows by Prob. 11-3-118. Now suppose that u is range closed. $(c) \Rightarrow (b)$. By Lemmas 11-1-7 and 11-3-2, $Nu'' = (Ru')^\perp = (Nu)^{\perp\perp} =$ weak $*$ closure of $Nu = Nu$ by Prob. 11-3-118. $(b) \Rightarrow (a)$. Let $h \in X''$, $u''h = y \in Y$. By Prob. 11-1-107, $y \in Ru$, say $y = ux$. Then $h - x \in Nu'' = Nu$ so $h \in x + Nu \subset X$.

6. Lemma If Y is reflexive and there exists a tauberian map from X to Y, then X is reflexive.

PROOF This is trivial.

7. Corollary Let X be a Banach space which has a closed subspace S such that S and X/S are both reflexive. Then X is reflexive.

PROOF Let $q: X \to X/S$ be the quotient map. It is tauberian, by Theorem 11-4-5, since it is onto and $Nq = S$. The result follows by Lemma 11-4-6.

PROBLEMS

In this list X, Y are Banach spaces, $u \in B(X, Y)$, and D is the unit disc in X.

101 If either X or Y is reflexive, show that u must be weakly compact.

102 If u is weakly compact and range closed, show that its range must be reflexive. [If u is onto, so is u''.]

103 Show that the inclusion map: $l^2 \to c_0$ is weakly compact and has dense range.

104 Show that a (matrix) map $A: c_0 \to c_0$ is weakly compact if and only if $\sum_k |a_{nk}| \to 0$ as $n \to \infty$ [Example 3-3-9].

105 Show that $a \otimes f$ (Definition 11-3-7) is weakly compact (a) by Theorem 11-4-2 and (b) by checking Definition 11-4-1 [Prob. 11-3-130].

106 Let $0 \neq f \in X'$ with X infinite dimensional. Show that X contains a closed bounded set S such that $f[S]$ is not closed. (Problem 11-4-201 generalizes this.)

107 Show that u is weakly compact if and only if $u[B]$ is weakly relatively compact for all bounded B.

108 u is called *compact* if $\overline{u[D]}$ is compact. Show that if u is compact its range must be separable, and if u is range closed and compact, its range must be finite dimensional. Compare Prob. 11-4-102. Further information may be found in [37] (pp. 485, 547–548), [82] (8B, 21A, B, C, D), and [116] (Chap. 8).

109 Show that a weakly compact operator whose range is in l must be compact [Example 9-5-5].

110 If X is reflexive, show that u is tauberian. Hence give an example of a tauberian operator which is not range closed.

111 Let u be one to one and u'' not one to one (Prob. 11-1-105). Show that u satisfies (c) but not (b) in Theorem 11-4-5.

112 Let A be the matrix whose nth row is δ^1 for $n = 1$, $\delta^{n-1} - \delta^n$ for $n > 1$. Show that $A: c_0 \to c_0$ satisfies (b) but not (a) in Theorem 11-4-5.

113 Let R be a reflexive Banach space. Show that the projection from $X \times R$ onto X provides an example of a range-closed tauberian operator with infinite-dimensional null space.

114 Call u *almost range closed* if $(Ru'') \cap Y = Ru$. Show that a range-closed operator is almost range closed. [See (b) $\Rightarrow$ (a) in Theorem 11-4-5.]

115 Show that every dual operator is almost range closed. (Hence a compact operator with infinite-dimensional range may be almost range closed and cannot be range closed [Prob. 11-4-108].)

116 Show that u is tauberian if and only if it is almost range closed and $Nu'' = Nu$.

117 Call u *nearly range closed* if the norm and weak $*$ closures of Ru' are the same. Show that a range-closed map is nearly range closed [Theorem 11-3-3].

118 Show that Nu is weak $*$ dense in Nu'' if and only if u is nearly range closed. Deduce that $Nu'' = Nu$ if and only if u is nearly range closed and Nu is reflexive [Prob. 11-3-118].

119 Show that neither of almost, nearly, range closed implies the other.

120 Show that u is almost range closed if and only if $\overline{u[D]} \subset Ru$.

121 Call u *disc closed* if $u[D]$ is closed. Show that every dual map is disc closed. This shows that the two parts of Prob. 11-3-124 are the same. Compare Prob. 11-4-201 which deals with a similar but not identical property [Theorem 9-1-12].

122 Show that every disc closed map is almost range closed [Prob. 11-4-120].

123 Show that a disc closed map need not be range closed [Prob. 11-4-121].

124 Show that a range closed map need not be disc closed [Prob. 6-2-104].

125 Let $u: l \to c_0$ be the inclusion map. Show that u is disc closed but not nearly range closed.

126 Show that u is tauberian if and only if $Nu'' = Nu$ and u is disc closed. Deduce that the matrix of Prob. 11-4-112 is not disc closed.

127 Let S be a vector subspace of X. Show that if $u[S]$ is closed, then $S + Nu$ must be closed, and that the converse holds if u is range closed. [Consider $u^{-1}[uS]$; in the latter case $u: X \to Ru$ is a quotient map.]

128 Improve Prob. 11-1-124 by showing that if $j''[X''\backslash X]$ and $j_2[X'']$ are disjoint. [j is tauberian by Theorem 11-4-5.]

129 Suppose that X is of codimension 1 in X'' (Remark 3-2-7). Show that the set of weakly compact operators is a maximal subspace of $B(X)$. (This is false for $X = c_0$ or l^p. Indeed, $B(X)$ has no ideal of codimension 1. See [56].)

130 Accepting the result of Prob. 11-4-301, suppose u is weakly compact and $v: Z \to Y$ has $v[Z] \subset R_u = u[X]$. Show that v is weakly compact. [Assuming X reflexive, give R_u the quotient topology; $v: Z \to R_u$ is continuous since it has closed graph; and $v[D]$ is weakly relatively compact in R_u, hence in Z.]

131 The inclusion map: $l \to c_0$ is weakly compact [Prob. 11-4-130; $X = l^2$].

---

201 Let $u \neq 0$. Show that u maps closed bounded sets onto closed sets if and only if u is a *semi-fredholm* map, that is, u is range closed and has finite-dimensional null space. [See [158], Theorem 2.1.]

202 Show that u maps closed vector subspaces onto closed sets if and only if u is range closed and Nu has finite dimension or codimension. [See [158], theorem 2.3, [47], theorem IV.1.10.]

203 Show that a weakly compact map from c_0 to c_0 must be compact [Prob. 11-4-109 and an analogue of Theorem 11-4-4].

---

301 If u is weakly compact, there exist a reflexive Banach space Z and maps $v: X \to Z$, $w: Z \to Y$ with $u = w \circ v$. The converse is trivial. [See [25], Corollary 1.]

TWELVE

COMPLETENESS

Many completeness theorems have the following form. Let $\mathscr{F}$ be a family of functions on a set X and $\mathscr{A}$ a collection of subsets of X. Then $\mathscr{F}$ is complete in the topology of uniform convergence on the sets of $\mathscr{A}$ if $f \mid A$ continuous for every $A \in \mathscr{A}$ implies $f \in \mathscr{F}$. As a first example, consider Theorem 8-6-5 which says that if every sequentially continuous function is continuous, the dual is complete in the topology of uniform convergence on the null sequences. Here $\mathscr{F} = X'$ and $\mathscr{A}$ is the class of null sequences. As a second example, let X be a topological space, $\mathscr{F} = C(X)$, $\mathscr{A} = $ compact sets. The resulting statement is a theorem about k spaces. See Prob. 5-1-201. Finally, Corollary 12-2-15, below, has this form.

We begin with an admissible topology, T°, which coincides with the weak * topology on equicontinuous sets. In Sec. 12-2, a less-restricted topology, aw^*, with this property is introduced. Various forms of completeness are related to compatibility of these topologies, as pointed out in Theorem 12-1-5, Prob. 12-1-113, and Remark 12-2-17.

> **Remark** Constant use will be made (without citation) of the fact that a set in the dual of a normed space is equicontinuous if and only if it is norm bounded. This is Example 9-1-6.

12-1 PRECOMPACT CONVERGENCE

Vast numbers of admissible topologies can be defined according to all the possible ways of designating suitable subsets of a member of a dual pair. These are rich fields for exploration and amusement, but not all of equal importance. We have selected, and shall continue to select, those which play a significant role in the development of the general theory. Here we concentrate on three such topologies, two of which were studied briefly in earlier chapters.

Let (X, Y) be a dual pair and T an admissible topology for X. The three families: (a) the set of T null sequences, (b) the set of T compact sets, (c) the set of T totally bounded sets, are obviously admissible; and the three resulting topologies on Y are written T_{c_0}, T_K, T°. The last-mentioned topology is called the *topology of precompact convergence*, a custom to which we bow rather than insisting on totally bounded convergence. The three topologies depend on T.

1. Lemma $\sigma(Y, X) \subset T_{c_0} \subset T_K \subset T^\circ \subset \beta(Y, X)$.

2. Example *For every dual pair* $\sigma^\circ(Y, X) = \beta(Y, X)$, where $\sigma^\circ(Y, X)$ stands for $[\sigma(X, Y)]^\circ$. Let $U \in \mathcal{N}(\beta)$, say $U = B^\circ$ with B a bounded set in X. Then B is $\sigma(X, Y)$ totally bounded. $[\sigma(X, Y)$ is a BTB topology as pointed out in Prob. 8-2-6.$]$ Thus $U = B^\circ \in \mathcal{N}(\sigma^\circ)$.

3. Example *It is possible that* $[\beta(X, Y)]_{c_0} \neq \sigma(Y, X)$, *a fortiori* $\beta^\circ(Y, X)$ *need not be* $\sigma(Y, X)$. Let N be an infinite-dimensional null sequence in a Banach space Y. Then $U = N^\circ \in \mathcal{N}[\beta_{c_0}(Y', Y)]$, where $\beta_{c_0}(Y', Y) = [\beta(Y, Y')]_{c_0}$. But $U \notin \mathcal{N}[\sigma(Y', Y)]$. $[$If not, $U \supset F^\circ$ with F finite and so $N \subset F^{\circ\circ}$, a finite-dimensional set. See also Prob. 12-1-106.$]$

It is trivial that equality may hold for the topologies mentioned in Example 12-1-3; indeed there are dual pairs (X, Y) with only one admissible topology for X $[$Prob. 8-5-7$]$.

4. Theorem Let T be an admissible topology for X such that (X, T) has the convex compactness property. Then T_K (hence also T_{c_0}) is compatible with (Y, X).

PROOF A special case (T compatible) of this result was given in Theorem 9-2-12. Here a little extra effort is needed. Let $U \in \mathcal{N}(T_K)$, say $U = A^\circ$ where A is a compact set in (X, T), and let B be the absolutely convex T closure of A. (This need not be $A^{\circ\circ}$ $[$Prob. 8-3-106$]$.) Then B is T compact, hence $\sigma(X, Y)$ compact. Since it is also absolutely convex, $B^\circ \in \mathcal{N}[\tau(Y, X)]$. Since $A^\circ \supset B^\circ$ the same is true of $U = A^\circ$.

Theorem 12-1-4 fails for T° $[$Prob. 14-2-104$]$ but can be obtained with a stronger assumption. (See also Prob. 12-1-113.)

5. Theorem Let T be an admissible topology for X such that (X, T) is boundedly complete. Then T° is compatible with (Y, X).

PROOF In the preceding proof replace A compact by A totally bounded. Again B is T compact and the rest of the proof is the same.

Theorem 12-1-5 may be compared with the fact that if $T = \sigma(X, Y)$ and (X, T) is boundedly complete then $\beta(Y, X)$ is compatible. This was proved in Theorem 10-2-4. It is a special case of Theorem 12-1-5 by Example 12-1-2.

6. Lemma Let (X, Y) be a dual pair, T an admissible topology for X, and $y = (y^\delta : D)$ a T equicontinuous net in Y. [*Note:* $Y \subset (X, T)'$.] Suppose that $y^\delta \to 0$ in $\sigma(Y, X)$. Then $y^\delta \to 0$ uniformly on each totally bounded set $B \subset (X, T)$.

PROOF Let $\varepsilon > 0$, $U = y^\circ \in \mathcal{N}(T)$ since y is equicontinuous. Then $B \subset F + \varepsilon U$ with F finite. Choose $\alpha \in D$ such that $\delta \geq \alpha$ implies $|y^\delta(f)| < \varepsilon$ for $f \in F$. Then $\delta \geq \alpha$ implies, for all $x \in B$, $|y^\delta(x)| = |y^\delta(f + \varepsilon u)| \leq \varepsilon + \varepsilon|y^\delta(u)| \leq 2\varepsilon$. Here $f \in F$ and $u \in U$ depend on x.

A consequence of this result is further evidence, Corollary 12-1-7, of the smallness of equicontinuous sets. It does not say anything about smallness of T° as illustrated by Example 12-1-2.

7. Corollary Let (X, Y) be a dual pair, T an admissible topology for X, and E a T equicontinuous set in Y. Then T_{c_0}, T_K, T°, and $\sigma(Y, X)$ all induce the same relative topology on E.

PROOF It is sufficient by Lemma 12-1-1 to show that $T^\circ|_E \subset \sigma(Y, X)|_E$. Let y be a net in E with $y \to 0$, $\sigma(Y, X)$. It is immediate from Lemma 12-1-6 that $y \to 0$ in T°.

8. Example Let (X, T) be an infinite-dimensional Banach space. By Corollary 12-1-7, $T^\circ = w^*$ on each disc, where $w^* = \sigma(X', X)$. This is because each disc in X', being norm bounded, is equicontinuous. Paradoxically, $T^\circ \neq w^*$. [See Example 12-1-3; $T = \beta(X, X')$ since X is barrelled.] Thus the identity map $i : (X', w^*) \to (X', T^\circ)$ has the paradoxical property that $i|D$ is continuous for every disc D (no matter how large) but i, itself, is not continuous.

9. Lemma Let T and T_1 be lc topologies for a vector space X, B an absolutely convex set in X with $T|_B = T_1|_B$. Then if B is T totally bounded, it is T_1 totally bounded.

PROOF It is shown in Prob. 12-1-107 that the assumption of absolute convexity cannot be omitted. Let $U \in \mathcal{N}(T_1)$. Then $U \cap B \supset V \cap B$ with $V \in \mathcal{N}(T)$. Now $\frac{1}{2}B \subset F + V$ with F a finite subset of $\frac{1}{2}B$ [Prob. 6-4-6]. It follows that $\frac{1}{2}B \subset F + V \cap B$. [$\frac{1}{2}b = f + v$ with $f \in F$, $v \in V$. Then $v = \frac{1}{2}b - f \in \frac{1}{2}B - \frac{1}{2}B = B$, so $v \in V \cap B$.] It follows that $\frac{1}{2}B \subset F + U$.

10. Theorem For any admissible topology T, $T^{\circ\circ} \supset T$.

PROOF Here $T^{\circ\circ} = (T^\circ)^\circ$. Say that T is a topology on X admissible for (X, Y) and let $U \in \mathcal{N}(T)$. We may assume $U = B^\circ$ with B absolutely convex and bounded in Y. Then B is $\sigma(Y, X)$ totally bounded [as in Example 12-1-2]. Since $T^\circ|_B = \sigma(Y, X)|_B$ by Corollary 12-1-7, it follows from Lemma 12-1-9 that B is T° totally bounded; hence $U = B^\circ \in \mathcal{N}(T^{\circ\circ})$.

11. Example $T^{\circ\circ} \neq T$. This may happen with $T = \sigma(X, Y)$, as shown in Examples 12-1-2 and 12-1-3. However, $\beta^{\circ\circ} = \beta$ always, since the former is admissible and larger.

12. Remark If T is compatible, Theorem 12-1-10 is much easier. Then U° is $\sigma(Y, X)$ compact and so it is T° compact by Corollary 12-1-7. Thus $U = U^{\circ\circ} \in \mathcal{N}[(T^\circ)_K]$ so we have the better result $(T^\circ)_K \supset T$.

An important method of relating topologies follows. The easy result of Prob. 12-1-1 will be used.

13. Theorem Let (X, Y) be a dual pair and T_X, T_Y admissible topologies for X, Y respectively. The following are equivalent:
(a) Every set in X which is T_Y equicontinuous is T_X totally bounded.
(b) $T_X^\circ \supset T_Y$.
(c) $T_Y|_E = \sigma(Y, X)|_E$ for each T_X equicontinuous set $E \subset Y$.
$(a)'$, $(b)'$, $(c)'$ are the same as (a), (b), (c) with X, Y interchanged.

PROOF It is obviously sufficient to prove $(a) \Rightarrow (b) \Rightarrow (c) \Rightarrow (a)'$.
$(a) \Rightarrow (b)$: Let $U \in \mathcal{N}(T_Y)$, say $U = B^\circ$ so that B is T_Y equicontinuous. By hypothesis B is T_X totally bounded and so $U = B^\circ \in T_X^\circ$. $(b) \Rightarrow (c)$: $T_Y|_E \subset T_X^\circ|_E = \sigma(Y, X)|_E$ by Corollary 12-1-7. The opposite inclusion holds since T_Y is admissible. $(c) \Rightarrow (a)'$: Let E be T_X equicontinuous. We may assume that E is absolutely convex since $E^{\circ\circ}$ is also T_X equicontinuous. Then E is $\sigma(Y, X)$ bounded, hence $\sigma(Y, X)$ totally bounded. By hypothesis and Lemma 12-1-9, E is T_Y totally bounded.

We shall isolate the most important part of this theorem and state it with a slightly different emphasis.

14. Theorem: Grothendieck interchange theorem Let (X, Y) be a dual pair and $\mathcal{A}, \mathcal{B}$ admissible families of subsets of X, Y respectively. Then each $A \in \mathcal{A}$ is $T_{\mathcal{B}}$ totally bounded if and only if each $B \in \mathcal{B}$ is $T_{\mathcal{A}}$ totally bounded.

PROOF This is just $(a) = (a)'$ in Theorem 12-1-13. A different proof is sketched in Prob. 12-1-117.

15. Corollary Let (X, Y) be a dual pair and T an admissible topology for X. Then T° is the largest admissible topology for Y which induces the same relative topology as $\sigma(Y, X)$ on each T equicontinuous set in Y.

PROOF The result is immediate from the equivalence of (b) and (c) in Theorem 12-1-13.

In Prob. 12-1-103 it is shown that "admissible" cannot be omitted (or replaced by "lcs"). See also Prob. 12-3-109.

We give an application of the interchange theorem, less for its intrinsic interest than to illustrate the ideas.

16. Corollary With X, Y, T as in Corollary 12-1-15, $T^{\circ\circ}$ has the same totally bounded sets, the same compact sets, and the same convergent sequences as T. Further, $T^{\circ\circ}$ is the largest admissible topology with the same totally bounded sets as T.

PROOF Each $T^{\circ\circ}$ equicontinuous set is T° totally bounded by definition. Hence each T° equicontinuous set is $T^{\circ\circ}$ totally bounded. But each T totally bounded set is T° equicontinuous by definition. Next, if K is T compact it is $T^{\circ\circ}$ totally bounded and complete since $T^{\circ\circ}$ is admissible and larger [Prob. 8-5-5]. Thus it is $T^{\circ\circ}$ compact [Theorem 6-5-7]. Third, if N is a T convergent sequence it is T compact; hence $T^{\circ\circ}$ compact. Thus $T^{\circ\circ}|_N = T|_N$ [Prob. 1-6-14]. So N is $T^{\circ\circ}$ convergent. Finally, let T_1 be admissible and with the same totally bounded sets as T. Every T° equicontinuous set is T_1 totally bounded. [It is T totally bounded.] Thus every T_1 equicontinuous set is T° totally bounded, which means $T^{\circ\circ}$ equicontinuous. Thus $T_1 \subset T^{\circ\circ}$.

17. Example Let $T = \sigma(l, l^\infty)$. Then $\beta(l, l^\infty) = \| \cdot \|_1$ has the same compact sets as T [Example 9-5-5], but is not $T^{\circ\circ}$ [Prob. 12-1-2]. Thus $T^{\circ\circ}$ *need not be the largest admissible topology with the same compact sets as T.*

PROBLEMS

In this list (X, Y) is a dual pair and T, T_1 are admissible topologies for X.

1 If $T \supset T_1$ then $T^\circ \subset T_1^\circ$—the same for T_{c_0} and T_K.

2 Let $\sigma = \sigma(X, Y)$. Show that $\sigma^{\circ\circ}$ is a BTB topology; hence cannot be normable if X is infinite dimensional [Corollary 12-1-16 and Prob. 8-2-6].

3 In the proof of Theorem 12-1-10 assume that $U \in \mathcal{N}[\beta(X, Y)]$. This would seem to give $T^{\circ\circ} \supset \beta(X, Y)$. Which step fails?

4 Suppose that $T_1 \supset T$ and T, T_1 have the same totally bounded sets, or, more generally, that each T compact set is T_1 totally bounded. Show that they have the same compact sets. The converse of the first part is false [proof of Corollary 12-1-16 and also Example 9-5-5].

5 Let (X, T) be a BTB space. Show that $T^\circ = \beta(X', X)$ [Example 12-1-2]. In particular [Example 6-4-14], if T is the Fréchet topology for ω, $T^\circ = \beta(\varphi, \omega)$, the largest lc topology for φ (Example 8-5-20).

---

101 Show that $T^{\circ\circ\circ} = T^\circ$ [Prob. 12-1-1 and Theorem 12-1-10].

102 Let (X, T) be a Mazur space. Show that X' is complete with T_K and T° [Theorem 8-6-5 and Prob. 8-5-5].

103 Show that all separated vector topologies for Y induce the same topology on $\sigma(X, Y)$ equicontinuous sets. Hence show that "admissible" cannot be omitted in Corollary 12-1-15 [Prob. 9-1-104].

104 Show, with $X = R$, that "totally bounded" cannot be omitted in Lemma 12-1-6.

105 Show, with $T = \sigma(X, Y)$, that "equicontinuous" cannot be omitted in Lemma 12-1-6.

106 Let Y be an infinite-dimensional reflexive Banach space. Show that $\beta(Y', Y)_{c_0}$ is complete while $\sigma(Y, Y')$ is not. Hence they are unequal. [See Theorem 8-6-5 and Prob. 8-3-3.]

107 Let $X = l$, $T = \sigma(l, \varphi)$, $T_1 = \|\cdot\|_1$, $B = \{n\delta^n\}$. Show that T, T_1 induce the same (discrete) topology on B, and B is T totally bounded but not T_1 bounded.

108 Show that every lcs separated topology coincides with its weak topology on its totally bounded sets. [$T^{\circ\circ}$ does this since T totally bounded implies T° equicontinuous.] For a boundedly complete space this is a trivial consequence of Prob. 1-6-14.

109 Show that (X, Y) is sequentially barrelled if and only if $[\sigma(Y, X)]_{c_0}$ is compatible with (X, Y).

110 Let B be a T totally bounded set. Show that $B^{\circ\circ}$ is T totally bounded. [See the proof of Corollary 12-1-16; $B^{\circ\circ}$ is T° equicontinuous.]

111 Show that T admissible cannot be omitted in Prob. 12-1-110. [Let $f \in X^{\#}\backslash X'$, $B = f^\perp \cap$ unit disc.]

112 Prove the equivalence of the following:

    (a) T° is compatible.

    (b) (Equivalence program). Every totally bounded set in (X, T) is $\sigma(X, Y)$ relatively compact.

    (c) Every totally bounded closed set in (X, T) is $\sigma(X, Y)$ compact.

113 Show that T° is compatible if and only if (X, T) is N complete (Prob. 6-5-107).

114 Let (X, T) be barrelled. Show that, in X', T° and weak $*$ have the same convergent sequences [Example 12-1-2 and Corollary 12-1-16]. (By Corollary 12-1-7, the same is true if X is only sequentially barrelled.)

115 Show that $n\delta^n \not\to 0$ in (l, n°), where n is the norm topology for $(\varphi, \|\cdot\|_\infty)$. [Consider $\{(1/n)\delta^n \subset \varphi.\}$ This shows that the assumption in Prob. 12-1-114 cannot be dropped.

116 Let $A \subset X$, $B \subset Y$, $B \subset F + A^\circ$ with F finite and A small of order F°. Show that A is small of order $3B^\circ$.

117 Let A be a bounded set in X, $B \subset Y$, $B \subset F + A^\circ$ with F finite. Show that A is a finite union of sets which are small of order $3B^\circ$. Deduce Theorem 12-1-14. [A is $\sigma(X, Y)$ totally bounded. Use Prob. 12-1-116.]

118 If α is a Hellinger–Toeplitz topology, show that α° is also. (Compare Prob. 11-2-102.)

119 Let $\sigma = \sigma(X, Y)$, $\tau = \tau(Y, X)$. Show that σ and τ_K have the same absolutely convex compact sets. [K is τ equicontinuous; hence τ_K totally bounded by Theorem 12-1-13. Apply Prob. 12-1-4.]

120 Let a be the family of absolutely convex compact sets in (X, T). Show that

    (a) T_a is compatible with (Y, X).

    (b) $T_a = T_K$ if and only if (X, T) has the convex compactness property.

    (c) If T is compatible then $T_{aa} \supset T$.

    (d) Part (c) fails for T admissible.

[For (c), U° is $\sigma(Y, X)$ compact; hence T_a complete by Prob. 8-5-5. It is T_a totally bounded by Theorem 12-1-13. For (d), take $T = \beta$ and apply (a).]

121 Prove the equivalence of the following:

    (a) $\sigma(X, Y)$ and $\tau(X, Y)$ have the same absolutely convex compact sets.

(b) $\sigma(Y, X)$ and $\tau(Y, X)$ have the same absolutely convex compact sets.

(c) $[\tau(X, Y)]_a = \tau(Y, X)$.

⟦(c) ⇒ (b) by Prob. 12-1-119 since $\sigma \subset \tau_a \subset \tau_K$.⟧

122 Let X be a Banach space such that the weak and norm topologies have the same compact sets (for example, l). Show that $\sigma(X', X)$ and $\tau(X', X)$ have the same compact sets. In particular, the unit disc of X' is $\tau(X', X)$ compact. ⟦See Prob. 12-1-121; $\sigma(X', X)$ is boundedly complete by Theorem 9-3-11.⟧

---

201 Show that $[\beta(Y, X)]_a$ and $\sigma(X, Y)$ have the same compact sets. ⟦See [112], Theorem 3.3.⟧

---

301 "Absolutely convex" cannot be omitted in Prob. 12-1-119; indeed, τ_K and τ_a need not have the same compact sets. ⟦See [112], Example 3.2.⟧

12-2 *aw**

Recall Corollary 12-1-15 which says that T° is the largest admissible topology for $(X, T)'$ which coincides with the weak * topology on T equicontinuous sets. The next definition attempts to define the largest such topology without any requirement such as admissibility. Whether or not the attempt actually succeeds turns out to be irrelevant. The reader should simply learn the meaning of the phrase "*aw** closed" without considering what sort of topology the collection of *aw** closed sets might form.

1. Definition Let (X, T) be a lcs space. A set $S \subset X'$ is said to be *aw** closed (almost weak * closed) if $S \cap U^\circ$ is weak * compact for each $U \in \mathcal{N}(X)$.

2. Example *A weak * closed set is aw* closed*, by the Alaoglu–Bourbaki theorem.

3. Lemma A set S is *aw** closed if and only if $S \cap E$ is weak * closed in E for each equicontinuous $E \subset X'$.

PROOF ⇐: Each U° is equicontinuous, so $S \cap U^\circ$ is weak * closed in U°; hence compact. ⇒: Let $U = E^\circ \in \mathcal{N}(X)$. Then $S \cap U^\circ$ is weak * closed. Finally, $S \cap E = S \cap (E \cap U^\circ) = (S \cap U^\circ) \cap E$, the intersection of E with a closed set.

4. Example *An aw* closed set which is not weak * closed* (in the dual of a normed space). Let $X = (\varphi, \| \cdot \|_\infty)$, $S = \{nP_n\}$, where $P_n(x) = x_n$. Then S is not weak * closed. ⟦For each x, $nP_n(x) = nx_n \to 0$, so $nP_n \to 0 \notin S$.⟧ However, S is *aw** closed. ⟦For each equicontinuous E is $\| \cdot \|_1$ bounded and so $E \cap S$ is finite, hence weak * closed.⟧

5. Example *Let (X, T) be a lcs space. Then T° closed implies aw* closed.* This is true since T° coincides with the weak * topology on equicontinuous sets ⟦Corollary 12-1-7⟧. The converse is false ⟦Prob. 12-2-113⟧.

6. Example *An aw* closed set which is not weak * closed* (in the dual of a Banach space). This improves Example 12-2-4. Consider Example 12-1-3 where, with $T =$ norm, $T° \neq$ weak $*$. Thus there is a $T°$ closed set which is not weak $*$ closed. The result follows by Example 12-2-5.

7. Lemma Any translate of an *aw** closed set is *aw** closed.

PROOF Let S be *aw** closed in X', $a \in X'$, $E \subset X'$ equicontinuous, and y a net in $(S + a) \cap E$ with $y \to e \in E$, weak $*$. Then $y - a \to e - a \in E - a$. Since $y - a \in S \cap (E - a)$ and $E - a$ is equicontinuous, it follows that $e - a \in S$ and so $e \in S + a$.

8. Definition Let X be a lcs space and f a linear functional on X'. Then f is called *aw** continuous if $f^\perp$ is *aw** *closed*.

Of course, if f is weak $*$ continuous [that is, $f : (X', \text{weak} *) \to Y$ is continuous], then f is *aw** continuous. This follows from Example 12-2-2.

9. Lemma Let $f \in X'^{\#}$. These are equivalent:
(a) f is *aw** continuous.
(b) $f | U°$ is weak $*$ continuous at 0 for each $U \in \mathcal{N}(X)$.
(c) $f(y) \to 0$ for each equicontinuous net y such that $y \to 0$, weak $*$.
(d) $f | E$ is weak $*$ continuous for each equicontinuous set $E \subset X'$.

PROOF $(a) \Rightarrow (b)$: Suppose $f | E$ is not weak $*$ continuous at 0 for a certain $E = U°$. There exists $\varepsilon > 0$ and a net $y = (y_\delta)$ in E with $y \to 0$, weak $*$, and $|f(y_\delta)| \geq \varepsilon$ for all δ. Now let $u \in X'$ with $f(u) = 1$, $v_\delta = u - y_\delta/f(y_\delta)$. Then $E_1 = \{u\} \cup \{v_\delta\}$ is equicontinuous since $|f(y_\delta)| \geq \varepsilon$, but $f^\perp \cap E_1$ is not weak $*$ closed in E_1. [$v_\delta \in f^\perp \cap E_1$, $v_\delta \to u \in E_1$, weak $*$, and $u \notin f^\perp$.] Thus f is not *aw** continuous. $(b) \Rightarrow (a)$: This is because $f^\perp \cap U° = [f | U°]^{-1} \{0\}$ is weak $*$ closed. The equivalence of (b) and (c) is trivial. Clearly, $(d) \Rightarrow (b)$. Finally, $(b) \Rightarrow (d)$. If e is a net in E with $e \to f$. The net $e - f$ is contained in $E - f$; hence is contained in $U°$ for some $U \in \mathcal{N}(X)$ since $E - f$ is equicontinuous.

10. Remark Let X be a lcs space, $f \in X'^{\#}$. We have the equivalence of: f is weak $*$ continuous, $f^\perp$ is weak $*$ closed, $f \in X$ [Theorems 8-1-7 and 4-5-10]. A remarkable result (Corollary 12-2-14) replaces these equivalent conditions by three weaker ones: f is *aw** continuous, $f^\perp$ is *aw** closed, $f \in \gamma X$, the completion of X. Of course, $f \in X$ implies f is *aw** continuous [f is weak $*$ continuous] and Corollary 12-2-16 says that the converse is true if and only if X is complete.

We first give a discussion of the completeness theorem for normed spaces. Examples 12-2-11 and 12-2-12 may be omitted since they are included in the immediately succeeding material; however, they may be read as a motivating and simpler treatment.

11. Example *Let X be a normed space. Then each aw* closed set is norm closed and each aw* continuous functional is in X''.* The second statement follows from the first since $f^\perp$ is norm closed [Theorem 4-5-10]. Suppose that S is not norm closed. Let $s_n \in S$ with $s_n \to y \notin S$. Let $E = \{s_n\} \cup \{y\}$. Since E is norm bounded it is equicontinuous. Now $S \cap E = \{s_n\}$ and $s_n \to y$, weak $*$; thus S is not weak $*$ closed in E. (Compare Example 12-2-4 where the given set is norm closed but not weak $*$ closed.)

But not every $f \in X''$ is *aw** continuous. It is a special case of the next example that if X is a Banach space, f is *aw** continuous if and only if $f \in X$.

12. Example Let X be a normed space, $f \in X'^{\,*}$. Then f is *aw** continuous if and only if $f \in$ norm closure of X in X''.

PROOF $\Leftarrow$: Let $g = (g_\delta)$ be an equicontinuous net in X' with $g \to 0$ weak $*$, $\varepsilon > 0$. Say $\| g_\delta \| < M$ for all δ. Choose $x \in X$ with $\| f - x \| < \varepsilon/M$. Then $|f(g_\delta)| < |f(g_\delta) - g_\delta(x)| + |g_\delta(x)| = |f(g_\delta) - \hat{x}(g_\delta)| + |g_\delta(x)| \leq \varepsilon + |g_\delta(x)| \to \varepsilon$. So $\limsup |f(g_\delta)| \leq \varepsilon$ and so $f(g_\delta) \to 0$. By Lemma 12-2-9, f is *aw** continuous. $\Rightarrow$: Let $v \in X'''$ with $v = 0$ on X. We shall show that $v(f) = 0$; the result follows by the Hahn–Banach Theorem 2-3-9, since we know by Example 12-2-11 that $f \in X''$. We may assume that $\| v \| = \| f \| = 1$. Then $v \in D^{\circ\bullet}$ where D is the unit disc in X', $D^\circ \subset X''$, $D^{\circ\bullet} \subset X'''$. Since D is absolutely convex, this means that $v \in \sigma(X''', X'')$ closure of D. Let a be a net in D with $a \to v$, $\sigma(X''', X'')$. Then $a \to 0$, $\sigma(X', X)$. [For each $x \in X$, $a(x) = \hat{x}(a) = \hat{a}(\hat{x}) \to v(\hat{x}) = 0$. Here $\hat{a}$ is the original net called a but considered in X'''.] Since a is a bounded net it is equicontinuous, so $v(f) = \lim \hat{a}(f) = \lim f(a) = 0$ by Lemma 12-2-9.

We now turn to the general case. By Example 12-2-12 the set of *aw** continuous linear functionals on a normed space X is a Banach space in which X is dense. An exactly analogous result holds for an arbitrary lcs space. It will be useful to prove a more general result. In the following theorem T is an admissible topology. Up till now it has been assumed compatible.

13. Theorem: Grothendieck's completion theorem Let (X, Y) be a dual pair and T an admissible topology for X; say $T = T_{\mathscr{A}}$ with $\mathscr{A}$ an admissible family of absolutely convex closed sets in Y. Let $G = \{f \in Y^{\,*}: f \,|\, A$ is $\sigma(Y, X)$ continuous for each $A \in \mathscr{A}\}$. Then G can be given a topology which makes it a completion of (X, T), that is, a complete lcs space which has X as a dense topological vector subspace.

PROOF First, G is clearly a vector space. Also $X \subset G$ since each x is $\sigma(Y, X)$ continuous on Y. Now $\mathscr{A}$ is admissible for (Y, G). [The only doubtful point is that the members of $\mathscr{A}$ are $\sigma(Y, G)$ bounded. To see this let $A \in \mathscr{A}$, $f \in G$, $\{a_n\} \subset A$. Then $(1/n)a_n \in A$ and $(1/n)a_n \to 0$, $\sigma(Y, X)$, since A is $\sigma(Y, X)$ bounded. Thus $(1/n)f(a_n) = f[(1/n)a_n] \to 0$ so $f[A]$ is bounded by Theorem 4-4-1.⟧ Let $\overline{T}$ be the resulting admissible topology for G. Then $\overline{T}|_X = T$. ⟦$A^\bullet \cap X = A^\circ$.⟧

Next, X is $\overline{T}$ dense in G. (It is obviously $\sigma(G, Y)$ dense since it is total over Y, but this is not good enough.) To prove this, let $v \in G'$, $v = 0$ on X. Then $\{f \in G : |v(f)| \leq 1\} \supset A^\bullet$ for some $A \in \mathscr{A}$. Considering $A \subset G'$, this says that v belongs to the bipolar of A in the dual pair (G, G'). This is just the $\sigma(G', G)$ closure of A, so there exists a net a in A with $a \to v$, $\sigma(G', G)$. Now $a \to 0$, $\sigma(Y, X)$. ⟦For $x \in X$, $[x, a] \to [x, v] = 0$ since $x \in G$ and $v = 0$ on X. We wrote $[x, v]$ for $v(x)$.⟧ For each $f \in G$, $v(f) = \lim a(f) = \lim f(a) = 0$ since f is $\sigma(Y, X)$ continuous on A. Thus $v = 0$ and X is dense as required.

It remains to prove that $(G, \overline{T})$ is complete. Let f_δ be a $\overline{T}$ Cauchy net in G. Then f_δ is $\sigma(G, Y)$ Cauchy and so $f_\delta \to f \in [Y^*, \sigma(Y^*, Y)]$ since the latter space is complete ⟦Prob. 8-1-5⟧. It is sufficient to prove that $f \in G$. ⟦If f_δ is $\overline{T}$ Cauchy and $\sigma(G, Y)$ convergent, it is $\overline{T}$ convergent by the usual argument involving F-linked topologies (Lemma 6-1-11 and Prob. 8-5-5).⟧ To this end let $A \in \mathscr{A}$ and let $a = (a_\gamma)$ be a $\sigma(Y, X)$ convergent net in A. We may assume that $a \to 0$. ⟦If $a \to b$, the net $(a_\gamma - b)$ is contained in some member of $\mathscr{A}$, by Theorem 8-5-15.⟧ Let $\varepsilon > 0$. There exists δ such that $\alpha \geq \delta$, $\beta \geq \delta$ implies $f_\alpha - f_\beta \in \varepsilon A^\bullet$. For each $a_\gamma \in a$ we have $|f(a_\gamma)| \leq |f(a_\gamma) - f_\delta(a_\gamma)| + |f_\delta(a_\gamma)| = \lim_\alpha |f_\alpha(a_\gamma) - f_\delta(a_\gamma)| + |f_\delta(a_\gamma)| \leq \varepsilon + |f_\delta(a_\gamma)|$. Also, $\lim_\gamma f_\delta(a_\gamma) = 0$ since $f_\delta \in G$ and so $\lim \sup |f(a)| \leq \varepsilon$, that is, $f(a) \to 0$.

The special case in which T is compatible has this appearance. It is a special case by Lemma 12-2-9.

14. Corollary Let X be a lcs space and G the set of aw^* continuous members of X'^*. Then G can be given a topology which makes it a completion of X.

15. Corollary: Grothendieck's completeness theorem Let (X, Y) be a dual pair and T an admissible topology for X; say $T = T_\mathscr{A}$ with $\mathscr{A}$ an admissible family of absolutely convex closed sets in Y. Then (X, T) is complete if and only if every $f \in Y^*$ such that $f|A$ is $\sigma(Y, X)$ continuous for each $A \in \mathscr{A}$ is also $\sigma(Y, X)$ continuous on Y, that is, belongs to X.

16. Corollary Let X be a lcs space. Then X is complete if and only if every aw^* continuous linear functional on X' is weak $*$ continuous (i.e., belongs to X) or, equivalently, if and only if every aw^* closed maximal subspace is weak $*$ closed.

17. Remark Corollary 12-2-16 is a success of the aw^* topology as compared with T°. (See the comparison between them at the beginning of this section.) We may interpret Corollary 12-2-16 as saying that X is complete if and only if aw^* is "compatible," i.e., if and only if $(X', aw^*)' = X$. (This is an abuse of notation since aw^* need not be a vector topology as far as we know. This point is considered in Example 12-3-12.) On the other hand, it is false that (X, T) must be complete if $T°$ is compatible. ⟦Let X be boundedly complete and not complete (Example 9-3-12) and apply Theorem 12-1-5.⟧ This is not unexpected since $T° \subset aw*$ ⟦Example 12-2-5⟧. See also Prob. 12-1-113.

18. Remark Let T be a topology on X' which is compatible with (X', X). Then a *convex set S is aw* closed if and only if $S \cap U°$ is T closed for each* $U \in \mathcal{N}(X)$. ⟦$\Rightarrow$: Since $T \supset$ weak *. $\Leftarrow$: By compatibility, $S \cap U°$ is weak * closed; hence weak * compact.⟧ Hence for $f \in X'$ *, *f is aw* continuous if and only if $f \mid E$ is T continuous for each equicontinuous set $E \subset X'$*. ⟦In Lemma 12-2-9, replace weak * by T.⟧

19. Corollary Let Y be a lcs space and $\mathcal{A}$ a family of absolutely convex closed sets which is admissible for (Y, Y'). Then $(Y', T_{\mathcal{A}})$ is complete if and only if every $f \in Y$ * such that $f \mid A$ is continuous for each $A \in \mathcal{A}$ is continuous, i.e., belongs to Y'.

PROOF This is precisely Corollary 12-2-15 with $\sigma(Y, X)$, in its first occurrence, replaced by the topology of Y, as allowed by Remark 12-2-18.

20. Example Let Y be a lcs space and let $\mathcal{A}$ be the set of absolutely convex bounded closed sets in Y so that $\beta(Y', Y) = T_{\mathcal{A}}$. An application of Corollary 12-2-19 is: *the strong dual of Y is complete if and only if every $f \in Y$ * such that $f \mid A$ is continuous for each bounded set $A \subset Y$ is continuous*. A special case is that the strong dual of a bornological space is complete (Corollary 8-6-6), for if $f \mid A$ is continuous (A bounded), f is bounded on A, as was proved in the course of Theorem 12-2-13, and so $f \in Y'$ ⟦Prob. 8-4-5⟧.

The next example was given a different treatment in Prob. 3-3-109.

21. Example: Sufficiency of the Silverman–Toeplitz conditions Let A be a matrix with $\sup_n \sum_k |a_{nk}| < \infty$. Let a^n be the nth row, that is, $(a^n)_k = a_{nk}$. Then $a^n \in l$ and $\{a^n\}$ is equicontinuous since it is $\| \cdot \|_1$ bounded. Suppose $\lim_n a_{nk} = 0$ for each k; this says precisely that $a^n \to 0$, $\sigma(l, \varphi)$. Now $T = \| \cdot \|_\infty$ is a compatible topology for φ and $c_0 = \gamma(\varphi, T)$, the completion of φ. Thus each $x \in c_0$ is $aw*$ continuous ⟦Example 12-2-12 or Corollary 12-2-14⟧. It follows from Lemma 12-2-9 that $[x, a^n] \to 0$, that is, $\sum_k a_{nk} x_k \to 0$ for each $x \in c_0$.

PROBLEMS

In this list (X, T) is a lcs space.

1 Let X be a normed space, $S \subset X'$. Show that S is $aw*$ closed if and only if $S \cap D$ is weak * compact (a) for every disc D and (b) for every disc D centered at 0.

2 Show that an $aw*$ continuous functional is bounded on each equicontinuous set. ⟦See the proof of Theorem 12-2-13.⟧

3 Let X be separable. Show that each weak $*$ sequentially closed set in X' is aw^* closed. Deduce that if X is separable and complete, $(X', \text{weak } *)$ is a Mazur space. (Compare Prob. 12-2-102.) ⟦See Theorems 9-5-3 and 4-5-10.⟧ (Neither "separable" nor "complete" can be omitted, by Probs. 14-7-109 and 9-3-117.)

4 If T, T_1 are admissible for (X, Y), $T_1 \supset T$ and (X, T) is complete, then (X, T_1) is complete by Prob. 8-5-5. Deduce this result from Corollary 12-2-19.

---

101 Let $X = (\varphi, \|\cdot\|_\infty)$, $S = \{f \in X' : \sum f(\delta^n)/n = 0\}$. Show that, considering $X' = l$, S corresponds to $\{y \in l : \sum y_n/n = 0\}$ and, considering $X'' = l^\infty$, S corresponds to $z^\perp$ where $z = \{1/n\} \in l^\infty$. Since $z \in \overline{X} \backslash X$, S is aw^* closed but not weak $*$ closed. Prove this directly.

102 Let X be sequentially barrelled. Show that each aw^* closed set in X' is weak $*$ sequentially closed. Deduce that if X is sequentially barrelled and $(X', \text{weak } *)$ is Mazur, X is complete. This improves Prob. 9-3-117.

103 Let X be barrelled. Show that every aw^* closed convex body in $(X', \text{weak } *)$ is weak $*$ closed ⟦Probs. 12-2-102 and 6-1-201⟧.

104 Show that the completion of a barrelled space is barrelled ⟦Prob. 9-3-124⟧.

105 Show that a set S in X' is aw^* closed if and only if $S \cap U^\circ$ is T° closed for each $U \in \mathcal{N}(X)$ ⟦Corollary 12-1-7⟧.

106 Let S be a dense subspace of X. Show that $X \subset \gamma S$ in the sense that each member of X is aw^* continuous on S' ⟦Probs. 9-1-7 and 9-3-104⟧.

107 Let S be a dense barrelled (or just sequentially barrelled) subspace of X. If $\{f_n\} \subset X'$, $f_n \to 0$ on S, show that $f_n \to 0$ on X ⟦Prob. 12-2-106⟧. Compare Prob. 9-3-138.

108 Give an example in which $\gamma X \not\subset X''$ in the sense that not every aw^* continuous f belongs to X'' ⟦Prob. 10-3-301⟧.

109 Let X be quasibarrelled. Show that every aw^* continuous $f \in X'^*$ is bounded on $[X', \beta(X', X)]$, that is, $\gamma X \subset (X')^b$ ⟦Prob. 12-2-2 and Theorem 10-1-11⟧.

110 Suppose that X is quasibarrelled and (X', X'') is bornological. Show that $\gamma X \subset X''$ in the sense that every aw^* continuous $f \in X''$ ⟦Prob. 12-2-109⟧. Deduce that γX is the closure of X in the bidual ⟦Corollary 8-6-6 and Theorem 10-1-8⟧.

111 Let T_1 be compatible and $T_1 \supset T$. Show that each aw^* closed set in $(X, T_1)'$ is aw^* closed in $(X, T)'$ ⟦Prob. 9-1-2⟧.

112 Deduce Theorem 8-6-5 ⟦(X', T_{c_0}) is complete if X is a Mazur space⟧ from Corollary 12-2-19.

113 Let B be a normed space and N a linearly independent null sequence in B'. Let $X = (B, \text{weak})$. Show that $N \cap E$ is finite for each equicontinuous $E \subset X'$; hence N is aw^* closed but not T° ⟦$= \beta(X', X) = \text{norm}$⟧ closed ⟦Prob. 9-1-104⟧.

114 Let X be relatively strong and suppose that $f \in X^*$, $f \mid A$ continuous for each bounded set A implies $f \in X'$. Show that X need not be bornological ⟦Example 12-2-20 and Prob. 9-5-108⟧.

12-3 STRICT HYPERCOMPLETENESS

The first result of this section says that if X is a lc metric space, aw^* is a lc topology, i.e., there is a lc topology for X' whose closed sets are precisely the aw^* closed sets. This is mildly interesting—the excitement mounts when we consider the possibility that aw^* might be compatible with (X', X). Under these circumstances we get a very strong version of the form of completeness represented by the type of statement: $S \cap A$ closed in A for all $A \in \mathcal{A}$ implies S closed (as in Corollary 12-2-16), namely, for convex sets, aw^* closed implies weak $*$ closed ⟦since aw^* and weak $*$ are compatible⟧. This type of completeness is called *strict hyper-*

completeness. We postpone the formal definition until after the promised metric result.

1. Theorem Let (X, T) be a lc metric space. The following topologies on X' are equal: T_{c_0}, T_K, T°, and aw^*.

PROOF Recall that the first three are the topologies of uniform convergence on the null sequences, compact sets, and totally bounded sets in X. Their equality with aw^* means that a set is, for example, T° closed if and only if it is aw^* closed. Since T° closed sets are aw^* closed ⟦Example 12-2-5⟧, the whole result will follow when it is shown that each aw^* closed set S is T_{c_0} closed. To this end let $0 \notin S$ and we shall construct $U \in \mathcal{N}(T_{c_0})$ with $U \pitchfork S$. ⟦This will be sufficient since aw^* closed is translation invariant by Lemma 12-2-7.⟧ Let $\{U_n\}$ be a base for $\mathcal{N}(X, T)$. Now $S \cap U_1^\circ$ is weak $*$ closed so there is a finite set $F_1 \subset X$ with $F_1^\circ \pitchfork [S \cap U_1^\circ]$. Consider the following infinite collection: U_2°, $S \cap U_2^\circ$, F_1°, $\{x^\circ : x \in U_1\}$. Their intersection is empty ⟦it is $U_2^\circ \cap S \cap F_1^\circ \cap U_1^\circ \subset S \cap F_1^\circ \cap U_1^\circ = \phi$⟧ and one of them, U_2°, is weak $*$ compact; hence a finite subcollection must have empty intersection, i.e., there exists $F_2 \subset U_1$, F_2 finite, with $\phi = U_2^\circ \cap S \cap F_1^\circ \cap F_2^\circ = U_2^\circ \cap S \cap (F_1 \cup F_2)^\circ$. We may assume $F_2 \neq \phi$ since enlarging F_2 does not spoil the preceding identity. Similarly, there exists a nonempty finite set $F_3 \subset U_2$ with $U_3^\circ \cap S \cap (F_1 \cup F_2 \cup F_3)^\circ = \phi$ and, in general, $F_n \subset U_{n-1}$ with $U_n^\circ \cap S \cap A_n^\circ = \phi$, where $A_n = \cup\{F_i : 1 \leq i \leq n\}$. Now set $N = \cup A_n$. This is a null sequence in X since $F_n \subset U_{n-1}$. Thus $N^\circ \in \mathcal{N}(T_{c_0})$. Finally, $N^\circ \pitchfork S$. ⟦For every n, $S \cap N^\circ \subset S \cap A_n^\circ \pitchfork U_n^\circ$ so $S \cap N^\circ \pitchfork U_n^\circ$ for each n. But $X' = \cup U_n^\circ$.⟧

A consequence of Theorem 12-3-1 is the exact identification of the totally bounded sets.

2. Corollary Let X be a lc metric space. Then a set B is totally bounded if and only if it lies in the absolutely convex closure of a null sequence.

PROOF $\Leftarrow$: By Theorem 7-1-5. $\Rightarrow$: $B^\circ \in \mathcal{N}(T^\circ)$ and $T^\circ = T_{c_0}$.

3. Theorem: Krein–Smulian theorem Let X be a lc Fréchet space. Then every aw^* closed convex set in X' is weak $*$ closed.

PROOF Such a set is T° closed by Theorem 12-3-1. But T° is compatible with (X', X) since X is boundedly complete ⟦Theorem 12-1-5⟧.

4. Remark This shows that no convex set could have figured in Example 12-2-6. Moreover, that example shows that "convex" cannot be omitted in Theorem 12-3-3.

5. Remark A primitive form of this theorem was given in Prob. 12-2-103.

6. Theorem: Banach–Dieudonné theorem Let X be a Banach space and D the unit disc in X'. Then every vector subspace S of X' such that $S \cap D$ is weak $*$ closed is itself weak $*$ closed.

PROOF The condition implies that S is aw^* closed. [For any $R > 0$, $S \cap D_R = R(S \cap D)$ is weak $*$ closed.] By Theorem 12-3-3, S is weak $*$ closed.

7. Definition A lcs space X is called *strictly hypercomplete* if every aw^* closed convex set in X' is weak $*$ closed.

Thus every lc Fréchet space is strictly hypercomplete.

8. Example *A strictly hypercomplete space X which is not relatively strong and a strictly hypercomplete relatively strong space which is not metrizable.* Let (F, T) be a lc Fréchet space, $X = F'$, and T_1 a lc topology for X with $T_K \subset T_1 \subset \tau(F', F)$, where K is the family of compact sets in (F, T) [possible since T_K is compatible with (F', F) by Theorem 9-2-12 and Example 9-2-10]. *Then (X, T_1) is strictly hypercomplete.* Let S be a convex set in $X'(= F)$ which is aw^* closed. We shall show that S is T closed. Since it is convex, it will follow that it is weak $* = \sigma(X', X) = \sigma(F, F')$ closed. Let $\{x_n\} \subset S$, $x_n \to x$, $E = \{x_n\} \cup \{x\}$. Then E is a compact set in (F, T); hence T_K equicontinuous; hence T_1 equicontinuous. Thus $S \cap E$ is weak $* = \sigma(F, F')$ closed in E; hence T closed in E. Thus $x \in S$.

In particular, (X, T_K) and $[X, \tau(X, X')]$ are strictly hypercomplete. The latter need not be metrizable or even quasibarrelled [Prob. 10-1-3]. The former need not be relatively strong. [Let F be a reflexive Banach space. The unit disc in $F' \in \mathcal{N}(\tau) \backslash \mathcal{N}(T_K)$ by Example 10-2-6 and Theorem 6-4-2.]

If one knows k spaces (Prob. 5-1-201), Example 12-3-8 may be done thus. If S is any aw^* closed set in X', $S \cap K$ is T compact for all compact $K \subset (F, T)$; hence T closed. If convex it is weak $*$ closed.

9. Remark In connection with Example 12-3-8, keep in mind the result of Theorem 12-3-1. See also Prob. 12-3-101.

10. Theorem A lc metric space (X, T) has the convex compactness property if and only if it is complete.

PROOF $\Rightarrow$: If S is a convex aw^* closed set it is T_{c_0} closed by Theorem 12-3-1. But T_{c_0} is compatible with (X', X) [Example 9-2-13] so S is weak $*$ closed. By Corollary 12-2-16, X is complete. Another proof is to note that each Cauchy sequence is included in a compact, hence complete, set by Corollary 12-3-2, hence is convergent. This gives a better result (Prob. 12-3-110). $\Leftarrow$: By Example 9-2-10.

The main application of Theorem 12-3-11 is to give an example in which aw^* is not a lc topology. Let us say that "aw^* is a lc topology" means that there exists a lc topology T for X' such that the T closed and aw^* closed sets are the same.

11. Theorem If aw^* is a lc topology for X' and X is a complete lcs space, then X is strictly hypercomplete.

PROOF It is sufficient to prove this for a real vector space since convexity uses only real scalars and since Theorem 1-5-2 is easy to apply. Let S be a convex aw^* closed set in X' and let $g \in X' \backslash S$. By Theorem 7-3-4 there exists an aw^* continuous $\varphi \in X'^{\#}$ such that $\varphi(g) > \sup \{\varphi(s) : s \in S\}$. Since X is complete it follows from Corollary 12-2-15 that there exists $x \in X$ such that $\varphi(f) = f(x)$ for all $f \in X'$. Then $\{f \in X' : |f(x) - \varphi(g)| < \varepsilon\}$ is a weak $*$ neighborhood of g and is disjoint from S if ε is sufficiently small. Thus S is weak $*$ closed.

12. Example Thus aw^* is not a lc topology if X is complete and not strictly hypercomplete. Examples occur in Probs. 12-4-113 and 12-4-301 and Example 13-2-14.

PROBLEMS

In this list (X, T) is a lcs space.

1 Show that a strictly hypercomplete space is complete. [Corollary 12-2-16].

101 Let (X, T) be strictly hypercomplete and T_1 a larger compatible topology. Show that (X, T_1) is strictly hypercomplete [Prob. 12-2-111].

102 Let (X, T) be a lc metric space. Show that the family of totally bounded sets is the saturated hull of the family of null sequences.

103 Let (X, T) be a separable lc Fréchet space. Show that every weak $*$ sequentially closed convex set in X' is weak $*$ closed [like Prob. 12-2-3 which gives the same result for maximal subspaces].

104 Let X be a Banach space, $f \in X'' \backslash X$ with $\|f\| = 1$. Let $S = \{u \in X' : |f(u)| \leq 1\}$, $D_r = \{u \in X' : \|u\| \leq r\}$. Show that $B = S \cap D_r$ is weak $*$ closed if $0 < r \leq 1$ but (by Theorem 12-3-3) is not weak $*$ closed for some r.

105 Show that the Banach–Alaoglu Theorem 9-1-12 cannot be generalized by replacing the unit disc in X' by the unit disc in an equivalent norm p [Prob. 12-3-104]. Deduce that p is not a dual norm, that is, X cannot be given an equivalent norm q such that $(X, q)' = (X', p)$.

106 Let p' be the norm on (X', p) in Prob. 12-3-105 and $q = p' | X$. Show that the norm on X' which is dual to q is smaller than p but equivalent to it. [Either use the fact that X is quasibarrelled or that $B^{\circ\circ}$ is the unit disc of q and is absorbed by B.]

107 Deduce the results of Probs. 11-3-117 and 11-3-118 by means of the Banach–Dieudonné theorem.

108 Show that the dual of a reflexive Fréchet space is strictly hypercomplete [Example 12-3-8].

109 If X is metrizable, show that T° is the largest topology for X' which induces the weak $*$ topology on each equicontinuous set in X'. Compare Prob. 12-1-103. [See Theorem 12-3-1.]

110 Show that a lc metric space is complete if and only if it has SCC, Prob. 11-1-112.

111 Show that there exists a Banach–Mackey space which neither is sequentially complete nor has the convex compactness property [Theorem 12-3-10 and Prob. 9-3-2].

112 (J. H. Webb). Let (X, T) be separable barrelled and metrizable. Show that (X', T°) is sequential [Probs. 12-1-114 and 12-2-3 and Theorem 12-3-1]. Compare Prob. 12-3-103 which gives a similar result but only for convex sets.

113 Show that φ, with its largest lc topology, is sequential [Probs. 12-3-112 and 12-1-5]. It is not N sequential, by Prob. 8-4-123.

---

201 Show that B, Prob. 12-3-104, is weak * compact if and only if $r \le 1$.

---

301 Theorem 12-3-10 is true with "lc" omitted. [See [30], Theorem 1.]

302 If X is a lc metric nonnormed space, $\beta(X', X)$ is not N sequential. It may be sequential by Probs. 12-3-112 and 12-1-5. [See [146], proposition 5.5.]

12-4 FULL COMPLETENESS

Recall that a lcs X is complete if and only if each aw^* closed maximal subspace is weak * closed, Corollary 12-2-16. Replacing maximal subspace by convex set we obtain the definition of strict hypercompleteness, Definition 12-3-7. Other definitions can easily be given; here is the most important one.

1. Definition A lcs space is called *fully complete* if every aw^* closed vector subspace is weak * closed.

Other names are *B complete* and *Ptak space* in honor of V. Ptak. The Banach–Dieudonné Theorem 12-3-6 says precisely that a Banach space is fully complete.

2. Definition A lcs space X is called B_r *complete* if each aw^* closed weak * dense vector subspace of X' is weak * closed or, equivalently, is equal to X'.

Such spaces are also called *infra-Ptak spaces*.

3. Theorem Each of the following conditions implies the next: strictly hypercomplete, fully complete, B_r complete, complete.

PROOF Every vector subspace is convex and every weak * dense vector subspace is a vector subspace; so the first two implications are trivial. Finally, if X is not complete, there exists a maximal subspace of X' which is aw^* closed but not weak * closed [Corollary 12-2-16]. It follows from Prob. 4-2-5 that this subspace is weak * dense and so X is not B_r complete.

Some results on the failure of the converses in Theorem 12-4-3 are given in Example 13-2-14, Theorem 12-4-19, and Prob. 12-4-204. See also Prob. 12-4-301.

Full completeness has an inheritance property (Theorem 12-4-5) not shared by completeness.

In Lemma 12-4-4, $u: X \to Y$ is linear and continuous; X, Y are lcs spaces. To prove Theorem 12-4-5 it would be sufficient to take u to be open, but the extra strength of Lemma 12-4-4 is needed later on.

4. Lemma If u is almost open, u' preserves $aw*$ closed sets.

PROOF Let $S \subset Y'$ be $aw*$ closed, $V \in \mathcal{N}(X)$, $A = u'[S] \cap V^\circ \subset X'$. The set A can also be reached starting from V by the route $X \to Y \to Y' \to X'$ thus: $A = u'[(uV)^\circ \cap S]$. [Let $f \in A$, say $f = u'g$; $y \in uV$, say $y = u(x)$, $x \in V$. Then $|g(y)| = |g(ux)| = |(u'g)(x)| = |f(x)| \leq 1$ since $f \in V^\circ$, $x \in V$. Thus $g \in (uV)^\circ$. Conversely, let $f = u'(g)$ with $g \in (uV)^\circ \cap S$; $x \in V$. Then $|f(x)| = |g(ux)| \leq 1$, so $f \in V^\circ$.] Now $u\overline{V} \in \mathcal{N}(Y)$ and so $(uV)^\circ \cap S = (\overline{uV})^\circ \cap S$ is weak $*$ compact. Since u' is weak $*$ continuous by Theorem 11-1-6, A is weak $*$ compact; hence $u'[S]$ is $aw*$ closed.

5. Theorem Every separated quotient of a fully complete space is fully complete.

PROOF Let $u: X \to Y$ be a quotient map ($=$ linear, continuous, open) where X is fully complete, Y separated. Let S be an $aw*$ closed subspace of Y'. Then $u'S$ is $aw*$ closed [Lemma 12-4-4]; hence weak $*$ closed. Thus $(u')^{-1}[u'S]$ is weak $*$ closed since u' is weak $*$ continuous. This set is S since u' is one to one [by Corollary 11-1-8, since u is onto].

6. Example In particular, a separated quotient of a fully complete space is complete. The special case of a lc Fréchet space was noted earlier.

These completeness concepts allow significant extensions of results proved earlier for Banach and Fréchet spaces. These are Theorem 11-3-4 and the open mapping and closed graph theorems. (See Theorems 12-4-9, 12-4-18, and 12-5-7.)

7. Remark In the remainder of this section, X, Y are lcs spaces and $u: X \to Y$ is a continuous linear map, unless otherwise specified.

8. Lemma Suppose that u is onto and u' is weak $*$ range closed. Then the quotient topology Qu (for Y) is compatible with (Y, Y').

PROOF Write Q for Qu and T for the original topology of Y, and Y' for $(Y, T)'$ and Y'_Q for $(Y, Q)'$. Now $Q \supset T$ since Q is the largest vector topology for which u is continuous; hence $Y'_Q \supset Y'$. Next, define $v: X \to (Y, Q)$ by $v(x) = u(x)$ for $x \in X$. Thus $v': Y'_Q \to X'$ and v' is one to one. The fact that $Y'_Q \subset Y'$ (completing the proof) follows from these remarks and $Rv' \subset Ru'$. [Applying Lemma 11-1-7 twice, $Rv' \subset (Nv)^\perp = (Nu)^\perp = Ru'$.]

9. Theorem: V. Ptak's open mapping theorem A continuous linear map u from a fully complete space X onto a barrelled space Y must be open.

PROOF Any map onto a barrelled space is almost open ⟦Prob. 9-3-3⟧, so by Lemma 12-4-4 the hypotheses of Lemma 12-4-8 are satisfied. Since Y is relatively strong, its topology is larger than Qu and so u is open.

10. Remark It follows immediately from Theorems 12-4-5 and 12-4-9 that *if a barrelled space is the continuous image of a fully complete space, it must itself be fully complete.*

The obvious uniqueness of topology results can now be written. They also follow from closed graph theorems and are postponed to the next section where, as it happens, they can be given with weaker hypothesis and with variations (Corollaries 12-5-10 and 12-5-14). The weaker hypothesis (B_r complete) is sufficient since such results deal only with one-to-one maps. See Theorem 12-6-11.

Theorem 12-4-9 generalizes the earlier version of the open mapping Theorem 5-2-4 only in the lc case (see Prob. 12-4-106). A non-lc version is given in Prob. 12-4-107.

A basic step in the proof of the open mapping theorem for Fréchet spaces is the proof that an almost open map is (sometimes) open ⟦Lemma 5-2-3⟧. This is also the essence of the proof of Theorem 12-4-9. It is worth while to isolate this result (Theorem 12-4-14). If the assumptions that Y is relatively strong and u is onto were added in Theorem 12-4-14, it would be an immediate consequence of the arguments just given.

11. Lemma Suppose that u' preserves weak $*$ closed vector subspaces. Then u is range closed.

PROOF Let $y \in \overline{Ru}$, $y \neq 0$. Then $u'[y^\perp]$ is a proper subspace of Ru'. ⟦Let $g \in Y'$, $g(y) \neq 0$. Then $u'g \notin u'[y^\perp]$ since otherwise $u'g = u'h$ with $h(y) = 0$. Then $g - h \in Nu' = (Ru)^\perp$ by Lemma 11-1-7. Thus $g - h \in y^\perp$ so that $g(y) = h(y) = 0$.⟧ By hypothesis, it is also weak $*$ closed so there exists $x \in X$ with $f(x) = 0$ for all $f \in u'[y^\perp]$, but $f(x) \neq 0$ for some $f \in Ru'$. For every $h \in Y'$ such that $h(y) = 0$ it follows that $h(ux) = 0$ ⟦$h(ux) = (u'h)(x) = 0$, by definition of x⟧ and so $u(x)$ is a multiple of y, say $u(x) = ty$. Now $u(x) \neq 0$ ⟦since $x \notin (Ru')^\perp = Nu$ by Lemma 11-1-7⟧ and so $t \neq 0$. Thus $y = u[(1/t)x] \in Ru$.

12. Remark A consequence of Lemma 12-4-13 is that the identity map between compatible topologies cannot be almost open without being open.

13. Lemma Let X be a lcs space, Y a vector space, and $u: X \to Y$ linear and onto. Let T, T_1 be topologies for Y which are compatible with each other. Suppose that $u: X \to (Y, T)$ is continuous and open, $u: X \to (Y, T_1)$ is almost open. Then $T_1 \supset T$ and $u: X \to (Y, T_1)$ is open.

PROOF Let $V \in \mathcal{N}(Y, T)$ be convex and T closed. Then $u^{-1}[V] \in \mathcal{N}(X)$ and so $cl_{T_1} u[u^{-1}V] \in \mathcal{N}(Y, T_1)$. Since u is onto, $u[u^{-1}V] = V$. Moreover, V is T_1 closed [Corollary 8-3-6]; so, finally, $V \in \mathcal{N}(X, T_1)$.

14. Theorem Let X be fully complete, Y any lcs space, and $u : X \to Y$ linear continuous and almost open. Then u is an open map onto Y.

PROOF By Lemmas 12-4-4 and 12-4-11, u is range closed. Being almost open, it must thus be onto. A special case of Lemma 12-4-4 is that $Ru' = u'[Y']$ is weak $*$ closed, so by Lemma 12-4-8 Qu is compatible with (Y, Y'). The result follows by Lemma 12-4-13.

We now turn to a generalization, Theorem 12-4-18, of part of Theorem 11-3-4. Remark 12-4-7 still applies.

15. Lemma Let X, Y be sets, $u : X \to Y$, $A \subset X$, $B \subset Y$. Then $u[A] \cap B = u(A \cap u^{-1}[B])$.

Lemma 12-4-16 gives the conclusion of Lemma 12-4-4 under different hypotheses.

16. Lemma Let u have the property that $(u')^{-1}$ preserves equicontinuous sets, that is, $E \subset X'$ equicontinuous implies $(u')^{-1}[E] \subset Y'$ equicontinuous. Then u' preserves aw^* closed sets.

PROOF Let S be an aw^* closed set in Y', $V \in \mathcal{N}(X)$. Then $(u')^{-1}[V^\circ]$ is equicontinuous and weak $*$ closed [Theorem 11-1-6]; hence weak $*$ compact [Alaoglu–Bourbaki]; thus $(u')^{-1}[V^\circ] \cap S$ is weak $*$ compact [Lemma 12-2-3]. The image of this set by u' is also weak $*$ compact [Theorem 11-1-6], and by Lemma 12-4-15 this is $u'[S] \cap V^\circ$. Thus $u'[S]$ is aw^* closed.

Next we give sufficient conditions for the hypothesis of Lemma 12-4-16. One which is not of much use in the present quest is given in Prob. 12-4-103. Recall that strong $*$ refers to $\beta(X', X)$ and $\beta(Y', Y)$.

17. Lemma Let Y be quasibarrelled and u' a strong $*$ isomorphism (into). Then $(u')^{-1}$ preserves equicontinuous sets.

PROOF Let $E \subset X'$ be equicontinuous. Then E is strong $*$ bounded [Theorem 9-1-8]. Hence $(u')^{-1}[E]$ is strong $*$ bounded and so it is equicontinuous [Theorem 10-1-11].

18. Theorem Let X be fully complete and Y quasibarrelled. Then if u' is a strong $*$ isomorphism (into), u is an open map of X onto Y.

PROOF By Lemmas 12-4-17 and 12-4-16, u' preserves aw^* closed sets; hence preserves weak $*$ closed subspaces. By Lemma 12-4-11, u is range closed; hence onto. [Since u' is one to one the range of u is dense, by Corollary 11-1-8.] Also, $Ru' = u'[Y']$ is weak $*$ closed so Lemma 12-4-8 yields the result. [Y is relatively strong by Theorem 10-1-9; thus its topology is larger than Qu.]

It may be worth noticing that if this result were available the proof of Theorem 11-3-3 [$(b) \Rightarrow (a)$, case (1)] could be abbreviated by omitting the part after u' is an isomorphism.

Theorem 12-4-18 is a variation of Theorem 12-4-9 with a weaker hypothesis on Y and a stronger one on u. As in Remark 12-4-10, the conditions of Theorem 12-4-18 imply that Y is fully complete.

Theorem 12-4-18 generalizes part of Theorem 11-3-4. It applies, for example, if X is a lc Fréchet space [Theorem 12-3-3] and Y is a lc metric space [Example 10-1-10]. Further developments of this nature are given in Prob. 12-4-304 and [82], sec. 21.

Note. The remainder of this section may be omitted.

There is an interesting special case in which the conclusions of Theorem 12-4-9 and 12-4-18 hold without any special hypothesis on the range space, i.e., when the domain is an algebraic dual, Theorem 12-4-23. Recall that there is only one topology for $X^{\#}$ which is admissible for the dual pair $(X^{\#}, X)$ [Prob. 8-5-7]. In the sequel, $X^{\#}$ will be assumed to have this topology. The interest is heightened by the fact that such a space may be a Fréchet space, namely, $\varphi^{\#} = \omega$ [Prob. 8-2-4].

19. Theorem For any vector space X, $X^{\#}$ is fully complete.

PROOF Observe that $\tau(X, X^{\#})$ is the largest lc topology for X [Example 8-5-19] and so every subspace of X is closed in this topology [Prob. 7-1-9]; hence $\sigma(X, X^{\#})$ closed. But this topology is the weak $*$ topology $\sigma[(X^{\#})', X^{\#}]$ for $(X^{\#})' = X$.

Theorem 12-4-19 is also a special case of Prob. 12-4-108.

20. Theorem Every continuous linear map $u: X^{\#} \to Y$ is range closed, where Y is any lcs space.

PROOF This follows from Lemma 12-4-11 since, as pointed out in the preceding proof, every subspace of X is weak $* = \sigma(X, X^{\#})$ closed.

The important special case $X^{\#} = Y = \omega$ is given an elementary proof in [153], Prob. 6-4-29.

21. Corollary With u as in Theorem 12-4-20, if u' is one to one, then u is onto.

PROOF By Theorem 12-4-20 and Corollary 11-1-8.

22. Lemma Let $u: X^\# \to Y$ be continuous, linear, and onto, where (Y, T) is any lcs space. Then there exists a vector space Z such that $Y = Z^\#$, $T = \sigma(Z^\#, Z)$.

PROOF Let $Z = Y'$ and $v: Y \to Z^\#$ the natural embedding, $(vy)(g) = g(y)$ for $y \in Y$, $g \in Z$. Then v is $T - \sigma(Z^\#, Z)$ continuous. [If $y \to 0$, $(vy)(g) = g(y) \to 0$ for each $g \in Z$.] Thus $v \circ u: X^\# \to Z^\#$ is continuous; hence range closed, by Theorem 12-4-20. It follows that v is range closed; hence onto. [[Y is dense in $Z^\#$ by Theorem 8-1-10.]] We now have $Y = Z^\#$, and since T is admissible for $(Z^\#, Z)$ it must be $\sigma(Z^\#, Z)$ since there is only one topology admissible for this dual pair, as pointed out above.

23. Theorem Let $u: X^\# \to Y$ be a continuous linear map where Y is any lcs space. Then these are equivalent:
(a) u is onto.
(b) u is an open map.
(c) u' is one to one.
Moreover, if these conditions hold, $Y = Z^\#$ with Z a subspace of X.

PROOF (a) and (c) are equivalent by Corollary 12-4-21. That (b) implies (a) is trivial. Finally, if (a) is true, $Y = Z^\#$ by Lemma 12-4-22. Lemma 12-4-8 may be applied because every subspace of X is $\sigma(X, X^\#)$ closed [[see the proof of Theorem 12-4-19]] so Qu is compatible with $(Y, Y') = (Z^\#, Z)$. Since this dual pair has only one admissible topology it must be Qu, so u is open. The last statement follows by (c) since $u': Z \to X$.

24. Corollary The topology of $X^\#$ is minimal, i.e., there is no lcs topology for $X^\#$ which is strictly smaller than its natural topology $\sigma(X^\#, X)$.

PROOF The special case $X = \varphi$, $X^\# = \omega$ was given in Prob. 8-2-107.

PROBLEMS

In this list X, Y are lcs spaces and $u: X \to Y$ a continuous linear map.
1 Let X be B_r complete, Y any lcs space, and $u: X \to Y$ one to one, linear, continuous, and almost open. Show that u is open; hence a linear homeomorphism. [[The proof of Theorem 12-4-14 applies, taking account of Corollary 11-1-8.]]

---

101 Show that the condition of Lemma 12-4-16 implies that u' is one to one.
102 Let (X, T) be fully complete and T_1 a larger compatible topology. Show that (X, T_1) is fully complete [[Prob. 12-2-111]].

103 If u is open, show that $(u')^{-1}$ preserves equicontinuous sets $[[(u')^{-1}E]^\circ \supset u[E^\circ]]$.

104 Show that in Remark 12-4-10, "barrelled" cannot be replaced by "quasibarrelled" $[$Example 5-2-8$(a)]$.

105 What does Theorem 12-4-14 say about the inclusion map from l to c_0? Prove it directly.

106 Show that Theorem 12-4-9 fails if "fully complete" is replaced by "Fréchet" $[$Prob. 9-3-116$]$. Thus local convexity is crucial here.

107 (Wendy Robertson's open mapping theorem). Let X be a Fréchet space, (Y, T) ultrabarrelled, and $u: X \to Y$ continuous, linear, and onto. Prove that u is open. $[$Let T_1 (on Y) be generated by $\{u[V] : v \in \mathcal{N}(X)\}$. Then $T_1 \subset T$ so u is almost open. See Lemma 5-2-3.$]$

108 If X is weakly complete it is fully complete in any topology compatible with (X, X'). $[\tau(X', X)$ is the largest lc topology for X', by Prob. 8-3-4 and Example 8-5-19. The result follows by Prob. 7-1-9.$]$

109 Let X, Y be Banach spaces and $u: X \to Y$ range closed but not onto. Show that (with the weak $*$ topologies on Y', X') u' is range closed but does not preserve all closed vector subspaces.

110 Show that every separated quotient (more generally every separated linear continuous image) of ω is (isomorphic to) ω or is finite dimensional $[$Theorem 12-4-23$]$.

111 Show that every continuous linear map from ω to a normed space has finite-dimensional range. Generalize to X^* $[$Probs. 12-4-110 and 4-4-111$]$.

112 Let S be a vector subspace of X' which with the relative topology of $\beta(X', X)$ is semireflexive. Show that S is aw^* closed. This generalizes Prob. 11-3-116. $[U^\circ \cap S$ is strong $*$ $[$hence $\sigma(S, S')]$ bounded and closed in S; hence $\sigma(S, S')$ compact.$]$

113 Let Y have a reflexive dense proper subspace S $[$e.g., Prob. 10-3-301$]$. Let $X = Y'$ with $\tau(Y', Y)$. Show that S satisfies the conditions of Prob. 12-4-112 but is not weak $*$ closed. Deduce that if a space is reflexive and not complete, its completion is not fully complete. $[$Note that Y is barrelled by Prob. 12-2-104.$]$

114 With S as in Prob. 12-4-112 suppose also that S is total and X is B_r complete. Show that (X, X') is reflexive. This generalizes Prob. 11-3-117 $[$Probs. 12-4-112 and 10-3-104$]$.

115 Let Y be a normed space and suppose that $u': Y' \to X'$ is almost open. Show that u' is open (any compatible topology on X' and norm on Y').

116 Let A be a row-finite matrix. Show that $A[\omega] \cap c$ is a closed subspace of c $[$Theorem 12-4-20$]$.

117 Let A be a row-finite one-to-one matrix. Show that c_A is a BK space and β (Prob. 7-2-121) can be taken to be 0. $[$In Lemma 5-5-11, take $Y = A[\omega] \cap c$, Prob. 12-4-116.$]$

118 If A is a reversible row-finite matrix, $A[\omega] = \omega$ $[$Theorem 12-4-20$]$.

119 Let $a_{nk} = n^k$. Show that A is onto ω, that is, $A[\omega_A] = \omega$. $[$Let f be an entire function with $f(n) = y_n.]$

120 Let A be a one-to-one matrix which is onto ω. Show that A must be row finite. Problem 12-4-119 shows that "one-to-one" cannot be omitted. $[$Let $u: \omega \to \omega_A$ be the inverse. By Theorem 12-4-20, $\omega_A = \omega.]$

121 Show that if A is a matrix mapping a BK space (such as c_0) into φ, then A has finite rank, i.e., for some m, $a_{nk} = 0$ for $n > m$. $[A$ is continuous with $\sigma(\varphi, \omega)$ by Theorems 1-6-10 and 3-3-13. Apply Prob. 12-4-111.$]$

---

201 Show that every closed subspace of a fully complete space is fully complete. $[$See $[58]$, p. 299.$]$

202 Write out the details of the following proof that there exists a continuous real function on $[0, 1]$ which is nondifferentiable on a set of positive measure. Assume this is false. Define $(\varphi_n f)(t) = n\{f[t + (1/n)] - f(t)\}$. Then $\varphi_n(f) \to f'$ almost everywhere. Define $\varphi: C \to \mathcal{M}$ by $\varphi(f) = f'$. Then φ is continuous. $[$Apply Prob. 12-4-107 and an appropriate form of Theorem 9-3-7.$]$ Let $f_n(t) = (1/n) \sin nt$. Then $f_n \to 0$ in C but $\varphi(f_n) \nrightarrow 0$ in $\mathcal{M}$.

203 In the dual of X^*, all absolutely convex aw^* closed sets are weak $*$ closed. This extends Theorem 12-4-19. $[$See $[82]$, 18 H.$]$

204 If $X = \mathscr{M}$, $X^{\#}$ is not strictly hypercomplete. Compare Theorem 12-4-19. [See [82], 18 H.]

205 Every infinite-dimensional closed subspace S of ω is linearly homeomorphic with ω. It follows as in Prob. 9-6-110 or from the form of S that S is complemented. [Let $N = \{n_k\}$ be a set of positive integers maximal with respect to $\sum\{\alpha_i x_i : i \in N\} = 0$ implies $\alpha_i = 0$ for all i. Let $u : S \to \omega$ be $u(x) = \{x_{n_k}\}$. Then $u[S]$ is dense since if $g = 0$ on it $g(x) = \sum \alpha_i x_i$ and $\alpha = 0$; $u[S]$ is closed since for $x \in S$, $x = \sum \alpha_k x_{n_k}$ and so if $y^r = u(x^r) \to y$ we have $x_{n_k}^r = y_k^r \to y_k$ and $x_m^r = \sum \alpha_k x_{n_k}^r = \sum \alpha_k y_k^r \to \sum \alpha_k y_k$. So $\lim x^r$ exists. Thus u is onto. Finally, u is clearly one to one.]

---

301 Let H be a completely regular Hausdorff space which is not normal. Then $C(H)$ has a non-complete quotient. [See [156], Prob. 13.2.110.] If, in addition, H is a k space, $C(H)$ is complete [Prob. 6-1-205]. By Theorem 12-4-5 this furnishes an example of a complete space which is not fully complete. [For H, see [156], example 6-7-3.]

302 If a locally bounded space has the property given for $X^{\#}$ in Theorem 12-4-20, it must be finite dimensional even if Y is restricted to be locally bounded. [See [63].]

303 Every lc Fréchet space which is not a Banach space has ω as a quotient. [See [88], 31.4.]

304 If X, Y are lc Fréchet spaces, u is range closed if and only if u' is (weak $*$) range closed. [See [82], theorem 21.9.]

12-5 CLOSED GRAPH THEOREMS

The basic closed graph Theorem 5-3-1 says that a linear map with closed graph between Fréchet spaces is continuous. Generalizations of this theorem have gone in many directions, such as maps between metric spaces ([156], prob. 9.1.118) and groups ([156], prob. 12.2.122). More detailed references are [47], [60], and [61]. For TVS maps a definitive reference is [116], especially the Appendix to the second edition which describes spaces with *webs*, a recent concept which leads to useful closed graph theorems of various forms. See also Prob. 13-1-301.

Note. In contrast with the preceding section, u is not assumed to be continuous.

1. Definition Let X, Y be lcs spaces and $u : X \to Y$ a linear map. Define $u' : Y' \to X^{\#}$ by $u'(g)(x) = g(ux)$. Then $\Delta = (u')^{-1}[X']$.

Thus Δ is the domain of u' in the sense that $u' : \Delta \to X'$. Whenever we can prove that $\Delta = Y'$ we shall have weak continuity of u [Lemma 11-1-1] and the resulting continuity in any Hellinger–Toeplitz pair of topologies. A sample theorem: if $\Delta = Y'$ and X is relatively strong, then $u : X \to Y$ is continuous [Prob. 11-2-1].

2. Lemma If u has closed graph, Δ is total over Y.

PROOF The converse is also true but is not needed. Let $y \in Y$, $y \neq 0$. Then $(0, y) \notin G$ (the graph of u) so there exist absolutely convex $V \in \mathscr{N}(X)$, $W \in \mathscr{N}(Y)$ with $(0, y) + V \times W \not\subset G$. It follows that $y + W \not\subset u[V]$. [Let $w \in W$, $v \in V$. Then $(v, y + w) = (0, y) + (v, w) \in (0, y) + V \times W$ so $(v, y + w) \notin G$. Hence $y + w \neq u(v)$.] Thus $y \notin \overline{u[V]}$. There exists $g \in Y'$ with $|g(y)| > 1$, $|g| \leq 1$ on $u[V]$

⟦Theorem 7-3-5⟧. Since $g(y) \neq 0$ the proof is concluded by showing that $g \in \Delta$. Now $u'g$ is bounded on V ⟦$|(u'g)(v)| = |g(uv)| \leq 1$⟧ and so $u'g \in X'$ ⟦Theorem 4-5-9⟧; hence $g \in \Delta$.

This result, together with Theorem 8-1-10, implies that Δ is weak $*$ dense in Y' if u has closed graph, so it is natural to look for conditions which imply that Δ is $aw*$ closed. (Sample theorem: if Δ is $aw*$ closed, Y is B_r complete, and u has closed graph, then u is weakly continuous.) Three such conditions are given in Lemmas 12-5-5 and 12-5-12 and Prob. 14-1-105.

3. Definition A linear map $u: X \to Y$ is called almost continuous if $\overline{u^{-1}[V]} \in \mathcal{N}(X)$ whenever $V \in \mathcal{N}(Y)$.

4. Example *A linear map defined on a barrelled space must be almost continuous,* for $\overline{u^{-1}[V]}$ is a barrel if V is.

This is reminiscent of Prob. 9-3-3 which says that maps onto a barrelled space are almost open. A similar duality is exhibited in the next result. Lemma 12-4-4 says, in particular, that if u is almost open the range of u' is $aw*$ closed. Here we replace "open" by "continuous" and "range" by "domain."

5. Lemma Let u be almost continuous; then Δ (Definition 12-5-1) is $aw*$ closed.

PROOF Let $V \in \mathcal{N}(Y)$. Consider the following four sets in Y, X, X', $Y^\#$ respectively: V, $u^{-1}[V]$, $A = (u^{-1}[V])^\circ$, $B = (u')^{-1}[A]$. Now $A = (\overline{u^{-1}[V]})^\circ$ so A is weak $*$ compact ⟦Alaoglu–Bourbaki theorem⟧. Thus B is weak $*$ closed in $Y^\#$ ⟦u' is weak $*$ continuous by Theorem 11-1-6⟧ and so $B \cap V^\circ$ is weak $*$ compact. The result follows from the equation $B \cap V^\circ = \Delta \cap V^\circ$ which is proved thus. If $g \in B \cap V^\circ$, then $g \in Y'$ and $u'(g) \in A \subset X'$ so $g \in \Delta$. If $g \in \Delta \cap V^\circ$, let $x \in u^{-1}[V]$. Then $|(u'g)(x)| = |g(ux)| \leq 1$ since $g \in V^\circ$ and $ux \in V$. Hence $u'g \in A$ and so $g \in B$.

6. Theorem Let $u: X \to Y$ be almost continuous and with closed graph and assume that Y is B_r complete. Then u is weakly continuous.

PROOF By Lemma 12-5-2, Δ is a weak $*$ dense vector subspace of Y'. By Lemma 12-5-5, it is weak $*$ closed. Thus $\Delta = Y'$ and the result follows by Lemma 11-1-1.

7. Theorem: V. Ptak's closed graph theorem Let X be barrelled and Y be B_r complete. Then each linear map with closed graph from X to Y is continuous.

PROOF By Theorem 12-5-6 and Lemma 12-5-4, u is weakly continuous. By the Hellinger–Toeplitz theorem (Corollary 11-2-6) u is continuous since X is relatively strong [Theorem 9-3-10] and the topology of Y is smaller than (or equal to) $\tau(Y, Y')$.

The way that Theorems 12-5-7 and 12-4-9 go can easily be remembered by realizing that the assumption that a space is barrelled is that its topology is large. Thus maps to a barrelled space are open and maps from a barrelled space are continuous.

As was done in Theorem 12-4-14 for almost open, we isolate the dual result for almost continuity, Theorem 12-5-9. Lemma 12-5-8 implies that the identity map between compatible topologies cannot be almost continuous without being continuous. Compare Remark 12-4-12.

8. Lemma Let X, Y be vector spaces and $u: X \to Y$ linear. Let T_1, T_2, (T_3, T_4) be lcs topologies for X, (for Y) which are compatible with each other. Suppose that u is $T_1 - T_3$ continuous and $T_2 - T_4$ almost continuous. Then u is $T_2 - T_4$ continuous.

PROOF Let $V \in \mathscr{N}(Y, T_4)$ be convex and T_4 closed. Then V is T_3 closed so $u^{-1}[V]$ is T_1 closed; hence T_2 closed. Thus $u^{-1}[V] \in \mathscr{N}(X, T_2)$.

9. Theorem Let Y be B_r complete, X any lcs space, and $u: X \to Y$ linear, closed graph, and almost continuous. Then u is continuous.

PROOF By Theorem 12-5-6, u is weakly continuous. By Lemma 12-5-8, it is continuous.

Theorem 12-5-7 can be deduced directly from this without use of the Hellinger–Toeplitz theorem.

10. Corollary Let (X, T) be barrelled, $(X, T_1)B_r$ complete, $T_1 \supset T$. Then $T_1 = T$.

PROOF $i: (X, T) \to (X, T_1)$ has closed graph since its inverse does. [It is continuous.] By Theorem 12-5-7, i is continuous.

11. Remark The hypothesis on X in Theorem 12-5-7 is precisely that X is relatively strong and $\sigma(X', X)$ is boundedly complete [Theorem 9-3-13]. This cannot be relaxed to sequentially complete [Prob. 12-5-105]; hence the interest of Theorem 12-5-13.

12. Lemma Let $(X', \text{weak} *)$ be sequentially complete and Y separable. Then Δ (Definition 12-5-1) is $aw*$ closed.

PROOF By Prob. 12-2-3 it is sufficient to show that Δ is weak $*$ sequentially closed. Let $\{g_n\} \subset \Delta$ with $g_n \to g \in [Y', \sigma(Y', Y)]$. Since u' is weak $*$ continuous, $u'g_n \to u'g \in [X^*, \sigma(X^*, X)]$. By hypothesis, $u'g \in X'$ and so $g \in \Delta$.

13. Theorem: N. J. Kalton's closed graph theorem Let X be relatively strong and with $(X', \text{weak} *)$ sequentially complete; let Y be separable and B_r complete. Then each linear map with closed graph from X to Y is continuous.

PROOF By Lemmas 12-5-2 and 12-5-12, $\Delta = Y'$ so u is weakly continuous by Lemma 11-1-1. The rest follows as in the proof of Theorem 12-5-7.

Corollary 12-5-10 can also be relaxed by replacing "bounded completeness" by "sequential completeness" and paying the price of separability. (See Remark 12-5-11.)

14. Corollary Let (X, T) be relatively strong and with $(X', \text{weak} *)$ sequentially complete; let (X, T_1) be separable and B_r complete and suppose $T_1 \supset T$. Then $T_1 = T$.

PROOF Use Theorem 12-5-13 instead of 12-5-7 in the proof of Corollary 12-5-10.

This leads to an interesting criterion for reflexivity of a Banach space B. Recall that B is reflexive if and only if it is weakly boundedly complete ⟦Theorem 10-2-4⟧. Weak sequential completeness is not enough ⟦Remark 10-2-7⟧ but once again it suffices to introduce a separability condition. This result was also noted in Prob. 10-2-104.

15. Example *A weakly sequentially complete Banach space B with a separable dual must be reflexive.* Let $X = B'$, $T = \tau(B', B)$, $T_1 = \beta(B', B) = \text{norm}$ ⟦Example 8-5-6⟧. By Corollary 12-5-14, $T_1 = T$.

16. Remark The property "C barrelled" was introduced and shown to imply weak $*$ sequential completeness in Probs. 9-4-103 and 9-4-107. Thus this property would serve as a hypothesis in Lemma 12-5-12 and the succeeding results. See also Prob. 12-6-202.

At the end of Sec. 12-4 some open mapping theorems were given for the algebraic dual of a vector space. In the same spirit, there is a closed graph theorem with no special assumptions about the domain. The key remark is, as before, that every subspace of Y is $\sigma(Y, Y^{\#})$ closed.

17. Theorem Every closed graph linear map $u: X \to Y^{\#}$ is continuous, where X is any lcs space.

PROOF By Lemma 12-5-2, Δ is weak $* = \sigma(Y, Y^{\#})$ dense in Y. By the preceding remark, it is closed, so $\Delta = Y$. As in the proof of Theorem 12-5-6, u is $\sigma(X, X') - \sigma(Y^{\#}, Y)$ continuous. Since the topology of X is larger than (or equal to) $\sigma(X, X')$, u is continuous.

PROBLEMS

1 If X is an infinite-dimensional lc Fréchet space, show that (X, T) is not B_r complete, where T is the largest lc topology ⟦Corollary 12-5-10 and Prob. 7-1-8⟧

2 Deduce Corollary 12-5-10 from Prob. 12-4-1.

3 Let $(X', \text{weak} *)$ be sequentially complete. Show that each closed graph linear map from X to a separable B_r complete space is weakly continuous.

---

101 Deduce from Example 12-5-15 that l^∞ is not separable ⟦Prob. 8-1-11⟧.

102 Deduce from Example 12-5-15 that c_0 is not weakly sequentially complete. Hence show that l^∞ is not weakly sequentially complete. ⟦If a space is weakly sequentially complete, so also are its closed subspaces, by Prob. 11-1-3.⟧ It follows that the inclusion in Prob. 9-1-116 must be proper.

103 Deduce from Lemma 12-5-2 that if u has closed graph, it still has closed graph if the topologies of X, Y are replaced by any other compatible topologies (a not unexpected result since the graph is convex).

104 Let X be a nonreflexive Banach space with separable dual. Let $T = \tau(X', X)$, $n = $ norm on X'. Show that the identity map $(X', T) \to (X', n)$ has closed graph but is not continuous. Consequently, which hypothesis in Theorem 12-5-13 cannot be dropped?

105 Let $X = [l^\infty, \tau(l^\infty, l)]$. Show that the identity map from X to l^∞ has closed graph but is not continuous, and that this illustrates Remark 12-5-11.

106 Let (X, T_1) be barrelled, (X, T) be B_r complete, and suppose that there exists a Hausdorff topology for X smaller than T_1 and T. Show that $T_1 \supset T$ ⟦like Prob. 5-5-106⟧.

107 Let T, T_1 be lcs topologies for a vector space X. Show that $i : (X, T) \to (X, T_1)$ has closed graph if and only if $(X, T)' \cap (X, T_1)'$ is total over X ⟦like Lemma 12-5-2⟧.

108 Deduce that X^* is minimal (Corollary 12-4-24) from Theorem 12-5-17.

109 Prove this variation of Theorem 12-4-14. If X is fully complete and $u : X \to Y$ is linear, closed graph, almost open and onto, then u is open. ⟦X/Nu is fully complete by Theorem 12-4-5 and Prob. 5-3-106 and $\bar{u} : X/Nu \to Y$ also has closed graph by Lemma 12-5-2; apply Theorem 12-5-9 to its inverse.⟧

110 A barrelled space may have a strictly larger Fréchet topology; compare Corollary 12-5-10 ⟦Prob. 9-3-116⟧.

111 Deduce from Probs. 12-5-1 and 12-3-108 that no Banach space may have countably infinite Hamel dimension.

112 Show that a linear map with closed graph from an ultrabarrelled space to a Fréchet space must be continuous ⟦like Prob. 12-4-107⟧.

113 Let b be a cardinal number such that $b^{\aleph_0} = b$ and X a vector space with Hamel dimension b. Show that X, with its largest lc topology, is not B_r complete ⟦Probs. 12-5-1 and 3-1-303⟧. For $b = \aleph_0$, $X = \varphi$ is strictly hypercomplete ⟦Prob. 12-3-108⟧.

12-6 CONVERSE THEOREMS

This section is concerned with the necessity of the various assumptions for the open mapping and closed graph theorems. To see, for example, that X must be barrelled in the closed graph Theorem 12-5-7, one could produce a nonbarrelled space for which the theorem fails. However, a more important question is whether barrelled is the "right" assumption, i.e., cannot be weakened in any way. This is the content of Theorem 12-6-3. This represents a solution of the *converse problem* for the closed graph theorem. The following are some earlier examples:

(*a*) The uniform boundedness theorem was proved for Banach spaces. The solution of the converse problem is that uniform boundedness holds if and only if the space is barrelled [Theorem 9-3-4].
(*b*) Bounded linear maps are continuous on normed spaces. The solution of the converse problem is that this is true if and only if the space is bornological [Prob. 8-4-106].
(*c*) The natural embedding of a normed space is a homeomorphism. The solution of the converse problem is that this is true if and only if the space is quasi-barrelled [Theorem 10-1-8].

Historically, the solution of the converse problem in Theorem 12-6-3 seems to have been the first of the results in this section. It was given by M. Mahowald in 1961.

As we have seen, it is very easy for a function to have closed graph; see, for example, Prob. 12-5-103. However, continuity of a function requires that the topology on its domain be large. Thus a closed graph theorem will hold only for spaces with large topologies. An extreme example is given in Prob. 12-6-1.

1. Lemma Let X be a lcs space and H a weak $*$ bounded set in X'. Give H the (relative) weak $*$ topology. Define $u: X \to C^*(H)$ by $(ux)(h) = h(x)$ for $x \in X$, $h \in H$. Then u is linear and has closed graph.

PROOF Clearly u is continuous when $C^*(H)$ has the pointwise topology, that is, $f \to 0$ if and only if $f(h) \to 0$ for each $h \in H$. [If $x \to 0$, $(ux)(h) = h(x) \to 0$ for each $h \in X'$, a fortiori for each $h \in H$. Thus $ux \to 0$ pointwise.] The result follows by Lemma 5-5-1.

2. Lemma The map in Lemma 12-6-1 is continuous if and only if H is equicontinuous.

PROOF u is continuous if and only if $u^{-1}[D] \in \mathcal{N}(X)$ where D is the unit disc in $C^*[H]$. But $u^{-1}[D] = \{x : |h(x)| \leq 1 \text{ for all } h \in H\} = H^\circ$, and $H^\circ \in \mathcal{N}(X)$ if and only if H is equicontinuous.

3. Theorem Let X be a lcs space. These are equivalent:
(*a*) X is barrelled.
(*b*) Every closed graph linear map from X to a B_r complete space must be continuous.
(*c*) Every closed graph linear map from X to a Banach space is continuous.
(*d*) Every closed graph linear map from X to $C^*(H)$ is continuous where H is an arbitrary Hausdorff space.

PROOF See also Prob. 12-6-101. That (*a*) implies (*b*) is Theorem 12-5-7. That (*b*) implies (*c*) is by Theorem 12-3-3. That (*d*) implies (*a*) is by Lemmas 12-6-1 and 12-6-2 and Theorem 9-3-4.

The converse problem for Kalton's closed graph Theorem 12-5-13 follows.

4. Theorem Let X be a lcs space. These are equivalent:
(a) $(X', \text{weak } *)$ is sequentially complete.
(b) Every closed graph linear map from X to a separable B_r complete space is weakly continuous.
(c) Every closed graph linear map from X to c is weakly continuous.
If X is relatively strong, omit "weakly" in (b) and (c).

PROOF That (a) implies (b) is given in the proof of Theorem 12-5-13. (c) $\Rightarrow$ (a). Let $\{f_n\}$ be a weak $*$ Cauchy sequence in X'. Define $u: X \to c$ by $u(x) = \{f_n(x)\}$. Then u is continuous when c has the relative topology of ω, that is, $y \to 0$ if and only if $y_n \to 0$ for each n. [If $x \to 0$, $f(x) \to 0$ for each $f \in X'$ so $f_n(x) \to 0$ for each n.] By Lemma 5-5-1, $u: X \to (c, \|\cdot\|_\infty)$ has closed graph. Thus u is weakly continuous. Define $f \in X^*$ by $f(x) = \lim_{n\to\infty} f_n(x)$ for $x \in X$. Then $f \in X'$ [$f = \lim \circ u$ and $\lim: c \to \mathcal{K}$ is weakly continuous]. Finally, $f_n \to f$ weak $*$ since $f_n(x) \to f(x)$ for each x. The last part follows by the usual application of the Hellinger–Toeplitz theorem.

The solution of the converse problem for Theorem 12-4-14 is Theorem 12-6-10.

5. Remark The notation to be used is as follows. Let X be a lcs space and S a subspace of X' which will always be assumed to have the topology $\sigma(X', X)|_S$. Let $Y = S'$, a vector space. The dual pair (S, Y) will use polars denoted by $\bullet$, while $\circ$ will be used for polars in the dual pair (X, X'). Finally, define $u: X \to Y$ by $u(x) = \hat{x}|S$, that is, $(ux)(s) = s(x)$ for $s \in S$.

6. Lemma With the notation of Remark 12-6-5, let $B \subset X$. Then $(u[B])^\bullet = B^\circ \cap S$.

PROOF Let $s \in B^\circ \cap S$, $b \in B$. Then $|(ub)(s)| = |s(b)| \le 1$ so $s \in (u[B])^\bullet$. The opposite inclusion is trivial.

7. Lemma With the notation of Remark 12-6-5, let $\mathcal{A}$ be the collection of subsets of S, each of which is an equicontinuous subset of X'. Then $\mathcal{A}$ is admissible for the dual pair (S, Y).

PROOF It is easy to check Definition 8-5-1; e.g., each $A \in \mathcal{A}$ is $\sigma(X', X)$ bounded [this is a very special case of the Alaoglu–Bourbaki theorem]; hence bounded in S. Thus $\mathcal{A}$ is a polar family. It is also obvious that $\mathcal{A}$ covers S, so by Remark 8-5-17, $\mathcal{A}$ is admissible.

8. Lemma With the notation of Lemma 12-6-7, $u: X \to (Y, T_\mathcal{A})$ is onto and continuous.

PROOF Let $y \in Y$. Extend y to $F \in [X', \sigma(X', X)]'$ by the Hahn–Banach theorem. Since the weak $*$ topology is compatible with (X', X), $F \in X$, that is, there exists x such that $F(g) = g(x)$ for all $g \in X'$. Thus $y(s) = s(x)$ for $s \in S$ and so $u(x) = y$. Thus u is onto. Next, let $V \in \mathcal{N}(Y, T_{\mathscr{A}})$, say $V = A^\bullet$. Then $A^\circ \in \mathcal{N}(X)$ and $u(A^\circ) \subset A^\bullet$ [if $x \in A^\circ$, $a \in A$, $|(ux)(a)| = |a(x)| \leq 1$]; thus u is continuous.

9. Lemma If in Lemma 12-6-7, S is aw^* closed in X', then $\mathscr{A}$ is compatible with the dual pair (S, Y) and u (Lemma 12-6-8) is almost open.

PROOF Let $A \in \mathscr{A}$. Then $A \subset A^{\circ\circ} \cap S$ which is $\sigma(X', X)$ compact; hence $\sigma(S, Y)$ compact. [$\sigma(S, Y) \subset \sigma(X', X)|_S$ by definition.] Thus A is included in an absolutely convex weakly compact set. By Corollary 9-2-4, $\mathscr{A}$ is compatible. Next let $V \in \mathcal{N}(X)$ be absolutely convex. Then $\overline{u[V]} = (u[V])^{\bullet\bullet} = (V^\circ \cap S)^\bullet$ [Lemma 12-6-6]. Since $V^\circ \cap S \in \mathscr{A}$ it follows that u is almost open.

10. Theorem Let X be a lcs space. These are equivalent:
(a) X is fully complete.
(b) Every continuous linear almost open map from X onto any lcs space Y is open.

PROOF $(a) \Rightarrow (b)$ by Theorem 12-4-14. $(b) \Rightarrow (a)$: Let S be an aw^* closed subspace of X'. By Lemmas 12-6-8 and 12-6-9 the map u of Remark 12-6-5 is open. Thus $T_{\mathscr{A}}$ is the quotient topology by u. Now if we replace S by its weak $*$ closure in X', Y, u will be unchanged. Since $\overline{S}$ is also aw^* closed, u is an open map by the preceding argument so $T_{\mathscr{B}}$ is also the quotient topology by u, where $\mathscr{B}$ is the collection of equicontinuous sets in $\overline{S}$. Hence $T_{\mathscr{A}} = T_{\mathscr{B}}$ and it follows that $S = \overline{S}$ since, by Lemma 12-6-9, $(Y, T_{\mathscr{A}})' = S$, $(Y, T_{\mathscr{B}})' = \overline{S}$. Thus S is weak $*$ closed.

Next is given the solution of the converse problem for Prob. 12-4-1.

11. Theorem Let X be a lcs space. These are equivalent:
(a) X is B_r complete.
(b) Every one-to-one continuous linear almost open map from X onto any lcs is open.

PROOF $(a) \Rightarrow (b)$ is Prob. 12-4-1. $(b) \Rightarrow (a)$: Let S be a total aw^* closed subspace of X'. Then the map u of Remark 12-6-5 is one to one and the rest of the proof of Theorem 12-6-10 is unchanged; it yields the result that S is weak $*$ closed.

Additional converse theorems and discussion may be found in Probs. 12-6-202, 12-6-301, 13-3-201, [116] (Chap. 6), and [75].

PROBLEMS

In this list X is a lcs space; all maps are linear.

1 Suppose that every closed graph map from X to an arbitrary lcs space is continuous. Show that X has its largest lc topology.

---

101 Add to the list in Theorem 12-6-3:

(e) Every closed graph linear map from X to $C^*(H)$ is continuous where H is an arbitrary $T_{3\frac{1}{2}}$ space.

(f) Same as (e) with H an arbitrary compact Hausdorff space [Theorem 4-5-7 and Prob. 9-6-102].

102 Show that $(Y, T_{\mathscr{A}})$ (Lemma 12-6-9) need not be relatively strong even if X is fully complete. [Take $S = X'$.]

103 Show that X' contains an equicontinuous set which is separating over X if and only if there exists a compact Hausdorff space H such that $X \subset C(H)$ with continuous inclusion [Lemma 12-6-2].

104 Let X be fully complete. Show that $(Y, T_{\mathscr{A}})$ (Lemma 12-6-9) is $X/S^\perp$.

105 Suppose that S (Remark 12-6-5) is metrizable and separable. Show that $X/S^\perp$ is separable [Prob. 12-6-104 and Example 9-5-4].

---

201 Does Theorem 12-6-10 hold if Y is assumed to be relatively strong?

202 Show that every closed graph linear map from X to c is continuous if and only if X is C barrelled. [See [75], Theorem 3.1.]

---

301 These are equivalent:

(a) Every continuous linear map of X onto a barrelled space is open.

(b) In $(X', \text{weak } *)$ boundedly complete subspaces are closed.

[See [60].] Compare Theorem 10-2-4 and Prob. 12-4-112.

302 A TVS X is ultrabarrelled if and only if every linear closed graph map from X to a Fréchet space is continuous. [See [143], p. 15.]

THIRTEEN

INDUCTIVE LIMITS

Two basic ways of making new topologies are by means of weak topologies on domains of maps and the quotient topology on the range of a map. The inductive limit is an extension of the quotient idea to many maps. The weak (sometimes called projective) topologies have as special cases product and sup topologies; the inductive topologies have as special cases quotient and inf topologies. The two concepts are dual to each other in the usual sense.

An important application of inductive limits is to the theory of distributions, which is shown in Sec. 13-3.

13-1 INDUCTIVE LIMITS

Suppose given a family $\{X_\alpha : \alpha \in \mathscr{A}\}$ of topological vector spaces, not necessarily locally convex, and, for each α a linear map $u_\alpha : X_\alpha \to Y$ where Y is a fixed vector space. Finally, assume that the union of the ranges of the u_α spans Y. This means that for each $y \in Y$, there exist $\alpha_1, \alpha_2, \ldots, \alpha_n \in \mathscr{A}$, $x_i \in X_{\alpha_i}$, with $y = \sum u_{\alpha_i}(x_i)$. $[\![y = \sum t_i u_{\alpha_i}(x_i) = \sum u_{\alpha_i}(t_i x_i).]\!]$

1. Definition A vector topology T for Y will be called a *test topology* if $u_\alpha : X_\alpha \to (Y, T)$ is continuous for each $\alpha \in \mathscr{A}$.

For example, the indiscrete topology is always a test topology so the set of test topologies is not empty.

2. Example Let $\mathscr{A}$ be the positive integers, $X_n = \mathscr{K}$ for each n, $Y = \varphi$, and $u_n(x) = x_n \delta^n$ for each n. Every vector topology for Y is a test topology [Prob. 6-3-3].

3. Example With $\mathscr{A}$, Y as in Example 13-1-2, take $X_n = \mathscr{K}^n$ and $u_n(x) = \sum_{k=1}^{n} x_k \delta^k$. As in Example 13-1-2, every vector topology for Y is a test topology.

4. Definition The unrestricted inductive limit topology $\mathfrak{T}_u$ for Y is $\vee \{T : T \text{ is a test topology for } Y\}$. The inductive limit topology $\mathfrak{T}_i$ for Y is $\vee \{T : T \text{ is a lc test topology for } Y\}$. The space $(Y, \mathfrak{T}_i)$ is called the inductive limit of $\{X_\alpha\}$ and $(Y, \mathfrak{T}_u)$ is called the unrestricted inductive limit.

More precisely, the maps u_α should also be mentioned.
It is immediate that $\mathfrak{T}_u \supset \mathfrak{T}_i$.

5. Example In Examples 13-1-2 and 13-1-3, $\mathfrak{T}_u(\mathfrak{T}_i)$ is the largest vector (lc) topology for φ. It is pointed out in Prob. 13-1-109 that they are equal.

6. Theorem Each $u_\alpha : X_\alpha \to (Y, T)$ is continuous where $T = \mathfrak{T}_u$ or $\mathfrak{T}_i$. Further, $\mathfrak{T}_u(\mathfrak{T}_i)$ is the largest vector (lc) topology for Y for which this is true.

PROOF The first statement is immediate from the continuity criterion of Theorem 1-6-8. The second statement is true because if u_α is continuous to (Y, T), T is a test topology; so $T \subset \mathfrak{T}_u$ (or $T \subset \mathfrak{T}_i$ in the lc case).

7. Theorem Let Z be a TVS and $f : (Y, \mathfrak{T}_u) \to Z$ linear. Then f is continuous if and only if $f \circ u_\alpha : X_\alpha \to Z$ is continuous for each $\alpha \in \mathscr{A}$.

PROOF $\Rightarrow$: By Theorem 13-1-6. $\Leftarrow$: σf is a test topology for Y [by the last part of Theorem 1-6-10 and Definition 4-1-10]; hence $\mathfrak{T}_u \supset \sigma f$. The result follows since $f : (Y, \sigma f) \to Z$ is surely continuous.

8. Theorem Let Z be a lc space and $f : Y \to Z$ linear. These are equivalent:
(a) $f : (Y, \mathfrak{T}_u) \to Z$ is continuous.
(b) $f : (Y, \mathfrak{T}_i) \to Z$ is continuous.
(c) $f \circ u_\alpha : X_\alpha \to Z$ is continuous for each α.

PROOF Equivalence of (a) and (c) is by Theorem 13-1-7. (b) $\Rightarrow$ (a) because $\mathfrak{T}_u \supset \mathfrak{T}_i$. Finally, (c) $\Rightarrow$ (b) by the argument of Theorem 13-1-7 since σf is locally convex.

9. Corollary $(Y, \mathfrak{T}_u)' = (Y, \mathfrak{T}_i)'$.
This also follows from Probs. 13-1-101 and 7-1-103.

10. Example Let $\mathscr{A}$ have one member α; write $X_\alpha = X$, $u_\alpha = u$. Then $u: X \to Y$ is onto and $\mathfrak{T}_u$ is the quotient topology [Remark 6-2-4]. If X is locally convex, $\mathfrak{T}_u = \mathfrak{T}_i$ [Prob. 7-1-2].

11. Theorem Let $\mathscr{B} = \{V \subset Y : V$ is absolutely convex and $u_\alpha^{-1}[V] \in \mathscr{N}(X_\alpha)$ for each $\alpha \in \mathscr{A}\}$. Then $\mathscr{B}$ is a local base for $\mathscr{N}(Y, \mathfrak{T}_i)$.

PROOF We first show (by checking Theorem 7-1-4) that $\mathscr{B}$ generates a lc topology for Y. Each $V \in \mathscr{B}$ is absorbing. [Let $y \in Y$, $y = \sum_{i=1}^n u_{\alpha_i}(x_i)$. Let $t > 0$ have the property that $tx_i \in (1/n)u_{\alpha_i}^{-1}[V]$ for $i = 1, 2, \ldots, n$. Then $nty \in V + V + \cdots + V = nV$ so $ty \in V$.] Also, $\mathscr{B}$ is clearly a filterbase. Finally, $\mathscr{B}$ is additive since $V \in \mathscr{B}$ implies $\frac{1}{2}V \in \mathscr{B}$. (See Lemma 7-1-3.) Now let T be the topology generated by $\mathscr{B}$. It is a lc test topology so $T \subset \mathfrak{T}_i$. But if $V \in \mathscr{N}(Y, \mathfrak{T}_i)$ is absolutely convex, $V \in \mathscr{B}$ by Theorem 13-1-6, so $\mathfrak{T}_i \subset T$.

12. Example Let Y be a vector space and $\{T_\alpha : \alpha \in \mathscr{A}\}$ be a family of vector topologies for Y, let $X_\alpha = (Y, T_\alpha)$, and take each u_α to be the identity map. Then $\mathfrak{T}_u = \wedge \{T_\alpha : \alpha \in \mathscr{A}\}$, the largest vector topology which is smaller than each T_α. (This is not $\cap T_\alpha$ in general [Prob. 5-5-107].)

13. Theorem An inductive limit of barrelled spaces is barrelled if it is separated. The same is true with bornological instead of barrelled.

PROOF Let B be a barrel in $(Y, \mathfrak{T}_i)$. For each α, $u_\alpha^{-1}[B]$ is a barrel in X_α; hence a neighborhood of 0. By Theorem 13-1-11, $B \in \mathscr{N}(Y, \mathfrak{T}_i)$. The same proof, using Lemma 4-4-8, yields the second part.

14. Example In particular, *a quotient of a barrelled space by a closed subspace is barrelled; the same is true with bornological instead of barrelled.* A special case is the fact that *a complemented subspace of a barrelled (or bornological) space is barrelled (or bornological).* [The projection is a quotient map by Remark 6-2-16.]

15. Theorem Every bornological space (Y, T) is an inductive limit of normed spaces; if sequentially complete, it is an inductive limit of Banach spaces.

PROOF Let $\mathscr{A}$ be the set of absolutely convex bounded closed sets in Y. For each $A \in \mathscr{A}$ let X_A be its span and p_A its gauge (Theorem 2-2-1). If Y is sequentially complete, X_A is a Banach space by Theorem 6-1-17. Take $u_A : (X_A, p_A) \to Y$ to be inclusion for each A. Since $p_A \supset T | X_A$ for each A [Lemma 4-5-1], T is a test topology so $T \subset \mathfrak{T}_i$. Conversely, let $V \in \mathscr{N}(Y, \mathfrak{T}_i)$ be absolutely convex, B a bounded set in (Y, T), and A the absolutely convex closure of B in (Y, T). Then by Theorem 13-1-6, $u_A^{-1}[V] \in \mathscr{N}(X_A)$ so this set includes εA for some $\varepsilon > 0$; hence $\varepsilon B \subset \varepsilon A = u_A[\varepsilon A] \subset V$. Thus V is an absolutely convex bornivore; hence $V \in \mathscr{N}(Y, T)$ and so $\mathfrak{T}_i \subset T$.

16. Corollary A lcs space is bornological if and only if it is the inductive limit of a set of normed spaces.

PROOF By Theorems 13-1-13 and 13-1-15.

Another corollary of the same two theorems is the special case of Remark 10-4-13 that a sequentially complete bornological space is barrelled.

17. Theorem An inductive limit of relatively strong spaces is relatively strong if separated.

PROOF Let $(Y, T)' = (Y, \mathfrak{T}_i)'$ with T locally convex. Then each $u_\alpha: X_\alpha \to Y$ is continuous when X_α and Y have their weak topologies. By the Hellinger–Toeplitz theorem, Corollary 11-2-6, each $u_\alpha: (X_\alpha, T_\alpha) \to (Y, T)$ is continuous so T is a test topology; hence $T \subset \mathfrak{T}_i$.

PROBLEMS

The notation at the beginning of the section is used in this list.

1 Show that an inductive limit of separated spaces need not be separated. ⟦See Example 13-1-10. An example in which the null spaces are closed is given in Prob. 5-5-107.⟧

2 Let $\mathscr{A}$ have one member α, $X_\alpha = (X, T)$. Take $Y = X$, $u = i$. Show that $\mathfrak{T}_u = T$, $\mathfrak{T}_i = T_c$, the associated lc topology. (See Prob. 7-1-103. Thus Corollary 13-1-9 extends that result.)

3 Let Y be a vector space, $\mathscr{A}$ the set of one-dimensional subspaces, each with its euclidean topology, and $u_A: A \to Y$ the inclusion map for each A. Show that $\mathfrak{T}_u(\mathfrak{T}_i)$ is the largest vector (lc) topology for Y ⟦as in Example 13-1-5⟧.

4 Show that an inductive limit of quasibarrelled spaces is quasibarrelled if separated ⟦as in Theorem 13-1-13⟧.

5 Show that for a countable set of lc spaces, $\mathfrak{T}_i = \mathfrak{T}_u$. ⟦Let $V_n \in \mathscr{N}(Y, \mathfrak{T}_u)$ be balanced, $V_n + V_n \subset V_{n-1}$, $n = 1, 2, \ldots$. Let $W_n \in \mathscr{N}(X_n)$ be absolutely convex with $u_n[W_n] \subset V_n$, $H =$ absolutely convex hull of $\{u_n[W_n] : n = 1, 2, \ldots\}$. Then $H \subset V_0$.⟧ Show how to modify the proof for a finite set.

6 Let Y be a (sequentially complete) bornological space with a countable cobase for its bounded sets. Show that Y is the inductive limit of a sequence of normed (Banach) spaces. ⟦Choose A from the cobase in Theorem 13-1-15.⟧

7 Let X be a lc metric space such that X' is bornological. Show that X' is the inductive limit of a sequence of Banach spaces. ⟦See Prob. 13-1-6. $\{U_n^\circ\}$ is a cobase if $\{U_n\}$ is a base for $\mathscr{N}$ since B bounded implies $B^\circ \in \mathscr{N}(X)$ by Theorem 10-1-11; X' is complete by Corollary 8-6-6. This result also follows from Probs. 13-1-110 and 8-4-105.⟧

---

101 Show that $\mathfrak{T}_i = (\mathfrak{T}_u)_c$ in general. (See Prob. 13-1-2.)

102 Write versions of Theorems 13-1-13 and 13-1-17 without the separation assumption.

103 Show that a base for $\mathscr{N}(\mathfrak{T}_i)$ is given by the set of absolutely convex hulls of unions of sets of the form $u_\alpha[V_\alpha]$, $V_\alpha \in \mathscr{N}(X_\alpha)$.

104 Omit the assumption that $\cup\{u_\alpha[X_\alpha]\}$ spans Y and prove Theorem 13-1-11 with the additional assumption that each V is absorbing.

105 Continue Prob. 13-1-104 by showing that $Y = E \oplus F$ with E the inductive limit in the ordinary sense and F having its largest vector topology.

106 Show that the inductive limit of countably many separable spaces is separable. Deduce that φ is separable with its largest lc topology. This also follows from Example 9-5-4.

107 Let (X, T) be a lcs space and form $\mathfrak{T}_i$ as in Theorem 13-1-15. Show that $\mathfrak{T}_i$ is the associated bornological topology of Prob. 8-4-113.

108 Use Prob. 13-1-3 to give an example where each X_α is lc but $\mathfrak{T}_i \neq \mathfrak{T}_u$ ⟦Prob. 7-1-106⟧.

109 Show that the largest vector topology and the largest lc topology for φ are the same and that φ is the only infinite-dimensional vector space for which this is true ⟦Probs. 13-1-5 and 13-1-108⟧.

110 Let X be quasibarrelled, $\mathscr{A} = \{U^\circ : U \in \mathscr{N}(X)\}$. Form $\mathfrak{T}_i$ for $Y = X'$ as in Theorem 13-1-15. Show that $\mathfrak{T}_i \supset \beta(X', X)$ and that these two topologies have the same bounded sets. ⟦$\mathscr{A}$ is a cobase for the β bounded sets as in Prob. 13-1-7; $\mathfrak{T}_i \supset \beta$ as in Theorem 13-1-15; a β bounded set B is bounded in some X_A since $B^\circ \supset$ some U as in Prob. 13-1-7; hence $u_A[B]$ is $\mathfrak{T}_i$ bounded.⟧

111 In Theorem 4-4-6 it was shown that in a metric vector space (∗) every bornivore is a neighborhood of 0. Show that in contrast with Theorem 13-1-13, (∗) is not preserved under inductive limit, even countable ⟦Prob. 8-4-111⟧.

112 Suppose that every closed graph linear map $X_\alpha \to Z$ is continuous. Show that the same is true for each such map $(Y, \mathfrak{T}_i) \to Z$ ⟦Theorem 13-1-7 and Prob. 5-3-120⟧.

113 Show that the inductive limit of countably many lc metric spaces need not be metrizable, even if separated ⟦Prob. 7-1-8⟧.

114 Show that $\|\cdot\|_{1/2}$ defines a non-lc topology on φ which lies between two separated lc topologies ⟦Prob. 13-1-109⟧.

115 Suppose $\mathscr{A}$ has n members. Show that $\{\sum_{i=1}^n u_i[V_i] : V_i \in \mathscr{N}(X_i)$ is balanced for $i = 1, 2, \ldots, n\}$ is a local base for $\mathscr{N}(Y, \mathfrak{T}_u)$.

201 Show that $\cup\{l^n : n = 1, 2, \ldots\}$ with the inductive limit topology by the inclusion maps is not metrizable. ⟦See [154], p. 65.⟧

301 Let X, Y be separated with X an inductive limit of lc second-category spaces and Y an inductive limit of a sequence of fully complete spaces. The closed graph (open mapping) theorem holds for maps from X to Y (Y onto X). ⟦See [116], Chap. 6, supplement 2.⟧

13-2 DIRECT SUMS

Let $\{(X_\alpha, T_\alpha) : \alpha \in \mathscr{A}\}$ be a set of lcs spaces, $\pi = \pi\{X_\alpha : \alpha \in \mathscr{A}\}$. Each X_α may be regarded as a subspace of π by identifying $x_\alpha \in X_\alpha$ with $y \in \pi$ defined by $y_\alpha = x_\alpha$, $y_\beta = 0$ for $\beta \neq \alpha$. Let $\sum = \sum\{X_\alpha : \alpha \in \mathscr{A}\}$ be the span of $\cup\{X_\alpha : \alpha \in A\}$ in π. Then $y \in \sum$ if and only if $y_\alpha = 0$ for all but finitely many α. Moreover, $\sum = \pi$ if and only if $\mathscr{A}$ is finite. The *direct sum topology d* is defined to be the inductive limit topology by the inclusion maps, i.e., in Sec. 13-1, $Y = \sum$ and each $u_\alpha : X_\alpha \to \sum$ is inclusion. Thus $\sum$ has the largest lc topology whose restriction to each (X_α, T_α) is smaller than (or equal to) T_α.

1. Example Suppose that X is a lcs space and $X_\alpha = X$ for all α. In this case we write $\sum X_\alpha$ as $X^{(\mathscr{A})}$. Of course, $X^{\mathscr{A}} = \pi X_\alpha$. In particular, $\varphi = \mathscr{K}^{(N)}$, where N is the positive integers, and $\omega = \mathscr{K}^N$. The direct sum topology on φ is its largest lc topology ⟦Example 13-1-5 or Prob. 13-1-3⟧.

2. Theorem $(\sum X_\alpha, d)$ is a lcs space and $d \mid X_\alpha = T_\alpha$ for each α; moreover, d is the largest lc topology with this property.

PROOF It is immediate from Theorem 1-6-11 that the product topology induces T_α on each X_α. Thus d is larger than the product topology and the results follow.

3. Remark A special case of Theorem 13-1-11 is that a local base for $\mathscr{N}(d)$ is formed by the collection of absolutely convex sets V such that $V \cap X_\alpha \in \mathscr{N}(X_\alpha)$ for each α.

4. Theorem Every inductive limit is a quotient of a direct sum. Let u_α: $X_\alpha \to Y$, $Q:(\sum X_\alpha, d) \to (Y, \mathfrak{T}_i)$ be defined by $Q(x_\alpha) = u_\alpha(x_\alpha)$ for each $\alpha \in \mathscr{A}$ and extend Q by linearity. Then Q is a quotient map.

PROOF Note that for $z \in \sum X_\alpha$, $z_\alpha \in X_\alpha$ for all α and $z_\alpha = 0$ for all but finitely many α, so that $z = x_{\alpha_1} + x_{\alpha_2} + \cdots + x_{\alpha_n}$ where $x_{\alpha_i} \in \sum X_\alpha$ as explained above; then $Q(z) = \sum u_{\alpha_i}(x_{\alpha_i})$. Q is onto since $\{u_\alpha[X_\alpha]\}$ spans Y; Q is continuous since $Q \circ i_\alpha = u_\alpha$ is continuous for each α where $i_\alpha : X_\alpha \to \sum$ is the inclusion map [Theorem 13-1-7]. Finally, Q is open, for let $V \in \mathscr{N}(\sum)$ be absolutely convex; then $Q[V] \in \mathscr{N}(Y)$. $[u_\alpha^{-1}(Q[V]) \supset V \cap X_\alpha \in \mathscr{N}(X_\alpha)$ by Theorem 13-2-2. The result follows by Theorem 13-1-11.]

5. Corollary Let $(Y, \mathfrak{T}_i)$ be the inductive limit of lcs spaces X_α by maps u_α. Then $\mathfrak{T}_i$ is separated if and only if the null space of Q (Theorem 13-2-4) is closed in $(\sum X_\alpha, d)$.

PROOF By Theorems 13-2-4 and 6-2-8.

The continuity criterion for maps on a direct sum follows immediately from Theorem 13-1-7. It is the first part of Lemma 13-2-6.

6. Lemma Let Z be a lcs space and $f: \sum X_\alpha \to Z$ linear. Then f is continuous if and only if $f \mid X_\alpha$ is continuous for each α. In particular, the projection maps $P_\beta : \sum X_\alpha \to X_\beta$ are continuous.

PROOF In the second part, $P_\beta x = x_\beta$; $P_\beta \circ u_\alpha = 0$ if $\beta \neq \alpha$, while $P_\beta \circ u_\beta$ is the identity map on X_β.

7. Remark This shows that $d \supset \pi$ where π is the product topology restricted to $\sum X_\alpha$. [The latter is the smallest vector topology making each P_α continuous.]

A sequence of computations follows, leading to an important completeness result, Theorem 13-2-12.

8. Lemma Let $\mathscr{B}$ be a finite subset of $\mathscr{A}$. Then π and d induce the same topology on $\sum_{\mathscr{B}}$.

PROOF Let us denote the induced topologies by π and d. By Remark 13-2-7, $d \supset \pi$. Conversely, the identity map from $(\sum_{\mathscr{B}}, \pi)$ to $(\sum_{\mathscr{B}}, d)$ is continuous since it is a finite sum of continuous maps, namely, the inclusion maps from the X_α. [For example, if $\mathscr{B}$ has two members, $x = (x_1, x_2, 0, 0, \ldots) = (x_1, 0, 0, 0, \ldots) + (0, x_2, 0, 0, \ldots)$.] Thus $\pi \supset d$.

9. Lemma Let $U_\alpha \in \mathscr{N}(X_\alpha)$ for each α, let V be the absolutely convex closure in $(\sum X_\alpha, d)$ of $\cup U_\alpha$, and let $\mathscr{B}$ be a finite subset of $\mathscr{A}$. For $x \in \sum X_\alpha$ let x' be defined by $x'_\alpha = x_\alpha$ if $\alpha \in \mathscr{B}$; $x'_\alpha = 0$ otherwise. Then $x \in V$ implies $x' \in V$.

PROOF The map $x \to x'$ is continuous in the direct sum topology d by Lemma 13-2-6, since $x' = \sum \{P_\alpha(x) : \alpha \in \mathscr{B}\}$. Thus it is sufficient to prove the result for the absolutely convex hull of $\cup U_\alpha$. [If F is continuous and $F[V] \subset V$, then $F[\overline{V}] \subset \overline{F[V]} \subset \overline{V}$.] This is obvious since $0 \in U_\alpha$ for each α.

10. Lemma The set V in Lemma 13-2-9 is π closed.

PROOF Let $y \in cl_\pi V$, $0 \neq y \in \sum X_\alpha$. Let $\mathscr{B} = \{\alpha : y_\alpha \neq 0\}$, a finite set. It is sufficient to show that $y \in cl_\pi(V \cap \sum_{\mathscr{B}})$ [for $y \in \sum_{\mathscr{B}}$ and Lemma 13-2-8 will imply that $y \in cl_d(V \cap \sum_{\mathscr{B}}) \subset cl_d V = V$]. Let W be a product neighborhood of 0 in πX_α, which we may assume to be πW_α where $W_\alpha \in \mathscr{N}(X_\alpha)$ for each α. (Of course, $W_\alpha = X_\alpha$ for most α but we may ignore this.) Let $x \in (y + W) \cap V$. Define x' as in Lemma 13-2-9. Then $x' \in V$ and $x' \in y + W$ since x'_α differs from x_α only for values of α for which $y_\alpha = 0$. Now $x' \in \sum_{\mathscr{B}}$ and $x' \in (y + W) \cap V$, which is what we wished to prove.

11. Lemma d is F linked to π.

PROOF This concept was given in Definition 6-1-9. The result follows from Lemma 13-2-10 since such sets form a base for $\mathscr{N}(\sum X_\alpha, d)$ by Theorem 13-1-11.

12. Theorem A direct sum of complete lcs spaces is complete.

PROOF Let $x = (x^\delta : D)$ be a Cauchy net in $(\sum X_\alpha, d)$. By Remark 13-2-7, x is π Cauchy; hence $x \to a \in \pi X_\alpha$ in the product topology since the latter is complete by Theorem 6-1-7. It will be sufficient to show that $a \in \sum X_\alpha$ [by Lemmas 13-2-11 and 6-1-11]. For each α with $a_\alpha \neq 0$, choose U_α to be a closed neighborhood of 0 in X_α with $a_\alpha \notin U_\alpha$; if $a_\alpha = 0$ let $U_\alpha = X_\alpha$. Let V be the absolutely convex hull of $\cup U_\alpha$. There exists $\delta \in D$ such that $\gamma \geq \delta$ implies $x^\gamma - x^\delta \in V$; this in turn implies that $x^\gamma_\alpha - x^\delta_\alpha \in U_\alpha$ for each α. Since U_α is closed it follows that $a_\alpha - x_\alpha^\delta \in U_\alpha$ for all α. Now $x_\alpha^\delta = 0$ for all but finitely many α, and so $a_\alpha \in U_\alpha$ for all but finitely many α. By definition of the U_α this yields $a \in \sum X_\alpha$.

13. Example *Every vector space X is complete with its largest lc topology.* Let H be a Hamel basis for X. Every $x \in X$ can be written uniquely as a finite sum $x = \sum th$, where $t \in \mathcal{K}$, $h \in H$. In other words, X is isomorphic with $\mathcal{K}^{(H)}$. Since all separated topologies agree on $\mathcal{K}$, d is the largest lc topology. The result follows from Theorem 13-2-12.

14. Example *A complete space which is not B_r complete; hence not fully complete.* Such a space is indicated in Example 13-2-13, taking account of Prob. 12-5-1 or Prob. 12-5-113.

Theorems 13-2-15 and 13-2-17 show the duality between direct sums and products.

15. Theorem Let $\{X_\alpha : \alpha \in \mathcal{A}\}$ be a set of TVSs, $\pi = \pi X_\alpha$. For each $f \in \pi'$, there exist $X_1, X_2, \ldots, X_n$, chosen from the X_α, and $g_i \in X_i'$ for each i such that $f(x) = \sum g_i(x_i)$ for each $x \in \pi$.

PROOF Let $U = \{x : |f(x)| < 1\}$. Then $U \in \mathcal{N}(\pi)$, so for some finite $F \subset \mathcal{A}$, $U \supset \cap \{P_\alpha^\perp : \alpha \in F\}$ where $P_\alpha x = x_\alpha$ for each α. Let $Q : \pi \to \cap \{P_\alpha^\perp : \alpha \notin F\}$ be the projection; then $Q = \sum \{I_\alpha \circ P_\alpha : \alpha \in F\}$ where $I_\alpha : X_\alpha \to \pi$ is the inclusion map. For $\alpha \in F$ define $g_\alpha = f \circ I_\alpha \in X_\alpha'$; for $\alpha \notin F$ let $g_\alpha = 0$. Then $f = f \circ Q = \sum \{f_\alpha \circ I_\alpha \circ P_\alpha : \alpha \in F\} = \sum \{g_\alpha \circ P_\alpha : \alpha \in F\}$.

16. Remark Thus π' is a direct sum; its topology is discussed in Prob. 13-2-202. Theorem 13-2-15 could also be derived from the sup result (Theorem 7-2-16) [Prob. 13-2-118], but only in the lc case [Prob. 7-2-301].

17. Theorem Let $X = \sum X_\alpha$, each X_α a lcs space. Let each X_α' be given an admissible topology $T_{\mathcal{A}(\alpha)}$ and let X' have the topology $T_\mathcal{A}$ where $\mathcal{A}$ is the set of finite unions of sets chosen from $\{\mathcal{A}(\alpha)\}$. Then X' is linearly homeomorphic with $\pi X_\alpha'$.

PROOF Let $g = (g_\alpha) \in \pi X_\alpha'$. For each $x = \sum x_\alpha \in \sum X_\alpha$ set $f(x) = \sum g_\alpha(x_\alpha)$, a finite sum. In particular, $f(x_\alpha) = g_\alpha(x_\alpha)$ so $f \in X'$ by Lemma 13-2-6. This map $h : \pi X_\alpha' \to X'$ is clearly an algebraic isomorphism onto. [Given f, set $g_\alpha(x_\alpha) = f(x_\alpha)$. Then $g_\alpha \in X_\alpha'$ by Lemma 13-2-6.] Further, h is continuous. [Let $A^\circ \in \mathcal{N}(X')$ where $A = \cup \{A_i : i = 1, 2, \ldots, m\}$. Then $A^\circ = \cap A_i^\circ$, a finite intersection, so we may assume $A \subset X_\beta$. Then $g \in h^{-1}(A^\circ)$ if and only if $g_\beta \in A^\circ$, g_α unrestricted for $\alpha \neq \beta$. So $h^{-1}(A^\circ) \in \mathcal{N}(\pi)$.] Also, h^{-1} is continuous. [Using the criterion of Theorem 1-6-11, $P_\alpha \circ h^{-1}(f) = f | X_\alpha \in X_\alpha'$ by Lemma 13-2-6.]

PROBLEMS

The notation at the beginning of the section is used in this list.

1 Write the direct sum of X and Y as $X \oplus Y$. Show that X, Y are complementary in $X \oplus Y$ so that this agrees with the previous use of this symbol in Sec. 5-3 [Lemma 13-2-6].

---

101 For $\mathcal{B} \subset \mathcal{A}$ let $\sum_{\mathcal{B}} = \sum\{X_\alpha : \alpha \in \mathcal{B}\}$ with direct sum topology $d_{\mathcal{B}}$. Show that $d_{\mathcal{B}} = d_{\mathcal{A}}|\sum_{\mathcal{B}}$.

102 Let T_1, T_2 be vector topologies for a vector space X. Show that $T_1 \wedge T_2$ is separated if and only if $i : (X, T_1) \to (X, T_2)$ has closed graph [Example 13-1-12, Corollary 13-2-5, and Lemma 13-2-8].

103 With T_1, T_2 as in Prob. 13-2-102, suppose $T_1 \cap T_2$ is Hausdorff. Show that $T_1 \wedge T_2$ is separated.

104 Let (T_α) be a family of vector topologies for a vector space X. Show that $\wedge T_\alpha$ is separated if and only if $\{x : \sum x_\alpha = 0\}$ is closed in $\sum\{X, T_\alpha\}$ [Example 13-1-12 and Corollary 13-2-5].

105 Show that $\sigma(X, X^*)$ is complete if and only if X is finite dimensional [Prob. 8-3-4]. A special case follows from Prob. 12-4-108 with Example 13-2-14.

106 In contrast with Prob. 13-2-105, $\tau(X, X^*)$ is always complete [Examples 13-2-13 and 8-5-19].

107 Let X be an infinite-dimensional vector space and consult Example 13-2-13. Show that $\pi \subset \sigma(X, X^*) \subset d = \tau(X, X^*) = \beta(X, X^*)$ and neither inclusion can be equality [Probs. 13-2-105 and 13-2-106]. In particular, $d \neq \pi$. This also follows from Remark 13-2-3 and Example 4-1-13.

108 Let f be a linear functional on $\sum X_\alpha$ such that $f | X_\alpha$ is continuous for each α. Show that f need not be π continuous.

109 Show that a direct sum of fully complete spaces need not be fully complete [Examples 13-2-13 and 13-2-14].

110 Let X have its largest lc topology T and let S be a subspace. Show that the relative topology is the largest lc topology T_S for S. [Write $(X, T') = (S, T_S) \oplus S_1$; $T' = T$; apply Theorem 13-2-2.]

111 Let Y be the inductive limit of two lcs spaces X_1, X_2 by maps u_1, u_2. Suppose Y has a Hausdorff topology T_0 making u_1, u_2 continuous. Show that $(Y, \mathfrak{T}_i)$ is separated. This extends Prob. 13-2-103. [With Q as in Corollary 13-2-5, let $(x_1, x_2) \to (a_1, a_2)$, $(x_1, x_2) \in NQ$. Then $u_2(a_2) = \lim u_2(x_2) = -\lim u_1(x_1) = -u_1(a_1)$.]

112 Problem 13-2-111 is false for three spaces. [Let V be a vector space with lcs topologies T, T' such that $T \cap T'$ is not Hausdorff, e.g., Prob. 5-5-107. Let $X_1 = X_2 = (V, T)$, $X_3 = (V, T')$, $Y = V \times V$, $u_1(x) = (x, 0)$, $u_2(x) = (0, x)$, $u_3(x) = (x, x)$. Now let $T_0 = T$ on $u_1[V]$ and $u_2[V]$, T' on $u_3[V]$, and be discrete elsewhere.]

113 Show that Prob. 13-2-111 holds for three spaces if T_0 has closed graph addition. (See Prob. 5-3-121.)

114 Show that Prob. 13-2-111 holds for any finite number of spaces if T_0 has continuous addition. [$\mathfrak{T}_i \supset T_0$.]

115 Show that $\sum X_\alpha$ is dense in πX_α [Theorem 13-2-15 or Prob. 4-1-106].

116 Show that $X = \mathcal{K}^{(\mathcal{X})}$ is not separable (in contrast with $\mathcal{K}^{(N)}$ of Prob. 13-1-106 and $\mathcal{K}^{\mathcal{X}}$, [156], Prob. 6-7-201, which are both separable.) [If $f_n(x) = 0$ for $x \notin S_n$ (a finite set) then every $f_n(x) = 0$ for $x \notin \cup S_k$, a countable set. So $\{f_n\}$ is not dense. A more elegant proof is to note that l^∞ is algebraically isomorphic with X by Prob. 1-5-17, and so X has a smaller nonseparable topology by Prob. 13-1-3.]

117 Let X be a vector space of Hamel dimension $\geq \mathfrak{c}$. Show that X with its largest lc topology is not separable [second hint in Prob. 13-2-116].

118 Let X be a vector space, Y a TVS and $u : X \to Y$ a linear map. Let X have the topology σu. Show that $u' : Y' \to X'$ is onto. Hence deduce Theorem 13-2-15 for lc spaces from Theorem 7-2-16.

119 Why does Prob. 1-6-112 not allow the deduction of a general form of Theorem 7-2-16 from Theorem 13-2-15, contradicting Prob. 7-2-301?

120 Accepting the result of Probs. 13-2-205 or 13-2-206 show that a closed subspace of a relatively strong (even barrelled) space need not be relatively strong. [Let X be complete but not relatively strong; an example is given in Probs. 10-2-118 and 12-1-102. Apply Prob. 7-2-113.]

---

201 Show that the direct sum topology for φ is the same as the box topology.

202 In Theorem 13-2-15 let each X_α be a lcs space and give each X'_α some polar topology. Show that the direct sum topology for π' (Remark 13-2-16) is the polar topology using the products of the various members of the $\mathscr{A}_\alpha$. (Each X'_α has $T_{\mathscr{A}}$; assume each member of each $\mathscr{A}_\alpha$ to be absolutely convex.) [See [116], Chap. 5, Proposition 26.]

203 There exists a nonseparable space with a separable strong dual [Probs. 13-2-116 and 13-2-202]. This cannot happen for metrizable spaces [Prob. 9-5-4].

204 Let B be a bounded set in $\sum X_\alpha$. Show that $B \subset \sum \{X_\alpha : \alpha \in F\}$ for some finite $F \subset \mathscr{A}$. [See [116], Chap. 5, Proposition 24.]

205 Show that a product of barrelled spaces is barrelled [Probs. 13-2-202 and 13-2-204].

206 Show that a product of relatively strong spaces is relatively strong. [See [82], 18.11.]

13-3 STRICT INDUCTIVE LIMITS

We begin with an extension result for neighborhoods of 0 in a relative topology. If X is a subspace of Y and $U \in \mathscr{N}(X)$ it is clear that there exists $V \in \mathscr{N}(Y)$ with $V \cap X \subset U$; not so clear that we can make $V \cap X = U$, that is, U can be "extended" to V.

   1. Lemma Let Y be a lc space, X a subspace with the relative topology, and $U \in \mathscr{N}(X)$ absolutely convex. Then there exists absolutely convex $V \in \mathscr{N}(Y)$ with $V \cap X = U$.

   PROOF Choose $W \in \mathscr{N}(Y)$ with $W \cap X \subset U$ and let V be the absolutely convex hull of $W \cup U$. Clearly, $V \in \mathscr{N}(Y)$ and $V \cap X \supset U$. Finally, let $x \in V \cap X$. Then $x = rw + su$, $w \in W$, $u \in U$, $|r| + |s| \leq 1$. If $r = 0$, $x = su \in U$. If $r \neq 0$, $w = (x - su)/r \in X$; since also $w \in W$ it follows that $w \in U$ and so $x \in U$.

   2. Lemma In Lemma 13-3-1, let $y \in Y \backslash \bar{X}$. Then V can be chosen not containing y.

   PROOF Choose W as before and also such that $y + W \not\subset X$. Then if $y \in V$ we would have $y = rw + su$, $w \in W$, $u \in U$, $|r| + |s| \leq 1$. This implies that $y - rw = su \in (y + W) \cap U \subset (y + W) \cap X = \phi$.

A strict inductive limit is the inductive limit of a sequence of spaces of a special form. Let Y be a vector space, $\{X_n\}$ a strictly increasing sequence of subspaces with $Y = \cup X_n$. Let each X_n have a lcs topology T_n such that $T_{n+1} | X_n = T_n$ for each n. Give Y the inductive topology using the inclusion maps; then Y is called the *strict inductive limit* of the sequence $\{X_n\}$. The same topology would be obtained by using the unrestricted inductive limit [Prob. 13-1-5].

3. Example Let $Y = \varphi$, $X_n = \{x : x_i = 0 \text{ for } i > n\}$ and let T_n be the unique lcs (euclidean) topology for X_n. The strict inductive limit is the largest lc topology for Y [Example 13-1-5]. It is slightly paradoxical that more than one topology for Y could induce T_n on each X_n. Here this is obvious since all lcs topologies for Y induce the same topology on each X_n, and Y certainly has more than one lcs topology, for example $\|\cdot\|_\infty$ and $\|\cdot\|_1$. Neither of these is the strict inductive limit topology, by Theorem 13-3-8.

4. Theorem Let $(Y, \mathfrak{T})$ be the strict inductive limit of $\{(X_n, T_n)\}$. Then $\mathfrak{T} \mid X_n = T_n$ for each n.

PROOF That $\mathfrak{T} \mid X_n \subset T_n$ follows from Theorem 13-1-6. Conversely, let $U_n \in \mathcal{N}(X_n)$ be absolutely convex. Applying Lemma 13-3-1 repeatedly we obtain, for each $k > n$, absolutely convex $U_k \in \mathcal{N}(X_k)$, with $U_k \cap X_{k-1} = U_{k-1}$, $U_k \cap X_n = U_n$. Let $V = \cup \{U_k : k \geq n\}$. By Theorem 13-1-11, $V \in \mathcal{N}(\mathfrak{T})$. Since $V \cap X_n = U_n$ it follows that $U_n \in \mathcal{N}(\mathfrak{T} \mid X_n)$.

5. Lemma Let $y_n \in Y \backslash \overline{X}_n$ for each n. Then $y_n \not\to 0$.

PROOF By adding terms to $\{y_n\}$ we may assume that $y_n \in X_{n+1}$ for each n. It follows from Theorem 13-3-4 that y_n is not in the closure of X_n computed in X_{n+1}. By Lemma 13-3-2, choose absolutely convex $V_n \in \mathcal{N}(X_n)$ such that $V_{n+1} \cap X_n = V_n$, $y_n \notin V_{n+1}$. (Take $V_1 = X_1$.) Let $V = \cup V_n$. As in Theorem 13-3-4, $V \in \mathcal{N}(Y)$ but $y_n \notin V$ for all n.

6. Corollary If B is a bounded set in Y, $B \subset \overline{X}_n$ for some n.

PROOF If this is false, B contains a sequence $\{b_n\}$ such that $(1/n)b_n \notin \overline{X}_n$ for each n; possibly the b_n are not distinct. By Lemma 13-3-5, $(1/n)b_n \not\to 0$ and so B is not bounded [Theorem 4-4-1].

The more important examples of strict inductive limits have the additional property that each X_n is closed in X_{n+1}. This is true, for example, if each X_n is complete.

7. Theorem Each X_n is closed in X_{n+1} if and only if each X_n is closed in Y.

PROOF $\Leftarrow$: By Theorem 13-3-4. $\Rightarrow$: Let $x \in \overline{X}_n$, the closure taken in Y. Then $x \in X_k$ for some $k \geq n$, so by Theorem 13-3-4, $x \in \overline{X}_n$, the closure taken in X_k. Clearly, X_n is closed in X_k so $x \in X_n$.

8. Theorem Let Y be the strict inductive limit of a sequence $\{X_n\}$ of closed subspaces. Then each bounded set in Y is included in one of the X_n and Y cannot be metrizable.

PROOF The first part follows from Corollary 13-3-6. If Y is metrizable, choose $y_n \in X_{n+1} \backslash X_n$. Choose $\varepsilon_n > 0$ with $\| \varepsilon_n y_n \| < 1/n$. Then $\varepsilon_n y_n \to 0$, contradicting Lemma 13-3-5.

Two counterexamples in connection with Theorem 13-3-8 are indicated in Probs. 13-3-102 and 13-3-303.

9. Remark Suppose now that each X_n is complete. It is trivial that Y is sequentially, even boundedly, complete by Theorem 13-3-8. It turns out (Theorem 13-3-13) that Y is complete; the proof in a nutshell is that if y is a Cauchy net we may assume that it is contained in some X_n; hence is convergent. The phrasing for filters is: if $\mathscr{F}$ is a Cauchy filter we may assume that some X_n meets every member of $\mathscr{F}$, etc.

The first step is a uniform separation result.

10. Lemma Let $U_n \in \mathscr{N}(Y)$, $A_n \subset Y$ and $A_n + U_n \not\cap X_n$ for each n. Then there exists $V \in \mathscr{N}(Y)$ with $A_n + V \not\cap X_n$ for each n.

PROOF We may assume each U_n to be absolutely convex and $U_{n+1} \subset U_n$ for each n. Take V to be the absolutely convex hull of $\cup (U_n \cap X_n)$. Then $V \in \mathscr{N}(Y)$ by Theorem 13-1-11. Now suppose that $x \in (A_n + V) \cap X_n$ for some n; $x = a + \sum_{i=1}^m t_i x_i$, $a \in A_n$, $x_i \in U_i \cap X_i$, $\sum |t_i| \leq 1$. If $m \leq n$, A_n contains a, which is equal to $x - \sum t_i x_i \in X_n$ and so $A_n + U_n$ meets X_n. If $m > n$, X_n contains $x - \sum_{i=1}^n t_i x_i$ which is equal to $a + \sum_{i=n+1}^m t_i x_i \in A_n + U_m \subset A_n + U_n$.

For example, if each A_n is a singleton y_n, then $y_n \notin -V$ and so $\{y_n\}$ cannot converge. This is the result of Lemma 13-3-5.

11. Lemma Let $\mathscr{F}$ be a Cauchy filter in Y. Then there exists n such that $A + U$ meets X_n for each $A \in \mathscr{F}$, $U \in \mathscr{N}(Y)$.

PROOF If not, for each n, $A_n + U_n \not\cap X_n$ for some $A_n \in \mathscr{F}$, $U_n \in \mathscr{N}$. Choose V as in Lemma 13-3-10 and $S \in \mathscr{F}$ which is small of order V. A contradiction will result when it is proved that $S \subset A_n + V$ for all n since this will imply, by the choice of V, that $S \not\cap X_n$ for each n, contradicting $Y = \cup X_n$. In fact, $S \subset F + V$ for all $F \in \mathscr{F}$. [Let $s \in S$, $f \in F \cap S$. Then $s - f \in S - S \subset V$ so $s \in f + V$.]

12. Lemma Let $\mathscr{F}$ be a filter in a TVS. Then $\mathscr{F}$ is Cauchy or convergent to x if and only if $\mathscr{F} + \mathscr{N}$ is.

PROOF First, $\mathscr{F} + \mathscr{N} \subset \mathscr{F}$. [For $A \in \mathscr{F}$, $U \in \mathscr{N}$, $A + U \supset A$ so $A + U \in \mathscr{F}$.] So if $\mathscr{F} + \mathscr{N}$ is convergent or Cauchy, $\mathscr{F}$ is also. The converse is by continuity of addition.

For the next proof the reader will be required to know the fact that a space is complete if each Cauchy filter is convergent (Prob. 6-5-101).

13. Theorem The strict inductive limit Y of complete spaces X_n is complete.

PROOF Let $\mathcal{F}$ be a Cauchy filter. By Lemmas 13-3-11 and 13-3-12 we may assume that some X_n meets each member of $\mathcal{F}$. Let $\mathcal{B} = \{A \cap X_n : A \in \mathcal{F}\}$. Then $\mathcal{B}$ is a Cauchy filter in X_n; hence convergent, say $\mathcal{B} \to x$. It will follow that $\mathcal{F} \to x$ when we show that $x \in \bar{A}$ for each $A \in \mathcal{F}$ [Lemma 6-5-1]. So let $A \in \mathcal{F}$, $U \in \mathcal{N}(Y)$. Then $x + (U \cap X_n) \in \mathcal{B}$ since $\mathcal{B} \to x$; so this set meets $A \cap X_n$. Hence $x + U$ meets A and so $x \in \bar{A}$, as required.

Theorem 13-3-13 contrasts with the fact that an inductive limit of complete spaces, even a separated quotient of a complete space, need not be complete [Prob. 12-4-301].

Consider the problem of whether a given space has the strict inductive limit topology. The space would have to be of the following form.

14. Definition An S_σ subspace of a TVS X is the union of a strictly increasing sequence of closed subspaces of X. An S_σ space is a lcs space which is an S_σ subspace of itself.

An S_σ space need not have the strict inductive limit topology, as was pointed out in Example 13-3-3.

15. Theorem A barrelled S_σ space must have the strict inductive limit topology.

PROOF Let $(X, T) = \cup X_n$ and let I be the strict inductive limit topology. Clearly $I \supset T$, and since T is relatively strong it is sufficient to prove that $(X, T)' \supset (X, I)'$. Let $f \in (X, I)'$; then $f \mid X_n$ is continuous for each n [Theorem 13-3-4] and so it can be extended to $f_n \in (X, T)'$. Now $f_n \to f$ pointwise on X since each $x \in X_n$ for some n, and so, by the closure Theorem 9-3-7, $f \in (X, T)'$.

16. Corollary A barrelled S_σ space cannot be metrizable.

PROOF By Theorems 13-3-15 and 13-3-8.

17. Corollary Let X be a lcs space which is the union of a strictly increasing sequence $\{X_n\}$ of subspaces. Then X has at most one barrelled topology for which each X_n is closed.

18. Example *The space φ has exactly one barrelled topology*, namely, its largest lc topology, which is also the strict inductive limit topology as in Example 13-3-3. This is because its constituent finite-dimensional subspaces must be closed [Prob. 6-3-1]. Earlier results showed that φ has no Fréchet topology, e.g., Prob. 3-2-112. The result here is more specific.

For use in Chap. 15 we need a refinement of Theorem 13-3-15. Only the special case needed will be given. The more general result is Prob. 13-3-113.

19. Theorem Let (X, T) be barrelled and A a closed subspace of countable codimension. Then A is complemented; indeed, any algebraic complement B is a complement. Moreover, B has its largest lc topology.

PROOF We may obviously assume that the codimension of A is infinite [Theorem 6-3-4]. Let $(Y, T_1) = (A, T) \oplus (B, b)$ where b is the unique barrelled topology for B [Example 13-3-18]. Then $T_1 \supset T$ since T induces T on A, $T \subset b$ on B, and T_1 is the largest topology with these properties. We shall show that $Y = X$, that is, $T_1 = T$. Since T is relatively strong, it is sufficient to prove that $(X, T)' \supset (Y, T_1)'$. Let $f \in Y'$ and let $\{x_n\}$ be a Hamel basis for B. Let $f_n \in X'$ satisfy $f_n = f$ on $A_n = \text{span}\,(A, x_1, x_2, \ldots, x_n)$. [This is possible since A_n is closed by Theorem 6-3-3 and so $x_n \notin A_{n-1}$; hence f_n may be chosen so that $f_n = f$ on A_{n-1} and $f_n(x_n) = f(x_n)$ by Prob. 7-2-10. We are using the fact that $f \mid A$ is T continuous since $T_1 = T$ on A by Theorem 13-2-2.] Then $f_n \to f$ pointwise on X and so, by the closure Theorem 9-3-7, $f \in X'$. This completes the proof that $T_1 = T$ and so $T \mid B = b$ [Theorem 13-2-2]. Finally, $X = Y = A \oplus B$.

20. Example The strict inductive limit of a sequence of lc Fréchet spaces is called an (LF) space. Such spaces cannot be metrizable but share some important properties of Fréchet spaces (e.g., by Theorem 13-1-13, they are barrelled and bornological; see also Prob. 13-3-304). The motivation for their invention lies in the following examples.

21. Example: Test functions Let X_n be the set of real C^∞ functions f (i.e., differentiable any number of times) defined on R with $f(x) = 0$ for $|x| > n$. Let $p_k(f) = \| f^{(k)} \|_\infty$, where $f^{(k)}$ is the kth derivative. Then $(X_n, \sigma\{p_k : k = 0, 1, 2, \ldots\})$ is a Fréchet space. The fact that X_n contains more than one point is not completely obvious; it will show up shortly. Any function on R which lies in X_n for some n is called a *test function*. The set of test functions is always written $\mathcal{D}$ and it is an (LF) space, the strict inductive limit of $\{X_n\}$. In making this statement we are obligated to show that $X_{n+1} \neq X_n$ [Prob. 13-3-4].

22. Example: Distributions A member of the dual $\mathcal{D}'$ of the space of test functions is called a *distribution*. Each $F \in C(R)$ defines a distribution, also called F, by the formula $F(f) = \int_{-\infty}^{\infty} F(t) f(t)\, dt$ [Prob. 13-3-5]. Distributions defined in this way are called *functions*. Define $d : \mathcal{D} \to \mathcal{D}$ by $d(f) = f^{(1)}$ (the derivative). This is a continuous operator by Theorems 13-3-4 and 13-1-8.

Thus it has dual map d'. The operator $-d' : \mathscr{D}' \to \mathscr{D}'$ is called the *differentiation operator*; this means, writing $D = -d'$, if u is a distribution, Du is the distribution defined by $(Du)(f) = -(d'u)(f) = -u(df) = -u(f^{(1)})$. In particular, for a continuously differentiable function F, $(DF)(f) = -F[f^{(1)}] = -\int F(t)f^{(1)}(t)\,dt = \int F^{(1)}(t)f(t)\,dt$ [integration by parts]. Thus DF is the ordinary derivative of F. [See Prob. 13-3-6.] The differentiation operator is defined for all distributions and, in particular, every continuous function has a derivative which is equal to its ordinary derivative if the latter exists and is continuous.

Further information is contained in the problems and in texts such as [138].

PROBLEMS

1 Show that a strict inductive limit of separated spaces is separated. (Compare Prob. 13-1-1.) [See Theorems 13-3-4 or 13-2-4.]

2 Let Y be a strict inductive limit of closed subspaces X_n. Show that a set in Y is bounded if and only if it is a bounded subset of some X_n [Corollary 13-3-6 and Theorem 13-3-4].

3 Show that an (LF) space must be of first category. Deduce its nonmetrizability from this and Theorem 13-3-13.

4 Let $a > 0$, $f(t) = \exp\left[t^2/(t^2 - a^2)\right]$ for $|t| < a$, 0 for $|t| \geq a$. Show that if $a = n + 1$, $f \in X_{n+1} \setminus X_n$ in Example 13-3-21.

5 Show that F, Example 13-3-22, is a distribution [Theorem 13-1-8].

6 Show that the embedding of $C(R)$ into $\mathscr{D}'$ (Example 13-3-22) is an isomorphism (into). [Use f, Prob. 13-3-4.]

7 Let $H(f) = \int_0^\infty f$ (the *Heaviside distribution*) and $\delta(f) = f(0)$ (the *Dirac distribution*). Show that these are distributions and not functions [Prob. 13-3-4] and that δ is the derivative of H.

8 For a real number k, let k be the corresponding distribution $k(f) = k \int_{-\infty}^\infty f(t)\,dt$. Such distributions are called *constant*. The Lebesgue measure is the constant distribution 1. Show that a test function f is a derivative if and only if $1(f) = 0$. [$\Leftarrow$: Let $g(t) = \int_{-\infty}^t f$.] Deduce that the set of derivatives is a closed maximal subspace of $\mathscr{D}$.

9 Show that the derivative of a distribution u is 0 if and only if u is constant. [$\Rightarrow$: $u = 0$ on the set of derivatives; use Prob. 13-3-8.]

10 Show that a separable infinite-dimensional separated TVS has a dense proper S_σ subspace [Prob. 6-3-1].

---

101 Show that a Fréchet space cannot have a closed subspace of countably infinite codimension [Theorems 6-3-3 and 1-6-1].

102 Fill in the details of this example showing that "closed" cannot be omitted in Theorem 13-3-8. Let (Y, M) be a lc metric space with $Y = \cup X_n$, each X_n dense, $X_{n+1} \supset X_n$ as in Prob. 3-1-4. If $\mathfrak{T}$ is the strict inductive limit topology for Y, $\mathfrak{T} \supset M$ by Theorem 13-1-6 and $\mathfrak{T} \subset M$ since M is relatively strong and $(Y, \mathfrak{T})' = (X_1, \mathfrak{T})' = (X_1, M)' = (Y, M)'$.

103 Show that the differentiation operator, Example 13-3-22, is weak $*$ continuous.

104 Show that the set of functions is weak $*$ dense in the space of distributions.

105 Let H be a noncompact completely regular space which has a countable cobase $\{K_n\}$ for its compact sets. For $K \subset H$ let $X_K = \{f \in C(H) : f(t) = 0 \text{ for } t \notin K\}$. Let $X = \cup X_{K_n}$ and give X the strict inductive limit topology $\mathfrak{T}$ where each X_{K_n} has $\|\cdot\|_\infty$. Show that X is an (LF) space.

106 Take $H = R$ in Prob. 13-3-105. Let u be the topology given by $\|\cdot\|_\infty$ on X. Show that $\mathfrak{T} \supset u \supset$ topology of compact convergence, and both inclusions are strict. [For example, u is not complete.]

107 A member of $(X, \mathfrak{T})'$ (Prob. 13-3-106) is called a *Radon measure* on R. Show that $\mathscr{D} \subset X$ with continuous inclusion; hence each measure μ leads to a distribution $i'\mu$; indeed, $i'\mu(f) = \mu(f)$ for each $f \in \mathscr{D}$. Show also that $F(f) = f^{(1)}(0)$ leads to a distribution F which does not come from a measure.

108 Show that an (LF) space has a countable cobase for its bounded sets if and only if the constituent spaces are Banach spaces [Theorem 13-3-8 and Prob. 4-5-105].

109 Let A be a countably infinite-dimensional subspace of a lc metric space, B an algebraic complement, and $Y = A \oplus B$. Show that Y is not barrelled [Corollary 13-3-16 and Example 13-1-14].

110 Let X be a lc metric S_σ space. By Corollary 13-3-16, $(X', \text{weak} *)$ is not boundedly complete. Show that it is not even sequentially complete. [Let $x^n \in X_n \backslash X_{n-1}$, $x^n \to 0$. Let $f_n \in X'$, $f_n = f_{n-1}$ on X_{n-1}, $f_n(x^n) = 1$. Then $f_n \to f$ and $f(x^n) = 1$ for all n.]

111 Extend Theorem 13-3-15 by omitting "barrelled" and assuming that the dual is weak $*$ sequentially complete [the same proof].

112 Let a subspace A of a lcs space X be called a co-S_σ if A has an algebraically complementary subspace B which is an S_σ in X, say $B = \cup B_n$. Let (X, T) be relatively strong and such that $f \in X^\#$ must be continuous if $f \mid A$ and $f \mid B_n$ are continuous for all n. Show that $T \mid B$ is the strict inductive limit topology $\mathfrak{T}$ and that A, B are complementary, hence are both closed. [Let d be the topology of $A \oplus (B, \mathfrak{T})$. Then $d \supset T$ by Theorem 13-1-7. Conversely, $(X, d)' \subset (X, T)'$ so $T \supset d$.]

113 Let X be relatively strong with $(X', \text{weak} *)$ sequentially complete, and A a co-S_σ in X such that $A + B_n = A \oplus B_n$ for each n (notation of Prob. 13-3-112). Show that B has the strict inductive limit topology and A, B are complementary, hence are both closed. [The criterion of Prob. 13-3-112 holds by a proof very similar to the latter part of the proof of Theorem 13-3-19.]

---

201 Show by means of Theorem 12-4-9 that every continuous linear map from a fully complete space onto Y, Prob. 13-3-109, must be open. Thus "barrelled" is not the solution of the converse problem for this theorem. [See [64].]

---

301 (T. Shirai; R. M. Dudley). The space $\mathscr{D}$ of test functions is bornological [Example 13-3-20]; hence C sequential [Prob. 8-4-127], but is not sequential. [See [146], pp. 345, 362.]

302 A strict inductive limit of a sequence of fully complete (even lc Fréchet) spaces need not be fully complete. [See [116], Chap. 6, supplement (1).]

303 Theorems 13-3-4 and 13-3-8 fail for noncountably many spaces. [See [86] and [59], p. 517.]

304 The closed graph and open mapping theorems hold for maps between (LF) spaces. [See Prob. 13-2-301 for a more general result.]

13-4 FINITE COLLECTIONS OF METRIC SPACES

Throughout this section a collection of n spaces will be considered and n will always have this meaning.

1. Theorem The unrestricted inductive limit $\mathfrak{T}_u$ of finitely many paranormed spaces (X_i, p_i) is given by a paranorm.

PROOF With the notation of Sec. 13-1, for $y \in Y$ define $p(y) = \inf \{\sum_{i=1}^n p_i(x_i) : x_i \in X_i, y = \sum_{i=1}^n u_i(x_i)\}$. To prove $p(y^1 + y^2) \le p(y^1) + p(y^2)$, let $y^j = \sum u_i(x_i^j)$,

$\sum p_i(x_i^j) < p(y^j) + \varepsilon$, $j = 1, 2$. Then $p(y^1 + y^2) \le \sum p_i(x_i^1 + x_i^2) \le p(y^1) + p(y^2) + 2\varepsilon$. For continuity of multiplication, we prove first

$$p(y^k) \to 0, \ t_k \to t \text{ in } \mathcal{K} \text{ implies } p(t_k y^k) \to 0 \tag{13-4-1}$$

〚For each k choose $x_i^k \in X_i$ with

$$y^k = \sum_{i=1}^n u_i(x_i^k), \qquad \sum_{i=1}^n p_i(x_i^k) < p(y^k) + 2^{-k} \tag{13-4-2}$$

Then $x^k \to 0$ as $k \to \infty$ for each i, so $p(t_k y^k) \le \sum p_i(t_k x_i^k) \to 0$.〛 Next,

$$t_k \to 0 \text{ in } \mathcal{K} \text{ implies } p(t_k y) \to 0 \tag{13-4-3}$$

〚Let $y = \sum u_i(x_i)$. Then $p(t_k y) \le \sum p_i(t_k x_i) \to 0$.〛 Now if $t_k \to t$ and $p(y^k - y) \to 0$, we have $p(t_k y^k - ty) \le p[t_k(y^k - y)] + p[(t_k - t)y] \to 0$ by (13-4-1) and (13-4-3). Thus p is a paranorm.

To prove that $p = \mathfrak{T}_u$ note first that p is a test topology, i.e., each $u_i: X_i \to (Y, p)$ is continuous. 〚$u_i(x_i) = u_1(0) + u_2(0) + \cdots + u_i(x_i) + \cdots + u_n(0)$ so $p[u_i(x_i)] \le p_i(x_i)$.〛 This proves that $\mathfrak{T}_u \supset p$. Conversely, let $p(y^k) \to 0$. For each k choose $x_i^k \in X_i$ satisfying (13-4-2). Then $x_i^k \to 0$ as $k \to \infty$ for each i and so $u_i(x_i^k) \to 0$ in $(Y, \mathfrak{T}_u)$. Thus $y^k \to 0$ in $(Y, \mathfrak{T}_u)$ and so $p \supset \mathfrak{T}_u$.

2. Theorem If, in Theorem 13-4-1, each X_i is lc, so is $\mathfrak{T}_u$ and $\mathfrak{T}_u = \mathfrak{T}_i$.

PROOF By Probs. 13-1-5 or 13-1-115.

3. Theorem If, in Theorem 13-4-1, each X_i is seminormed, so is $\mathfrak{T}_u$ ($= \mathfrak{T}_i$).

PROOF Let p be as in Theorem 13-4-1, $y \in Y$, $t \in \mathcal{K}$, $\varepsilon > 0$. Let $y = \sum u_i(x_i)$ with $\sum p(x_i) < p(y) + \varepsilon$. Then $ty = \sum u_i(tx_i)$ so $p(ty) \le \sum p(tx_i) = |t| \sum p(x_i) \le |t| p(y) + \varepsilon |t|$. So $p(ty) \le |t| [p(y)]$. Replacing t by $1/t$ and y by ty yields the opposite inequality.

4. Theorem The unrestricted inductive limit of finitely many complete paranormed spaces is complete. The inductive limit of finitely many complete lc paranormed spaces is complete.

PROOF Let $\{y^k\}$ be a sequence in $(Y, \mathfrak{T}_u)$ with $\sum p(y^k) < \infty$, where p is given in Theorem 13-4-1. For each k choose $x_i^k \in X_i$ satisfying (13-4-2). Then, for each i, $p_i(x_i^k) < p(y^k) + 2^{-k}$ for all k and so $\sum_{k=1}^\infty x_i^k$ converges in X_i, say to x_i for each i. Then $\sum_{k=1}^m y^k = \sum_{i=1}^n u_i(\sum_{k=1}^m x_i^k) \to \sum_{i=1}^n u_i(x_i)$ as $m \to \infty$. Completeness of Y follows by Theorem 5-1-2.

5. Example Let X_1, X_2 be closed subspaces of a Fréchet space (X, q). Here q stands for both the paranorm and the topology of X. Note that Y need not be a closed subspace of X 〚Prob. 5-3-111〛 where $Y = X_1 + X_2$. Let $u_i: X_i \to Y$ be the inclusion map for $i = 1, 2$. Obviously, q is a test topology for Y and so $\mathfrak{T}_u \supset q$. In particular, $\mathfrak{T}_u$ is separated so, by Theorems 13-4-1 and 13-4-4, $(Y, \mathfrak{T}_u)$ is a Fréchet space. Moreover, it is continuously included in X, that is, the inclusion map is continuous [since $\mathfrak{T}_u \supset q$]. By Theorem 13-4-1, $\mathfrak{T}_u$ is given by $p(y) = \inf \{q(x_1) + q(x_2) : y = x_1 + x_2, \; x_i \in X_i \text{ for } i = 1, 2\}$. Some further properties of this example are spelled out in Probs. 13-4-103 to 13-4-106.

PROBLEMS

101 Show that the inductive limit of finitely many paranormed spaces is paranormed [Probs. 13-1-101 and 7-1-105].

102 Show that "unrestricted" cannot be omitted in the first sentence of Theorem 13-4-4, even if there is only one space. [Take $X = l^{1/2}$.]

103 Suppose that the sum S of finitely many closed subspaces of a Fréchet space X is a proper subspace of X. Show that S is a first-category subspace of X and if not closed is nonbarrelled. [See Example 13-4-5 and Theorem 12-5-7. Consider the identity map on S.]

104 In Example 13-4-5, assume that $X_1 \cap X_2 = \{0\}$. Show that $\mathfrak{T}_u$ is given by $p(x_1 + x_2) = q(x_1) + q(x_2)$.

105 In Example 13-4-5, show that p, q are equivalent on Y if and only if $X_1 + X_2$ is closed in X [Corollary 5-2-7 and Theorem 13-4-4].

106 Show that it is possible for two nonequivalent norms on a vector space to agree on each of two closed complementary subspaces [Probs. 13-4-104 and 5-3-111].

107 Show that in Theorem 13-4-3, p is the gauge of $\{y : y = \sum u_i(x_i), \sum \|x_i\| \leq 1\}$. What does this say if there is only one space?

108 Show that the sum of two FH spaces has the largest topology which makes addition continuous and is smaller on each space than its own topology.

109 Let Z be an FH space and X, Y algebraically complementary subspaces of Z which can be made into FH spaces (with possibly different topologies). Show that X, Y are complementary; hence closed. [Z is an FH space with $p_X + p_Y$, by Example 13-4-5. Now use Corollary 5-5-8.]

110 Let S be an algebraic complement of l in c_0. Show that S has no complete norm (or paranorm) larger than $\|\cdot\|_\infty$. [It would be an FK space, contradicting Prob. 13-4-109. Thus the new topology cannot even be larger than that induced by ω.]

111 Let X be a lc Fréchet space, $f \in X^* \backslash X'$. Show that $f^\perp$ has no larger complete norm (or paranorm). [As in Prob. 13-4-110. A special case follows from Prob. 3-1-105.]

112 Let $X \supset \varphi$ be a lc FK space with $1 \notin X$. Show that $X + [1]$ is a coregular FK space. [X is closed by Prob. 13-4-109.]

FOURTEEN

COMPACTNESS

14-1 WEAK COMPACTNESS

An important property of metric spaces is the equivalence of various forms of compactness. The purpose of this section is to extend this equivalence to a class of weak topologies (Theorem 14-1-9 and Example 14-1-10). The exposition follows that of [82].

A set B in a topological space C is called *conditionally countably compact* if every sequence x in B has a cluster point $c \in C$. This means that for every neighborhood U of c, $x_n \in U$ for infinitely many n.

1. Lemma Let E be a compact topological space, (M, d) a compact metric space, and give $C = C(E, M)$ the pointwise topology. Let B be a conditionally countably compact subset of C and $f \in \bar{B} \subset M^E$. Then $f \in C(E, M)$; indeed, f is continuous when E has the smaller topology wB, Theorem 1-6-10.

PROOF If possible, let f be not wB continuous at $x_0 \in E$. There is an open neighborhood V of $f(x_0)$ with $U = f^{-1}[V]$ not a wB neighborhood of x_0. By the assumption on f there exists $f_1 \in B$ with $d(f_1 x_0, f x_0) < 1$. Since f_1 is wB continuous there exists $x_1 \in E \backslash U$ with $d(f_1 x_1, f_1 x_0) < 1$. Next choose $f_2 \in B$ with $d(f_2 x_i, f x_i) < \frac{1}{2}$ for $i = 0, 1$ and $x_2 \in E \backslash U$ with $d(f_j x_2, f_j x_0) < \frac{1}{2}$ for $j = 1, 2$. This process yields sequences $\{x_i\} \{f_j\}$ in $E \backslash U$, B respectively with

$$d(f_j x_i, f x_i) < \frac{1}{j} \qquad \text{for } i = 0, 1, \ldots, j-1 \qquad (j = 1, 2, \ldots) \quad (14\text{-}1\text{-}1)$$

$$d(f_j x_i, f_j x_0) < \frac{1}{i} \qquad \text{for } j = 1, 2, \ldots, i \qquad (i = 1, 2, \ldots) \quad (14\text{-}1\text{-}2)$$

Since $M\backslash V$ is a compact metric space, $\{x_n\}$ has a subsequence $\{t_n\}$ with $f(t_n) \to y \in M\backslash V$. It follows that

$$\lim_i \lim_j f_j(t_i) = y, \qquad \lim_j \lim_i f_j(t_i) = f(x_0) \qquad (14\text{-}1\text{-}3)$$

Let x be a cluster point of $\{t_n\}$ in (E, T) and f_0 a cluster point of $\{f_n\}$ in C. Then, taking t_0 to be x_0,

$$f_0(t_i) = f(t_i) \qquad \text{for } i = 0, 1, 2, \dots \qquad (14\text{-}1\text{-}4)$$

$$f_j(x) = f_j(x_0) \qquad \text{for } j = 0, 1, 2, \dots \qquad (14\text{-}1\text{-}5)$$

These are true because, for each n, the map which takes g to $g(t_n)$ is a continuous map from C into M and so preserves cluster points. Now Eq. (14-1-3) implies that $y = \lim_i f_0(t_i) = f_0(x) = f(x_0)$ by (14-1-4) and (14-1-5). This contradicts the facts that $y \in M\backslash V$ and V is a neighborhood of $f(x_0)$.

2. Corollary Let E, M, C, B be as in Lemma 14-1-1. Then B is relatively compact.

PROOF The closure of B in M^E is compact by Tychonoff's theorem, and it lies in $C(E, M)$ by Lemma 14-1-1.

3. Lemma With E, M, C, B as in Lemma 14-1-1, let $\{f_n\} \subset B$, $A \subset E$. Suppose that $f_n \to f \in C$ pointwise on A; then $f_n \to f$ on $\bar{A}$.

PROOF If possible, let $a \in \bar{A}$, $f_n(a) \not\to f(a)$. For some V, a neighborhood of $f(a)$, $f_{n(k)}(a) \notin V$. Let $g \in C$ be a cluster point of $\{f_n\}$; then, for each $x \in E$, $g(x)$ is a cluster point of $\{f_{n(k)}(x)\}$. For $x \in A$ this means that $g(x) = f(x)$; hence $g(a) = f(a)$ and so $f(a)$ is a cluster point of $\{f_{n(k)}(a)\}$, which contradicts the definition of V.

4. Definition A topological space X is called N sequential if for every $A \subset X$, every $x \in \bar{A}$ is the limit of a sequence of points of A.

This terminology occurred also in Prob. 8-4-116.

5. Lemma In Lemma 14-1-1 assume that E is also separable. Then B is N sequential.

PROOF Let $S \subset B$, $f \in \bar{S}$, and let A be a countable dense set in E. Using the subscript A to denote restriction of functions to A, we have $f_A \in \bar{S}_A$ in the space $B_A \subset M^A$. Since M^A is a metric space [Prob. 2-1-12], f_A is a sequential limit point of S_A, that is, there exists $\{f_n\} \subset S$ with $f_n \to f$ pointwise on A. By Lemma 14-1-3, $f_n \to f$ on E.

The separability hypothesis is removed in Lemma 14-1-8.

6. Lemma With E, M, C as in Lemma 14-1-1, let $S \subset C$ and $f \in \overline{S} \subset C$. For every positive integer n and $\varepsilon > 0$, S contains a finite set $F = F(n, \varepsilon)$ which meets $\cap \{U(f, x_i, \varepsilon) : i = 1, 2, \ldots, n\}$ for every n-tuple $(x_1, x_2, \ldots, x_n)$ of points of E, where $U(f, x, \varepsilon)$ stands for $\{g \in C : d(gx, fx) < \varepsilon\}$.

PROOF For each $z = (x_1, x_2, \ldots, x_n) \in E^n$ choose $f_z \in S$ with $d(f_z x_i, f x_i) < \varepsilon$ for all i; let $V_z = \{x \in E : d(f_z x, fx) < \varepsilon\}$. Then for each z, $W_z = V_z^n$ is an open set in E^n and contains z. This cover of E^n may be reduced to a finite subcover $(W_{z_1}, W_{z_2}, \ldots, W_{z_n})$. Let $F = (f_{z_1}, f_{z_2}, \ldots, f_{z_n})$. For any $z = (x_1, x_2, \ldots, x_n)$, say $z \in W_{z_k}$. Then for each i, $f_{z_k} \in U(f, x_i, \varepsilon)$.

7. Lemma Under the conditions of Lemma 14-1-6, S has a countable subset F with $f \in \overline{F}$.

PROOF Let $F = \cup \{F(n, 1/m) : n, m = 1, 2, \ldots\}$ (Lemma 14-1-6). For any neighborhood U of f there exist $x_1, x_2, \ldots, x_n \in E$ and $\varepsilon > 0$ with $U \supset \cap \{U(f, x_i, \varepsilon) : i = 1, 2, \ldots, n\}$. For $m > 1/\varepsilon$, $F(n, 1/m)$ meets $\cap U(f, x_i, 1/m) \subset \cap U(f, x_i, \varepsilon) \subset U$. Thus F meets U.

8. Lemma With (E, T), M, C, B as in Lemma 14-1-1, B is N sequential.

PROOF This improves Lemma 14-1-5. Let $S \subset B$, $f \in \overline{S}$. By Lemma 14-1-7 we may assume that S is countable. Let $B_1 = S \cup \{f\}$, $w = wB_1$; then $w \subset T$ and so $E_1 = (E, w)$ is a compact pseudometric space; hence separable. By Corollary 14-1-2, B_1 is relatively compact in $C(E_1, M)$. By Lemma 14-1-5, B_1 is N sequential; thus f is the limit of a sequence in S.

9. Theorem: Eberlein–Smulian theorem Let E be a compact topological space, M a compact metric space, and $C = C(E, M)$ with the pointwise topology. A subset of C is compact if and only if it is countably compact and if and only if it is sequentially compact. The same is true for relatively compact, relatively countably compact, and relatively sequentially compact. All compact subsers are N sequential.

PROOF Recall that for any property P, S is called relatively P if $\overline{S}$ has property P. Any sequentially compact or compact set is countably compact. If B is countably compact it is relatively compact by Corollary 14-1-2 and closed by Lemma 14-1-8 applied to its closure; hence compact. Lemma 14-1-8 also shows that any compact set is sequentially compact. The second part follows by considering the closures.

10. Example *The equivalences of Theorem 14-1-9 hold for the weak topology of a normed space X (but not for the weak $*$ topology; see Prob. 14-1-1).* Let E be the unit disc of X' with the weak $*$ topology and M the unit disc of the complex plane. We shall show that the unit disc D of X, with w, the weak topology, is homeomorphic into $C(E, M)$. The map is $x \to \hat{x} \mid E$, where

$\hat{\infty}$ is the natural embedding. Thus $\hat{x}(e) = e(x)$ for $e \in E$, $x \in D$. $[\![$If $x \to a$ in (D, w) and $e \in E$, $\hat{x}(e) = e(x) \to e(a) = \hat{a}(e)$, that is, $\hat{x} \to \hat{a}$ pointwise. Conversely, if $\hat{x} \to \hat{a}$ pointwise then $e(x) \to e(a)$ for each $e \in E$, from which it follows easily that $x \to a$ weakly.$]\!]$ Thus (D, w) is covered by the Eberlein–Smulian theorem and the result follows since every countably compact set is bounded. $[\![$See the proof outlined in Prob. 6-4-106.$]\!]$

Let X be a Banach space. A sufficient condition that $(X', \text{weak} *)$ be a Mazur space was given in Problem 12-2-3, namely that X is separable. A condition which is more general $[\![$Problem 9-5-1$]\!]$ is that D, the unit disc in X', with its weak $*$ topology, is N sequential. (For example, this holds, by Example 14-1-10, if D is homeomorphic with some subset of a Banach space in its weak topology — such a set is called an *Eberlein compact*.) For if f is weak $*$ sequentially continuous on X', $f | D$ is continuous $[\![f[\overline{A}] \subset f(A)$ for each $A \subset D]\!]$ and so by Theorem 12-3-6, f is weak $*$ continuous.

It is clear that this proof applies to reflexive spaces also; but Problem 8-6-1 gives it more easily.

Some additional details about the results of this section may be found in the problems of [156], sec. 13.4. More general theorems relating the forms of compactness in arbitrary spaces may be found in [88], §24. See, for example, Prob. 14-1-302. Since the weak $*$ topology is included in this class, the best one can hope for is results for forms of relative compactness. (See Prob. 14-1-1.)

PROBLEMS

1 Show that the result of Example 14-1-10 is false for the weak $*$ topology of the dual of a Banach space $[\![$Theorem 9-6-2$]\!]$.

101 Suppose that a reflexive Banach space X has the property (as l does) that weakly convergent sequences are norm convergent. Show that it must be finite dimensional $[\![$Examples 14-1-10 and 10-2-6 and Theorem 6-4-2$]\!]$.

102 Let X be a reflexive infinite-dimensional Banach space. Show that X' contains a sequence $\{f_n\}$ with $\| f_n \| = 1$, $f_n \to 0$ weak $*$. $[\![$See Prob. 14-1-101. This is also a special case of Prob. 9-5-301.$]\!]$

103 Show that l has no reflexive infinite-dimensional subspace $[\![$Probs. 14-1-101 and 8-1-8$]\!]$.

104 Show that c_0 has no reflexive infinite-dimensional quotient $[\![$Prob. 14-1-103 and Theorem 11-3-4$]\!]$.

105 (A. G. McIntosh's closed graph theorem). Let X be a lcs space with $(X', \text{weak} *)$ sequentially complete. Show that every linear map with closed graph from X to a reflexive Banach space Y is continuous. $[\![$As in Lemma 12-5-12, Δ is weak $*$ sequentially closed. The unit disc D in Y is weak $*$ sequentially compact by the Banach–Alaoglu and Eberlein–Smulian theorems. Thus $\Delta \cap D$ is weak $*$ compact and so Δ is aw^* closed. Apply the Banach–Dieudonné theorem.$]\!]$

106 Let X be a lc metric space such that weakly convergent sequences are convergent. Show that weakly compact sets are compact. $[\![$By Example 14-1-10 you may use sequential compactness.$]\!]$ Problem 14-1-102 shows that X cannot be infinite dimensional and reflexive normed.

107 Let E, M be topological spaces, $B \subset M^E$. Say that B satisfies the ILC (*iterated limit condition*) if whenever x, f are sequences in E, B such that both iterated limits $\lim \lim f_n(x_m)$ exist they must be equal. Now let M be a compact metric space and $B \subset C(E, M)$. Show that if B satisfies the ILC, every $f \in \bar{B} \subset M^E$ is continuous and B is a (pointwise) relatively compact set in $C(E, M)$ [as in Lemma 14-1-1 and Corollary 14-1-2].

108 Let $E = M = R$, $f_n(x) = nx$. Show that $B = \{f_n\}$ satisfies the ILC but is not relatively compact in $C(E, M)$. Let $g_n(x) = nx$ for $|x| \leq n^{-1/2}$, $1/x$ for $|x| \geq n^{-1/2}$. Show that $B = \{g_n\}$ satisfies the ILC but $g_n \to g \notin C(E, M)$. Thus "compactness of M" cannot be dropped in either part of Prob. 14-1-107.

109 Suppose that S is a vector subspace of l which is closed in l^2. Show that S is finite dimensional. [$i: S \to l$ is weakly compact as in Prob. 11-4-131. Apply Prob. 14-1-103. An analogous result follows from Prob. 15-3-1 applied to c_0.]

110 Suppose that X is a Banach space and that there exists a reflexive Banach space Z and a continuous linear map u from Z onto a dense proper subspace of X. Show that the unit disc D in X' is weak $*$ sequentially compact. [For $\{f_n\} \subset D$, $\{u'(f_n)\}$ has a weak $*$ convergent subsequence $\{u(g_n)\}$ by Example 14-1-10. Then $\{g_n\}$ is weak $*$ convergent by Prob. 9-3-104.] The application to l^∞ is also given in Prob. 11-1-109.

111 Let X be a weakly sequentially complete Banach space such that the unit disc D in X'' is weak $*$ sequentially compact. Show that X is reflexive. This extends Example 12-5-15, which is a special case by Prob. 9-5-1. [The unit disc of X is weakly sequentially compact. Apply Examples 14-1-10 and 10-2-6.]

---

301 c_0 has no reflexive infinite-dimensional subspace. [See [8], p. 194, Theorem 1.]

302 Let X be a lcs space, $A \subset X$ complete [or just $\tau(X, X')$ complete], absolutely convex, and countably compact. Then A is compact. [See [88], 24.2 (1′).]

303 A bounded weakly closed set S in a complete lcs X is weakly compact if and only if every $f \in X'$ assumes its sup on S. [See [66], Theorem 6.]

304 The following are equivalent for $u \in B(X, Y)$ where X, Y are Banach spaces:

    (*a*) u is tauberian.

    (*b*), (*c*) $A \subset X$ is weakly relatively compact if it is bounded and $u[A]$ is (weakly) relatively compact.

Compare Prob. 11-4-107. [See [79].]

    Also equivalent are:

    (*a*) u is semifredholm.

    (*b*) $A \subset X$ is relatively compact if it is bounded and $u[A]$ is compact.

    (*c*) Every bounded sequence x in X with $ux_n \to 0$ has a convergent subsequence. [See [95].]

305 Let X, Y be Banach spaces, $u \in B(X, Y)$. Then u is semifredholm if and only if $u \mid Z$ is, for each reflexive subspace Z of X. [$\Rightarrow$: By Prob. 14-1-304. See [79].] It follows that tauberian and semifredholm are equivalent if X has no reflexive infinite-dimensional subspace. (See Probs. 14-1-103 and 14-1-301.) This generalizes the Berg–Crawford–Whitley theorem; see [44].

306 Let H be a compact Hausdorff space, $C = C(H)$, and X the span of H in C'. If a subset of C is weakly closed, bounded, and $\sigma(C, X)$ compact, it is weakly compact. [See [88], 24.5 (3).]

14-2 CONVEX COMPACTNESS

The main result of this section is a sufficient condition for the convex compactness property. This is a special case of a theorem of M. Krein (Prob. 14-2-301). We shall make use of the Lebesgue bounded convergence theorem in the following form.

1. Lemma Let H be a compact Hausdorff space. A sequence $\{g_n\}$ in $C(H)$ tends to 0 weakly if and only if $g_n \to 0$ pointwise boundedly, that is, $g_n(h) \to 0$ for each $h \in H$ and, for some M, $|g_n(h)| < M$ for all n, h.

PROOF A proof may be found in such texts as [57], Theorem 12.24; or [88], 24.5 (2).

In the remainder of this section K is a weakly compact set in a lcs space X, $C = C(K)$ where K is given the weak topology $\sigma(X, X')$, making it a compact Hausdorff space. The map $v: X' \to C$ is defined by $v(f) = f \mid K$. This is a continuous linear map when X' has $\beta(X', X)$ since $f \to 0$, $\beta(X', X)$, implies $f \to 0$ uniformly on bounded sets in X, in particular, $\|f\|_\infty \to 0$. Thus $v': C' \to X''$.

2. Lemma Let X be a separable complete lcs space. Then $v': C' \to X$.

PROOF This means that $F = v'(\varphi) \in X$ for $\varphi \in C'$, that is, that $F \in X''$ is $\sigma(X', X)$ continuous [Theorem 8-1-7]. It is sufficient to show that $F^\perp$ is aw^* closed [Corollary 12-2-16] and for this it is sufficient to show that $F^\perp$ is weak $*$ sequentially closed [Prob. 12-2-3]. Let $\{f_n\}$ be a sequence in $F^\perp$ with $f_n \to f$ weak $*$. Now $\{f_n\}$ is $\beta(X', X)$ bounded. [X is a Banach–Mackey space by Theorem 10-4-8 and the result follows by Theorem 10-4-5.] It follows that there exists M with $|f_n(x)| \le M$ for $x \in K$. [$K^\circ \in \mathcal{N}[\beta(X', X)]$ so $\{f_n\} \subset MK^\circ$.] Let $g_n = v(f_n - f)$ so that $g_n \to 0$ pointwise and boundedly. By Lemma 14-2-1 $\varphi(g_n) \to 0$. But $\varphi(g_n) = F(f_n - f) = -F(f)$ and so $f \in F^\perp$.

3. Theorem Let X be a separable complete, lcs space. Then X, with any compatible topology, has the convex compactness property.

PROOF It is sufficient to prove this for the weak topology [Theorem 9-2-11]. Let K be a weakly compact set in X. Let D be the unit disc in C' where $C = C(K)$, as in the earlier discussion. Now D is weak $*$ compact [Banach–Alaoglu] and $v': (C', \text{weak} *) \to [X, \sigma(X, X')]$ is continuous. [By Lemma 14-2-2 and Theorem 11-1-6 since $\sigma(X'', X')|_X = \sigma(X, X')$.] Thus $v'[D]$ is a weakly compact absolutely convex set in X and the result will follow when it is shown that $K \subset v'[D]$. Let $x \in K$. Define $\varphi \in C'$ by $\varphi(g) = g(x)$ for $g \in C$. Clearly, $\|\varphi\| = 1$ and $v'(\varphi) = x$. [For $f \in X'$, $v'(\varphi)(f) = \varphi(vf) = (vf)(x) = f(x) = x(f)$, considering $x \in X''$.]

4. Theorem Let X be a Banach space. Then X, with any compatible topology, has the convex compactness property.

PROOF As in Theorem 14-2-3 it is sufficient to prove this for the weak topology. Let K be weakly compact and K_0 its absolutely convex closure. It is sufficient to show that K_0 is weakly countably compact [Example 14-1-10] so let $\{x^n\} \subset K_0$. Since K_0 is the norm closure of the absolutely convex hull H

of K, for each n there is a sequence $\{y_k^n\}$ in H with $y_k^n \to x^n$ in norm; hence weakly. Each y_k^n is a finite linear combination of members of K; let S be the linear closure of this (countable) set of members of K. Then S is a separable Banach space and $K \cap S$ is a weakly compact set in S. [The weak topology of S is $\sigma(X, X')|_S$ and S is weakly closed in X, since norm closed and convex.] Each $x^n \in S$ and so, by Theorem 14-2-3, $\{x^n\}$ has a weak cluster point in the absolutely convex closure in S of $K \cap S$, *a fortiori* in K_0.

PROBLEMS

101 Let X be a Banach space or a separable complete lcs space and let K be the family of weakly compact sets in X. Show that $T_K = \tau(X', X)$. (See Prob. 14-2-301) [Theorems 14-2-3, 14-2-4, and 9-2-12.]

102 Show that c_0 with its weak topology has the convex compactness property but is not sequentially complete.

103 Show that the dual pair (l, c_0) is sequentially barrelled but not C barrelled [Probs. 9-4-107 and 9-4-108].

104 Show that the convex compactness property for a lcs space X is not sufficient to imply that T° is compatible for (X', X); see Theorem 12-1-4 [Example 12-1-2].

105 Let B be a bounded set in a quasibarrelled space X. Show that B is relatively compact in $[X'', \sigma(X'', X')]$. $[B^\circ \in \mathcal{N}[\beta(X', X)]$ and so $B^{\circ\circ}$ is weak $*$ compact.]

106 Let X be relatively strong and such that (X', T) has the convex compactness property for all admissible T. Show that X need not be barrelled. Compare Prob. 9-3-1, a converse result [Theorem 14-2-4 and Problem 8-4-6].

107 The topologies $\tau(l, c_0)$, $\sigma(l, l^\infty)$, $\|\cdot\|_1$ all have the same convergent sequences and compact sets. [See Probs. 14-2-102, 8-6-119, 9-2-4, and 14-1-106. This is also a special case of Prob. 14-2-302.]

108 Show that a relatively strong space need not be C sequential [Probs. 14-2-107 and 8-4-201].

201 A relatively strong Mazur space must be C sequential. [T^+ is comparible.]

301 (M. Krein's theorem). Let K be a weakly compact set in a lcs space X. Then if the absolutely convex closure of K is $\tau(X, X')$ complete it is weakly compact. [See [88] 24.5(4), 24.6; and [82] 17.12.] It follows that if X is boundedly complete, it has the convex compactness property in any topology compatible for X, X'. Thus, for example, Prob. 14-2-101 holds for any boundedly complete space.

302 (Dunford–Pettis property). Let H be a locally compact Hausdorff space and $C = C_0(H)$. Then $\tau(C', C)$ and $\sigma(C', C'')$ have the same convergent sequences and the same compact sets. [See [49], p. 153, Theorem 4; and [38], pp. 633–639.]

14-3 EXTREME POINTS

For $x, y \in X$, a vector space, $x \neq y$, let $(x, y) = \{tx + (1 - t)y : 0 < t < 1\}$, called an *open interval*. If $x = y$, take (x, y) to be empty. For $K \subset X$, a point $z \in K$ is called an *extreme point* of K if $x, y \in K$ implies $z \notin (x, y)$. For example, a vertex of a triangle or a point on a circle is an extreme point of the triangular or circular region in the plane. A set $S \subset K$ is called an *extreme set* if $x \in K$, $y \in K \backslash S$ implies $(x, y) \not\subset S$. Thus K and the empty set are extreme sets of K as is, for example, an interval containing a vertex in a side of the triangle if K is a triangular region.

1. Lemma Let X be a lcs space and K a compact set in X. Then each nonempty compact extreme set S of K contains an extreme point of K.

PROOF We may assume that X is a real space since the definitions use real scalars only. Let P be the collection of nonempty compact extreme sets of K, each of which is a subset of S; ordering P by inclusion, let C be a maximal chain and $E = \cap C$. Clearly, $E \in P$. [For example, $E \neq \phi$ since K is compact and each finite subset of C has a smallest member; hence has nonempty intersection.] Also E is minimal in P. It remains to show that E has only one point; if not, let $f \in X'$ take on two different values on E and let m be the maximum value of f on E.

By the assumption, $M = \{x \in E : f(x) = m\}$ is a proper subset of E, is compact, and is nonempty. It is also an extreme set of K, contradicting the minimality of E. [Let $x \in K$, $y \in K \backslash M$. If either x or y does not belong to E, $(x, y) \not\subset E$ a fortiori $(x, y) \not\subset K$. If $x, y \in E$; for each $z \in (x, y)$, $f(z) = tf(x) + (1 - t)f(y) < m$ so $z \notin M$.]

2. Remark The proof uses about X only the fact that X' is total; thus the result would apply to certain non-lc spaces such as $l^{1/2}$.

3. Theorem: Krein–Milman theorem Let X be a lcs space, K a compact set in X, and E the set of extreme points of K. Then the convex closure C of E includes K.

PROOF As in Lemma 14-3-1 we may assume that X is real. If possible, let $k \in K \backslash C$. By the Hahn–Banach Theorem 7-3-4, there exists $f \in X'$ with $f(k) > \sup \{f(x) : x \in C\}$. Let m be the maximum value of f on K, $M = \{x \in K : f(x) = m\}$. Then $M \subset K \backslash C$ and, as in Lemma 14-3-1, M is an extreme set of K. By Lemma 14-3-1, M contains an extreme point of K which contradicts the fact that it is disjoint from C.

4. Corollary In a lcs space, each compact convex set is equal to the convex closure of its set of extreme points.

This proof of the Krein–Milman theorem is due to E. Artin and J. L. Kelley. The theorem has many applications in abstract harmonic analysis. An extended account is given in [38], Chap. 10 and also p. 209. See also [88], 25.2–25.5.

5. Example The unit disc D in c_0 has no extreme points. [Let $z \in D$, $|z_n| < 1$ for some n. Let $x = z + \varepsilon \delta^n$, $y = z - \varepsilon \delta^n$ where $0 < \varepsilon < 1 - |z_n|$. Then $x, y \in D$, $z = (x + y)/2$.] Thus c_0 cannot be given a lcs topology which makes D compact. In particular (Prob. 14-3-1), c_0 *is not equivalent with any dual Banach space*. A better result is contained in Prob. 11-1-122.

PROBLEMS

1 Let X be a lcs space and $U \in \mathcal{N}$. Show that U° is the convex weak $*$ closure of its set of extreme points. In particular, U° has extreme points.

2 Show that the unit disc in a dual Banach space has extreme points.

301 Let X be a separable space which is the dual of a Banach space. Then each convex bounded norm closed set is the convex norm closure of its set of extreme points; "separable" cannot be omitted. [See [13] and [32].]

14-4 PHILLIPS' LEMMA

A valuable treatment of the results of this section may be found in [32]. Let $b = \text{bfa}(H, \mathcal{A})$ be the set of bounded finitely additive set functions defined on an algebra $\mathcal{A}$ of subsets of a set H. A special case was used in Examples 2-3-14 and 2-3-15. An *algebra* in this sense is a collection of subsets of H which contains H and is closed under complementation and finite unions.

For each $\mu \in b$, define $|\mu| \in b$ (called the *variation* of μ) by $|\mu|(e) = \sup \{\sum_{i=1}^n |\mu(e_i)| : e = \cup e_i, \ e_i \not\subset e_j \text{ for } i \neq j, \text{ each } e_i \in \mathcal{A}\}$ and $\|\mu\| = |\mu|(H)$. It follows from Lemma 14-4-2 that $\|\mu\| < \infty$ for $\mu \in b$; indeed, b is a normed space.

1. Lemma Let E be a finite set of complex numbers. There exists $F \subset E$ such that $|\sum \{z : z \in F\}| \geq \frac{1}{6} \sum \{|z| : z \in E\}$.

PROOF Divide E into three sets by trisecting the complex plane. For at least one of these three sets, call it F, we have $\sum \{|z| : z \in F\} \geq \frac{1}{3} \sum \{|z| : z \in E\}$. We may assume $-\pi/3 \leq \arg z \leq \pi/3$ for $z \in F$. Then $|\sum \{z : z \in F\}| \geq \frac{1}{2} \sum \{|z| : z \in F\}$.

2. Lemma Let $\mu \in b$. Then there exists $e \in \mathcal{A}$ with $|\mu(e)| \geq \frac{1}{7} \|\mu\|$.

PROOF Let H be a disjoint finite union $\cup s_i$ with $\sum |\mu(s_i)| \geq \frac{6}{7} \|\mu\|$. By Lemma 14-4-1 we may select $e_1, e_2, \ldots, e_m$ from among the s_i with $|\sum \mu(e_i)| \geq \frac{1}{6} \sum |\mu(s_i)|$. Set $e = \cup e_j$.

3. Lemma For $\mu \in b$, $e \in \mathcal{A}$, set $\bar{\mu}(e) = \sup \{|\mu(s)| : s \subset e, s \in \mathcal{A}\}$. Then for any disjoint sequence $\{e_k\} \subset \mathcal{A}$ we have $\sum \bar{\mu}(e_k) < \infty$.

PROOF Let $s_k \subset e_k$ with $|\mu(s_k)| > \bar{\mu}(e_k) - 2^{-k}$. Fix a positive integer m. By Lemma 14-4-1, choose $g_1, g_2, \ldots, g_n$ from among $s_1, s_2, \ldots, s_m$ so that $|\sum \mu(g_j)| \geq \frac{1}{6} \sum_{i=1}^m |\mu(s_i)|$. Let $g = \cup g_j$. Then $6 |\mu(g)| \geq \sum_{i=1}^m \bar{\mu}(e_i) - 1$. Since $|\mu(g)| < M$ for some M, depending only on μ, the result follows.

We now specialize to bfa(N), Example 2-3-14. Thus $\mathcal{A}$ is the collection of all subsets of N.

4. Lemma : R. S. Phillips' lemma For each $n \in N$, let $\mu_n \in \mathrm{bfa}(N)$. Suppose that $\mu_n(e) \to 0$ for each $e \subset N$. Then $\sum_{k=1}^{\infty} a_{nk} \to 0$ as $n \to \infty$, where $a_{nk} = |\mu_n(\{k\})|$.

PROOF The matrix A has null columns, i.e.,

$$a_{nk} \to 0 \quad \text{as} \quad n \to \infty \qquad \text{for each } k \tag{14-4-1}$$

This follows from the hypothesis by taking $e = \{k\}$. Also

$$\sum_k a_{nk} < \infty \qquad \text{for each } n \tag{14-4-2}$$

by taking $e_k = \{k\}$ in Lemma 14-4-3 applied to μ_n. Suppose now that the conclusion is false. By passing to a subsequence and multiplying by a constant we may express this assumption as

$$\sum_k a_{nk} > 1 \qquad \text{for all } n \tag{14-4-3}$$

Let $0 < \varepsilon < \frac{1}{20}$, $n_1 = N_1 = 1$ and (inductively) having chosen n_i, N_i, choose $N_{i+1} > N_i$ and $n_{i+1} > n_i$ so that

$$\sum_{k=N_{i+1}}^{\infty} a_{n_i k} < \varepsilon, \qquad \sum_{k=1}^{N_{i+1}-1} a_{n_{i+1} k} < \varepsilon \tag{14-4-4}$$

These choices are possible by (14-4-2) and (14-4-1). By Lemma 14-4-1, for each i, choose a set of integers $T_i \subset [N_i, N_{i+1})$ so that $|\mu_n(T_i)| \geq \frac{1}{6} \sum_{k=N_i}^{N_{i+1}-1} a_{nk}$, from which it follows, using (14-4-3) and (14-4-4) [take i instead of $i+1$ in the second part of (14-4-4)], that

$$|\mu_{n_i}(T_i)| > \frac{1 - 2\varepsilon}{6} \tag{14-4-5}$$

Now $\{T_n\}$ is a disjoint sequence of finite sets of integers. Divide $\{T_n\}$ up into infinitely many infinite sequences $\Delta_1^1, \Delta_2^1, \Delta_3^1, \ldots$. For example, we could take $\Delta_1^1 = (T_1, T_3, T_5, T_7, \ldots)$, $\Delta_2^1 = (T_2, T_6, T_{10}, T_{14}, \ldots)$, etc. Let $D_i^1 = \cup \Delta_i^1$, that is, D_i^1 is an infinite subset of N containing each integer which belongs to some $T_j \in \Delta_i^1$. Also, $D_1^1, D_2^1, \ldots,$ are disjoint.

By Lemma 14-4-3, there exists m_1 such that $\bar{\mu}_{n_1}(D_{m_1}^1) < \varepsilon$.

Divide $\Delta_{m_1}^1$ into infinitely many infinite sequences $\Delta_1^2, \Delta_2^2, \Delta_3^2, \ldots$, just as we did $\{T_n\}$; let $D_i^2 = \cup \Delta_i^2$. Again by Lemma 14-4-3, there exists m_2 such that $\bar{\mu}_{n_2}(D_{m_2}^2) < \varepsilon$. Continuing in this way we obtain $\Delta_{m_{i+1}}^{i+1} \subset \Delta_{m_i}^i$, $D_{m_{i+1}}^{i+1} \subset D_{m_i}^i$, $\bar{\mu}_{n_i}(D_{m_i}^i) < \varepsilon$.

Let E be the set (of subsets of N) whose kth member is the kth member of $\Delta_{m_k}^k$, say $E = (T_{q_1}, T_{q_2}, \ldots)$ and let $e = \cup E$. The required contradiction will be reached by showing that $\mu_n(e) \not\to 0$. Fix i with $T_i \in E$; there are infinitely many such i. Then $|\mu_{n_i}(e)| = |\mu_{n_i}(T_i) + \mu_{n_i}(e \backslash T_i)| > (1 - 2\varepsilon)/6 - |\mu_{n_i}(f_i)|$ where $f_i = e \backslash T_i$, by (14-4-5). The proof will be completed by showing that $|\mu_{n_i}(f_i)| < 3\varepsilon$, since this will imply that $|\mu_{n_i}(e)| > (1 - 2\varepsilon)/6 - 3\varepsilon > 0$ for infinitely many i. To prove this, write $f_i = u \cup v \cup w$ where $u = \cup \{T_j \in E : j < i\}$, $v =$

$\cup \{T_j \in E : i < j < q_i\}$, and w is the union of the remaining $T_j \in E$. Note that T_i figures in none of these three sets. Now $u \subset [1, N_i)$, so by the second part of (14-4-4) (with $i + 1$ replaced by i) we have $|\mu_{n_i}(u)| < \varepsilon$; v is finite and $v \subset [N_{i+1}, \infty)$, so by the first part of (14-4-4), $|\mu_{n_i}(v)| < \varepsilon$; all T_j contributing to w have $j = q_k$ with $k \geq i$; for such j, T_j is the kth member of $\Delta_{m_k}^k$ so $T_j \subset D_{m_k}^k \subset D_{m_i}^i$. This proves that $w \subset D_{m_i}^i$ and so $|\mu_{n_i}(w)| \leq \bar{\mu}_{n_i}(D_{m_i}^i) < \varepsilon$.

5. Corollary With the notation of Lemma 14-4-4, $\mu_n(\{n\}) \to 0$.

PROOF For $|\mu_n(\{n\})| = a_{nn} \leq \sum_k a_{nk}$.

6. Corollary For each $n \in N$, let $\mu_n \in \text{bfa}(N)$. Suppose that $\{\mu_n(e)\}$ is bounded for each $e \subset N$. Then $\{\mu_n(\{n\})\}$ is bounded.

PROOF It is sufficient to show that $\varepsilon_n \mu_n(\{n\}) \to 0$ whenever $\varepsilon_n \to 0$, and this follows immediately by applying Corollary 14-4-5 to $\{\varepsilon_n \mu_n\}$.

7. Example *The equivalence of norm and weak sequential convergence in* l (Prob. 8-1-8) *can be deduced from Lemma 14-4-4.* Let $x^n \to 0$ weakly. Considering $x^n \in (l^\infty)'$ we have, for $y \in l^\infty$, $[x^n, y] = \int y \, d\mu_n$ with $\mu_n \in \text{bfa}(N)$ [Example 2-3-15]. For $e \subset N$, let y be the characteristic function of e; the assumption on $\{x^n\}$ implies that $\mu_n(e) \to 0$. By Lemma 14-4-4, $\sum_k a_{nk} \to 0$; it remains to show that $\sum a_{nk} = \|x^n\|_1$, and this is because $x_k^n = [x^n, \delta^k] = \int \delta^k \, d\mu_n = \mu_n(\{k\})$.

8. Remark Schur's Theorem 1-3-2 is equivalent to the property of l mentioned in Example 14-4-7; hence it follows from Lemma 14-4-4. Conversely, a weak form of Lemma 14-4-4 can be deduced from Schur's theorem, namely, the same result assuming that each μ_n is countably additive. If x is any sequence of zeros and ones, let $e = \{i \in N : x_i = 1\}$. Then, with $b_{nk} = \mu_n(\{k\})$ we have $(Bx)_n = \mu_n(e) \to 0$. By Schur's theorem $\sum a_{nk} = \sum |b_{nk}| \to 0$. (We have used the stronger form resulting from Remark 15-2-3.)

9. Example c_0 *is not complemented in* l^∞. Let $P_n \in c_0'$ be given by $P_n(x) = x_n$. If the result is false, each P_n can be extended to $g_n \in (l^\infty)'$ with $g_n \to 0$ weak * [Prob. 5-3-5]. For $x \in l^\infty$ we have $g_n(x) = \int x \, d\mu_n$ with $\mu_n \in \text{bfa } N$ [Example 2-3-15]. Let $e \subset N$ and let x be the characteristic function of e. Then $\mu_n(e) = g_n(x) \to 0$. By Corollary 14-4-5, $g_n(\delta^n) = \mu_n(\{n\}) \to 0$. But $g_n(\delta^n) = P_n(\delta^n) = 1$.

Example 14-4-9 is improved by Example 14-7-8.
Phillips' lemma also implies a useful uniform boundedness result, Theorem 14-4-11.

10. Lemma Let α be a positive unbounded set function on an algebra $\mathscr{A}$ of subsets of a set H. Suppose that $\alpha(e \backslash f) \geq |\alpha(e) - \alpha(f)|$ and, when e, f are disjoint, $\alpha(e \cup f) \leq \alpha(e) + \alpha(f)$. Then $\mathscr{A}$ contains a disjoint sequence $\{e_n\}$ with $\alpha(e_n) \to \infty$.

PROOF We shall define $\{f_n\} \subset \mathscr{A}$ inductively in such a way that α is unbounded on each f_n. Take $f_1 = H$ and, when f_n is chosen, take $f \subset f_n$ with $\alpha(f) > n + 2\alpha(f_n)$. Now α is unbounded on either f or $f_n \backslash f$ so let f_{n+1} be one of these two sets with α unbounded on f_{n+1}. It follows that $\alpha(f_{n+1}) > n + \alpha(f_n)$. $[\![$If $f_{n+1} = f$ this is trivial; if $f_{n+1} = f_n \backslash f$, $\alpha(f_{n+1}) \geq \alpha(f) - \alpha(f_n) > n + \alpha(f_n).]\!]$ Now let $e_n = f_n \backslash f_{n+1}$. Then $\alpha(e_n) \geq \alpha(f_{n+1}) - \alpha(f_n) > n$.

A σ *algebra* is an algebra which is also closed under countable unions.

11. Theorem: Nikodym's theorem Let $B \subset \mathrm{bfa}(H, \mathscr{A})$ where $\mathscr{A}$ is a σ algebra. Suppose that $\sup \{|\mu(e)| : \mu \in B\} < \infty$ for each $e \in \mathscr{A}$. Then $\sup \{|\mu(e)| : \mu \in B, e \in \mathscr{A}\} < \infty$.

PROOF If the conclusion is false $\alpha(e) = \sup \{|\mu(e)| : \mu \in B\}$ obeys the hypotheses of Lemma 14-4-10 and so $\{e_n\}$ can be chosen as specified there. Choose $\mu_n \in B$ with $|\mu_n(e_n)| \to \infty$. For $T \subset N$ let $e_T = \cup \{e_n : n \in T\}$ and define $v_n \in \mathrm{bfa}(N)$ by $v_n(T) = \mu_n(e_T)$. For each $T \in N$, $\{v_n(T)\} = \{\mu_n(e_T)\}$ is bounded, so by Corollary 14-4-6 $\{v_n(\{n\})\}$ is bounded. But $v_n(\{n\}) = \mu_n(e_n)$.

12. Remark The "collecting technique" used in this proof is the standard method for reducing arguments to the case covered by Phillips' lemma. It occurs also in Theorem 14-6-10 and Probs. 14-6-103 and 14-6-112.

PROBLEMS

In this list $b = \mathrm{bfa}(H, \mathscr{A})$ with $\mathscr{A}$ a σ algebra of subsets of H.

1 Let $B \subset b$ and suppose that there exists M such that $|\mu(e)| < M$ for all $\mu \in B$, $e \in \mathscr{A}$. Show that B is a bounded set in the normed space b $[\![$Lemma 14-4-2$]\!]$.

2 Let $B \subset b$ and suppose that for each $e \in \mathscr{A}$ there exists M_e with $|\mu(e)| < M_e$ for all $\mu \in B$. Show that B is a bounded set in the normed space b $[\![$Theorem 14-4-11 and Prob. 14-4-1$]\!]$.

---

101 Deduce that c_0 is not linearly homeomorphic with a dual Banach space (Prob. 11-1-122) from Example 14-4-9 and Prob. 5-3-6.

102 Let S be an algebraic complement of c_0 in l^∞. Show that S cannot be given a complete norm larger than $\|\cdot\|_\infty$ (or even a lc Fréchet metric larger than that of ω—in short S cannot be made into a lc FK space). $[\![$See Prob. 13-4-109.$]\!]$

103 Let $i: c_0 \to l^\infty$ be inclusion so that $i': Y \to X$ where $Y = (l^\infty)'$, $X = l = c_0'$. Let Y be given its weak $*$ topology; then the quotient topology on X is also the weak $*$ topology $[\![$Prob. 8-3-201$]\!]$. So $X = Y/S$, with these topologies where $S = N i' = c_0^\perp$. Show that Y/S contains a null sequence $\{u_n + S\}$ such that for no selection $v_n \in u_n + S$, $v_n \in Y$, it is true that $v_n \to 0$. Thus the topology of a quotient cannot be defined by net convergence in this way. $[\![$Take $u_n(x) = P_n(x) = x_n$. Then for $x \in c_0$, $v_n(x) = P_n(x)$ so $v_n \nrightarrow 0$, as shown in Example 14-4-9.$]\!]$ Some further discussion is given at the end of sec. 6.5 in $[156]$; see also $[106]$.

104 In Lemma 14-4-1 replace $\geq \frac{1}{6}$ by $> 1/\pi$. This is the best extension. ⟦See [27].⟧

301 c_0 is quasicomplemented in l^∞. ⟦See [119], p. 185.⟧

14-5 THE SPACE L

The next two sections contain criteria for weak conditional sequential compactness, Theorems 14-5-5 and 14-6-7. They are applied to the construction of G spaces in Sec. 14-7.

The material of the next three sections is largely drawn from [6], [49], [130], and [148].

We shall make use of a few measure theoretic definitions and results as given, for example, in [51]. Let H be a set, $\mathscr{A}$ a σ algebra of subsets of H, and μ a finite positive measure such that $(H, \mathscr{A}, \mu)$ is a measure space. For $e, f \in \mathscr{A}$ let $d(e, f) = \mu[(e\backslash f) \cup (f\backslash e)]$. Then d is a pseudometric for $\mathscr{A}$.

1. Lemma If $\mathscr{A}$ is countably generated, $(\mathscr{A}, d)$ is separable.

PROOF This may be found in [51], Theorem 40B.

Next is a version of the Radon–Nikodym theorem.

2. Lemma With $H, \mathscr{A}, \mu$ as above, let v be a measure defined on the sets of $\mathscr{A}$ such that $\mu(e) = 0$ implies $v(e) = 0$. Then there exists $f \in L_1(H, \mathscr{A}, \mu)$ such that $v(e) = \int_e f \, d\mu$ for $e \in \mathscr{A}$. In case $\mu = |v|$, f can be chosen so that $|f(x)| = 1$ for all $x \in H$.

PROOF See [37], Theorem III.10.7; or [51], Theorem 31B. For the last part see [123], Theorem 6.12.

The special case of interest will be that in which H is a compact Hausdorff space, $\mathscr{A}$ is the collection of Borel sets in H, and μ is a finite positive regular measure.

3. Definition $L = L_1(H, \mathscr{A}, \mu)$, the Banach space of μ integrable complex functions on H; $\| f \|_1 = \int_H |f| \, d\mu$.

The dual of L is the space L^∞ of bounded measurable functions on H with $\| \cdot \|_\infty$. (See [37], Theorem IV.8.5.)

4. Lemma Let $\{f_n\} \in L$ satisfy $\int_e f_n \, d\mu \to 0$ for all Borel sets e. Then $f_n \to 0$ weakly.

PROOF Let $\mu_n(e) = \int_e f_n d\mu$ for each Borel set e. By Theorem 14-4-11 and Lemma 14-4-2 there exists M such that $\| \mu_n \| < M$ for all n. ⟦Note that

$|\mu_n(e)| \le \int_H |f_n| d\mu$ so each $\mu_n \in \text{bfa}(H, \mathscr{A})$ and $\|\mu_n\| = |\mu_n|(H)$ as in the preceding section.] Let $g \in L' = L^\infty$; $\varepsilon > 0$. There exists a simple function $h \in L^\infty$ (i.e., a function with finite range) such that $\|g - h\|_\infty < \varepsilon$. [Let $a < -\|g\|_\infty, b > \|g\|_\infty, a = y_0 < y_1 < y_2 < \cdots < y_m = b$ with $|y_i - y_{i-1}| < \varepsilon$ for each i, $e_i = \{x : y_{i-1} < g(x) \le y_i\}$. Choose h to be a suitable constant t_i on each e_i.] Then $\int_H g f_n d\mu = u_n + v_n$ where $|u_n| = |\int (g - h) f_n d\mu| < \varepsilon \int |f_n| d\mu = \varepsilon \|\mu_n\| < \varepsilon M$ and $v_n = \int h f_n d\mu = \sum t_i \mu_n(e_i) \to 0$.

5. Theorem Let $B \subset L$ be bounded and infinite. Suppose that for each $\varepsilon > 0$ there exists $\delta > 0$ such that $\mu(e) < \delta$ implies $\int_e |f| d\mu < \varepsilon$ for all $f \in B$. Then B contains a weakly convergent sequence of distinct points.

PROOF We may assume that B is countably infinite. Let $\{U_n\}$ be a basis for the open sets of the complex plane and $\mathscr{A}_1$ the σ algebra of subsets of H generated by $\{f^{-1}[U_n] : f \in B, n = 1, 2, \ldots\}$. Let $\{e_n\}$ be a dense countable subset of $(\mathscr{A}_1, d)$ (Lemma 14-5-1). For each n, $|\int_{e_n} f d\mu| \le \|f\|_1 < M$ for every $f \in B$ where M is a constant depending only on B. Thus the set $\{\int_{e_n} f d\mu : f \in B\}$ is a bounded infinite set of complex numbers for each fixed n, and so contains a convergent sequence of distinct points. By the diagonal process we can choose a fixed sequence $\{f_i\}$ of members of B such that $\lim_i \int_{e_n} f_i d\mu$ exists for each n. But then $\lim_i \int_e f_i d\mu$ exists for all $e \in \mathscr{A}_1$. [Let $e \in \mathscr{A}_1, \varepsilon > 0, \delta$ as in the statement. Choose e_n with $d(e, e_n) < \delta$. For any i, j, $|\int_e (f_i - f_j) d\mu| \le |\int_e f_i - \int_{e_n} f_i| + |\int_{e_n} (f_i - f_j)| + |\int_{e_n} f_j - \int_e f_j| = u + v + w$, say. Now $|u| \le \int_a |f_i| d\mu$ where $a = (e\backslash e_n) \cup (e_n\backslash e)$. So by hypothesis $|u| < \varepsilon$. Similarly, $|w| < \varepsilon$. Also, $|v| = |\int_{e_n} (f_i - f_j) d\mu| < \varepsilon$ for large i, j.] Set $v(e) = \lim_i \int_e f_i d\mu$ for each $e \in \mathscr{A}_1$. This defines a measure which has the property of Lemma 14-5-2 so there exists $f \in L_1(H, \mathscr{A}_1, \mu)$ with $v(e) \int_e f d\mu$ for $e \in \mathscr{A}_1$. By Lemma 14-5-4, $f_n \to f$ weakly in $L_1(H, \mathscr{A}_1, \mu)$ which is a subspace of L since $\mathscr{A}_1 \subset \mathscr{A}$. Thus $f_n \to f$ weakly in L [Prob. 11-1-3].

PROBLEMS

101 Show that each Borel set e leads naturally to $F_e \in L'$ with $\|F_e\| = \mu(e)/\|\mu\|$. Thus the collection of Borel sets in H is a subset of L'. [$F_e(f) = \int_e f d\mu$.]

102 Prove the converse of Lemma 14-5-4 [Prob. 14-5-101].

103 Take $H = [0, 1]$, $\mu =$ Lebesgue measure, $f_n(x) = \sin n\pi x$. Show that $f_n \to 0$ weakly in L.

104 In Prob. 14-5-103, if $\varepsilon_n \to 0$ it is false that $\int_e f_n d\mu = 0(\varepsilon_n)$ for every e. [With $\mu_n(e) = \varepsilon_n^{-1/2} \int_e f_n$, Theorem 14-4-11 contradicts $\varepsilon_n^{1/2} \|\mu_n\| = \|f_n\|_1 = 2/\pi$.]

14-6 THE SPACE $M(H)$

In this section let H be a compact Hausdorff space and $M(H)$ the Banach space of bounded regular Borel measures on H with $\|\mu\| = |\mu|(H)$. Thus for each $\mu \in M$, $(H, \mathscr{A}, \mu)$ is a measure space with $\mathscr{A}$ the collection of Borel sets in H.

1. Lemma For each positive $\mu \in M(H)$, there is an equivalence mapping of $L = L_1(H, \mathscr{A}, \mu)$ into $M(H)$ given by $f \to v$ where $v(e) = \int_e f \, d\mu$.

PROOF Recall that an equivalence is an isometric linear map. The only nontrivial part of the proof is that $\|v\| = \|f\|_1$. By Lemma 14-5-2 there exists $g \in L_1(H, \mathscr{A}, |v|)$ with $v(e) = \int_e g \, d|v|$ for all $e \in \mathscr{A}$, with $|g(x)| = 1$ for all $x \in H$. Thus $g \, d|v| = f \, d\mu$ and so $d|v| = \bar{g}f \, d\mu$. It follows that $\bar{g}(x)f(x) \geq 0$ for all x and so $\|v\| = |v|(H) = \int_H \bar{g}f \, d\mu = \int_H |\bar{g}f| \, d\mu = \int_H |f| \, d\mu = \|f\|_1$.

2. Lemma For each sequence $\{v_n\} \subset M(H)$, there exists a finite positive regular Borel measure μ on H such that the range of the map in Lemma 14-6-1 includes $\{v_n\}$.

PROOF Let $\mu(e) = \sum [|v_n|(e)]/(2^n \|v_n\|)$. By Lemma 14-5-2, there exists $f_n \in L_1(H, \mathscr{A}, \mu)$ for each n such that $v_n(e) = \int_e f_n \, d\mu$ for each $e \in \mathscr{A}$.

3. Remark Thus each countable set in $M(H)$ is contained in a closed subspace which is (of course) a Banach space and is equivalent with a space $L = L_1(H, \mathscr{A}, \mu)$. Thus sequential properties of $M(H)$ may be deduced from those of L; weak sequential properties are derived by examining $L' = L^\infty$ instead of $M(H)'$ [Prob. 11-1-3].

4. Example Let $\{v_n\} \subset M(H)$ satisfy $v_n(e) \to 0$ for each Borel set e. Then $v_n \to 0$ weakly. As in Lemmas 14-6-1 and 14-6-2, for each $e \in \mathscr{A}$, $\int_e f_n \, d\mu = v_n(e) \to 0$, so, by Lemma 14-5-4, $f_n \to 0$ weakly. By Remark 14-6-3, $v_n \to 0$ weakly. Now see Prob. 14-6-104.

The main compactness result is Theorem 14-6-7. Its hypothesis is motivated by Prob. 14-6-103.

5. Lemma Let $\{\mu_n\} \subset M(H)$ and suppose that for every disjoint sequence $\{U_n\}$ of open sets in H, $\mu_n(U_n) \to 0$. Then for every compact $K \subset H$ and $\varepsilon > 0$, K lies in an open set U such that $|\mu_n|(U \backslash K) < \varepsilon$ for all n.

PROOF This is a "uniform regularity" result. Suppose it is false for some ε, K. Choose v_1 (one of the μ_n) with $|v_1|(H \backslash K) > \varepsilon$. There exists compact K_1 disjoint from K such that $|v_1(K_1)| > \varepsilon/7$. [By Lemma 14-4-2 choose $e \subset H \backslash K$ with $|v_1(E)| > \varepsilon/7$, and, by regularity, compact $K_1 \subset e$ with $|v_1(K_1)| > \varepsilon/7$.] Enclose K_1, K in disjoint open sets U_1, V_1 such that $|v_1(U_1)| > \varepsilon/7$ [by regularity]. Choose v_2 (one of the μ_n) such that $|v_2|(V_1 \backslash K) > \varepsilon$. Ensure that $v_2 \neq v_1$ by making V_1 smaller if necessary. Similarly, $v_i \neq v_j$ in the following. As before, there exists compact $K_2 \subset V_1 \backslash K$ with $|v_2(K_2)| > \varepsilon/7$. Enclose K_2, K in disjoint open sets U_2, V_2 with $U_2 \cup V_2 \subset V_1$, $|v_2(U_2)| > \varepsilon/7$. Thus U_1, U_2 are disjoint. Continuing in this way we obtain $|v_n(U_n)| > \varepsilon/7$.

6. Lemma With the hypotheses of Lemma 14-6-5, for every open $V \subset H$, $\varepsilon > 0$, there exists a compact set $C \subset V$ with $|\mu_n|(V \setminus C) < \varepsilon$ for all n.

PROOF Let $K = H \setminus V$. Form U as in Lemma 14-6-5 and take $C = H \setminus U$.

7. Theorem Let $B \subset M(H)$ be bounded and infinite. Suppose that $\mu_n(U_n) \to 0$ whenever $\{U_n\}$ is a disjoint sequence of open sets and $\{\mu_n\} \subset B$. Then B contains a weakly convergent sequence of distinct points.

PROOF We may assume that $B = \{\mu_n\}$. As in Remark 14-6-3 we may assume that $B \subset L$, that is, $\mu_n(e) = \int_e f_n \, d\mu$, and it must be proved that $\{f_n\}$ has a weakly convergent subsequence in L. For this we shall prove the criterion of Theorem 14-5-5. For convenience, write $v_n = |\mu_n|$. The hypothesis of Theorem 14-5-5 becomes as follows.

For each $\varepsilon > 0$, there exists $\delta > 0$ such that for each Borel set e,

$$\mu(e) < \delta \quad \text{implies} \quad v_n(e) < \varepsilon \quad \text{for all } n \tag{14-6-1}$$

Suppose that (14-6-1) is false for some ε. Then for every $\delta > 0$ there exists a Borel set e with $\mu(e) < \delta$ but $v_n(e) \geq \varepsilon$ for some n. Indeed, there is an open set e with this property. [Let U be open, $U \supset e$ with $\mu(U) < \delta$, by regularity. Then $v_n(U) \geq v_n(e) \geq \varepsilon$.] Take $\delta = 2^{-m-1}$, $m = 1, 2, \ldots$, and obtain open U_m with $\mu(U_m) < 2^{-m-1}$, $v_{n(m)}(U_m) \geq \varepsilon$. Let $V_m = \cup\{U_i : i \geq m\}$. Then $\{V_m\}$ is a decreasing subsequence of open sets with $\mu(V_m) \leq 2^{-m}$, $v_{n(m)}(V_m) \geq \varepsilon$. By Lemma 14-6-6, for each m there exists compact $K_m \subset V_m$ with

$$v_i(V_m \setminus K_m) < \frac{\varepsilon}{2^{m+1}} \qquad \text{for all } i \tag{14-6-2}$$

Let $K = \cap K_m$. Then $\mu(K) \leq \mu(\cap V_m) = 0$ and it follows from the way that μ was defined in Lemma 14-6-2 that $v_i(K) = 0$ for all i. From this and Lemma 14-6-5, K lies in an open set U with $v_i(U) < \varepsilon/2$ for all i.

A contradiction will be reached when we show that $v_i(U) > \varepsilon/2$ for some i. Since all the K_n are compact and their intersection lies in U, some finite intersection $C_r = K_1 \cap K_2 \cap \cdots \cap K_r \subset U$. Take $i = n(r)$. Then $v_i(U) \geq v_i(C_r) = v_i(V_r) - v_i(V_r \setminus C_r) \geq \varepsilon - v_i(V_r \setminus C_r)$. Finally, $V_r \setminus C_r = \cup\{V_r \setminus K_j : j = 1, 2, \ldots, r\} \subset \cup\{V_j \setminus K_j : j = 1, 2, \ldots, r\}$; so, by (14-6-2), $v_i(V_r \setminus C_r) < \varepsilon/2$.

The preceding ideas may be applied to an important dual pair.

8. Definition The symbol (C, M) will stand for the dual pair $C = C(H)$, H a compact Hausdorff space, $M = M(H)$, $[f, \mu] = \int_H f \, d\mu$.

9. Definition A compact Hausdorff space is called *extremally disconnected* if the closure of every open set is open.

Examples will occur in the next section. Those wishing to pursue such spaces further may consult [156], secs. 14.1 and 14.2.

10. Theorem Let H be an extremally disconnected space. Then each $\sigma(M, C)$ convergent sequence $\{\mu_n\}$ in M is weakly convergent [in the Banach space $M(H)$].

PROOF We may assume $\mu_n \to 0$, $\sigma(M, C)$. It is clear [Prob. 14-6-1] that $M \subset [C(H), \|\cdot\|_\infty]'$ and the norm of $C(H)'$ is, on M, given by $\|\mu\| = |\mu|(H)$. We shall see in the next section that $M = C'$, but this is not needed now. It follows that $\{\mu_n\}$ is bounded in $M(H)$ by the uniform boundedness principle [Remark 3-3-12] and we show first that it has a weakly convergent subsequence by the criterion of Theorem 14-6-7. If this is false, there exists $\varepsilon > 0$ and a disjoint sequence $\{U_n\}$ of open sets in H with $|\mu_n(U_n)| > \varepsilon$. [Actually a subsequence of $\{\mu_n\}$ should be used, but it may be called $\{\mu_n\}$ since it is also $\sigma(M, C)$ convergent.] We may assume each U_n is open and closed. [Write $U = U_n$, $\mu = \mu_n$. Let $K \subset U$ be compact with $|\mu|(U \backslash K) < \varepsilon/2$. Let V be an open set with $K \subset V \subset \bar V \subset U$. Then $\bar V$ is open and closed, and $|\mu(\bar V)| = |\mu(U) - \mu(U \backslash \bar V)| > \varepsilon - \varepsilon/2$.] For each $T \subset N$ (the integers) let $e_T = \cup\{U_n : n \in T\}$ and $v_n(T) = \mu_n(e_T)$ so that $v_n \in \text{bfa}(N)$. For each $T \subset N$, $v_n(T) = \mu_n(E_T) = \int_H f \, d\mu_n \to 0$, where f is the characteristic function of e_T. [$f \in C(H)$ since e_T is open and closed.] By Phillips' lemma, Corollary 14-4-5, $v_n(\{n\}) \to 0$, that is, $\mu_n(U_n) \to 0$. This contradiction shows that Theorem 14-6-7 applies and so each $\sigma(M, C)$ convergent sequence has a weakly convergent subsequence. If the conclusion of Theorem 14-6-10 is false, $\{\mu_n\}$ has a subsequence with no subsequence converging weakly to 0. This subsequence cannot have any convergent subsequence since the weak topology is larger than $\sigma(M, C)$, and this contradicts what was just proved.

PROBLEMS

1 Show that each $\mu \in M(H)$ leads to a member $\hat\mu$ of $C(H)'$, namely, $f \to \int_H f \, d\mu$, and $\|\hat\mu\| = \|\mu\|$. (This map is actually onto, see Prob. 2-3-301.)

2 For $h \in H$, let $\delta_h(e) = 1$ if $h \in e$, 0 if $h \notin e$. Show that $\delta_h \in M(H)$ for each h. It is called the *point mass* at h.

101 Let δ be a point mass and $\mu_n = n\delta$. Show that $B = \{\mu_n\}$ obeys the conditions of Theorem 14-6-7 except for boundedness. Deduce that "bounded" cannot be omitted in Theorem 14-6-7.

102 Prove the converse of the result in Example 14-6-4 [like Prob. 14-5-102].

103 Let $\{\mu_n\} \subset M(H)$ and suppose that $\mu_n \to 0$ weakly. Show that for any disjoint sequence $\{e_n\}$ of Borel sets, $\mu_n(e_n) \to 0$. [Form v_n as in the proof of Theorem 14-4-11; now $v_n(\{n\}) \to 0$.]

104 Extend Prob. 14-6-103 to conclude that $|\mu_n|(e_n) \to 0$ [Lemma 14-4-2].

105 Let $H = [0, 1]$, $\lambda = $ Lebesgue measure. Show that λ lies in the convex weak $*$ closure of the set of point masses in the setting of Prob. 14-6-1. [If not, the Hahn–Banach theorem supplies $f \in C(H)$ with $\int f(t) \, dt > \sup f(h)$.]

106 Let X be the linear closure of the point masses in $M(H)$ with $H = [0, 1]$. Show that $\lambda \notin X$ where $\lambda = $ Lebesgue measure. [For $0 \le t_1 \le t_2 \le \cdots \le t_m \le 1$ choose f with $\int_0^1 |f| = 1$, $f(t_i) = 0$.]

107 Give X, Prob. 14-6-106, the weak $*$ topology, $\sigma(X, C)$, Prob. 14-6-1. Show that X does not have the convex compactness property [Probs. 14-6-105 and 14-6-106].

108 With X as in Prob. 14-6-106, show that $\beta(C, X) = \|\cdot\|_\infty \cdot [\![\|\cdot\|_\infty = \beta(C, C') \supset \beta(C, X)$. Conversely, let D be the unit disc in C. Then $(D^\circ \cap X)^\circ \subset D$ for if $f \in (D^\circ \cap X)^\circ$ and $0 \le t \le 1$ then $|f(t)| \le 1$ since $t \in D^\circ \cap X$. So $D \in \mathcal{N}[\beta(C, X)]$.]

109 With X as in Prob. 14-6-106, show that $\tau(X, C) = \tau(C', C)|_X$. [Let B be absolutely convex and $\sigma(C, X)$ compact. Then B is norm bounded by Corollary 9-2-5 and Prob. 14-6-108; hence $\sigma(C, C')$ compact by Prob. 14-1-306.]

110 (D. H. Fremlin). With X as in Prob. 14-6-106, $\tau(X, C)$ has the convex compactness property but $\sigma(X, C)$ does not. Compare Prob. 14-2-301 and Theorem 9-2-11. [The second part is by Prob. 14-6-107. Let (K) be the absolutely convex closure of the compact set K. By Probs. 14-6-109 and 14-2-302, K is $\sigma(C', C'')$ compact. By Theorem 14-2-4 and Prob. 14-2-302, (K) is $\tau(C', C)$ compact; hence $\tau(X, C)$ compact by Prob. 14-6-109.]

111 A compact Hausdorff space is called σ *stonean* (in honor of M. H. Stone) if the set of open and closed sets is a base for the topology and is a σ complete lattice. (Every sequence has a least upper bound.) Show that an extremally disconnected space is σ stonean.

112 Prove the result of Theorem 14-6-10 when H is σ stonean. [In the proof take e_T to be the least upper bound.]

14-7 *GB* AND *G* SPACES

In 1953 A. Grothendieck discovered (Example 14-7-7) that the dual of l^∞ has the property that its weak $*$ convergent sequences are weakly convergent. (See [49].) Recall that, in l, weakly convergent sequences are norm convergent (Prob. 8-1-8) and that in no infinite-dimensional dual Banach space can every weak $*$ convergent sequence be norm convergent (Prob. 9-5-301).

1. Definition A Banach space X is called a *GB* space if it has the property that every weak $*$ convergent sequence in X' is weakly convergent.

Thus Grothendieck's theorem implies that l^∞ is a *GB* space.

It is trivial that *a reflexive space is a GB space*—among separable spaces this is the only kind of *GB* space.

2. Theorem A separable Banach space X is a *GB* space if and only if it is reflexive.

PROOF Alternate proofs and extensions are given in Probs. 14-7-103, 14-7-108, and 14-7-111. $\Rightarrow$: Let D be the unit disc of X' so that $(D, \text{weak} *)$ is a compact metric space [Theorem 9-5-3]. Thus it is sequentially compact and the *GB* assumption implies that (D, weak) is sequentially compact. By the Eberlein–Smulian theorem, Example 14-1-10, D is weakly compact and so X' is reflexive [Example 10-2-6]; hence X is reflexive [Theorem 3-2-9].

The preceding remarks show that *GB* is not a hereditary property. [Consider $c_0 \subset l^\infty$.] It is, however, Q hereditary [Theorem 14-7-3]; thus, using the open mapping theorem, a Banach space must be a *GB* space if it is the image of a *GB* space under a continuous linear map.

3. Theorem A separated quotient of a *GB* space is again a *GB* space.

PROOF Let $u: X \to Y$, $g_n \to 0$ weak $*$ in Y'. Then $u': Y' \to X'$ is weak $*$ continuous [Theorem 11-1-6] so $u'(g_n) \to 0$ weak $*$ in X'; hence weakly, assuming that X is a *GB* space. Now u' is a norm isomorphism [Theorem 11-3-4] so $(u')^{-1}$ is weakly continuous [Example 11-1-4]; hence $g_n \to 0$ weakly. [Note that $(u')^{-1}$ is defined only on the range of u'.]

A *G space* is an infinite compact Hausdorff space H such that $C(H)$ is a *GB* space. The trivial case in which $C(H)$ is reflexive could arise only if H is finite [Prob. 10-2-120]. It is convenient to rule out this case in the definition. It is a nontrivial task to construct an example of a *G* space; such a space must have a certain degree of pathology, as indicated in Theorem 14-7-4; e.g., *a G space cannot be metrizable or even first countable* since such a space, if compact, would have to be sequentially compact.

4. Theorem Sequential convergence in a *G* space is trivial, i.e., a *G* space H cannot contain a convergent infinite sequence.

PROOF A direct proof is spelled out in Prob. 14-7-101. Suppose $x_n \to x$. Define $u: C(H) \to c_0$ by $u(f) = \{f(x_n) - f(x)\}$. This is onto. [Either enclose each x_n in an open set U_n not meeting U_m for $m \neq n$, or use Tietze's extension theorem ([156], Theorem 8-5-3) to extend f from $\{x_n\} \cup \{x\}$.] Thus c_0 is a quotient of $C(H)$ which contradicts the fact that c_0 is not reflexive, using Theorems 14-7-2 and 14-7-3.

To show that l^∞ is a *GB* space (Example 14-7-7) observe first that $l^\infty = C(\beta N)$, where N is the positive integers and βN is its Stone–Cech compactification. This was pointed out in Corollary 9-6-3 and Prob. 9-6-102. Thus we have to prove that βN is a *G* space. This is done in the next two theorems.

5. Theorem βN is extremally disconnected.

PROOF Let U be an open set, $V = U \cap N$. The characteristic function of V is continuous and bounded on N and thus has a continuous extension to $f \in C(\beta N)$. Since N is dense, V is dense in U; hence $f = 1$ on $\bar{U}$. Applying the same argument to $\beta N \backslash \bar{U} = W$, we obtain $f = 0$ on $\bar{W}$. Thus $\bar{U} = \{x: f(x) = 1\} = \{x: f(x) > \frac{1}{2}\}$ since f takes on only the values 0 and 1. Hence $\bar{U}$ is open.

We shall now need to use the Riesz representation theorem, a standard result from analysis. The statement and references are given in Prob. 2-3-301. [Since H is compact, $C_0(H)$ and $C(H)$ have the same meaning.] What it says is that $\|\cdot\|_\infty$ induces on $C(H)$ a topology compatible with the dual pair (C, M), Definition 14-6-8.

6. Theorem Every extremally disconnected space H is a G space, i.e., $C(H)$ is a GB space.

PROOF Let $\{u_n\} \subset C(H)'$, $u_n \to 0$, weak $*$. By Riesz representation we identify u_n with $\mu_n \in M(H)$, $\int_H f \, d\mu_n \to 0$ for each $f \in C(H)$. By Theorem 14-6-10, $\mu_n \to 0$ weakly.

7. Example l^∞ *is a GB space* since, as just noted, $l^\infty = C(\beta N)$.

8. Example c_0 *is not a quotient of* l^∞, that is, there is no continuous linear map from l^∞ onto c_0; it would be a quotient map by the open mapping theorem. If this were false c_0 would be a G space by Theorem 14-7-3 and reflexive by Theorem 14-7-2.

PROBLEMS

In this list H is an infinite compact Hausdorff space.

1 Let X be a Banach space such that every weak $*$ convergent sequence in X' has a weakly convergent subsequence. Show that X is a GB space.

101 Suppose $x_n \to x_0$ in H. Let δ_n be the point mass at x_n, $n = 0, 1, 2, \ldots$. Show that $\delta_n - \delta_0$ tends to 0 weak $*$ but not weakly. [See Prob. 14-6-102; consider $\{x_0\}$.]

102 Show that a closed subspace of a G space is a G space. [If $H \subset K$, $C(H)$ is a quotient of $C(K)$. See Theorem 14-7-3.]

103 Show that a GB space is reflexive if and only if the unit disc in X' is weak $*$ sequentially compact. This extends Theorem 14-7-2. Compare Example 14-1-11 [proof of Theorem 14-7-2].

104 If X is a GB space show that X' is weakly sequentially complete [Prob. 8-1-10 and Example 9-3-8].

105 Show that there exists no range-closed continuous linear map from l^∞ to l or c_0 [Theorem 14-7-3 and Probs. 14-1-103 and 14-1-301].

106 Let X be a nonreflexive GB space, w, w^* the weak and weak $*$ topologies on X'. Show that w, w^* do not have the same compact sets although they have the same convergent sequences. Compare Prob. 14-1-106 [Example 10-2-6].

107 With the notation of Prob. 14-7-106 show that a sequence may have different cluster points in w and w^*. [The unit disc in X' is not weakly countably compact, by Example 14-1-10.]

108 Let X be a GB space. Show that $[X', \sigma(X', X)]^s = X''$ where X^s denotes the set of sequentially continuous linear functions on X. Deduce Theorem 14-7-2 from this and Prob. 12-2-3.

109 If X is a nonreflexive GB space show that $(X', \text{weak} *)$ is not a Mazur space. This shows that the converse of the first part of Corollary 8-6-6 fails even if the space is relatively weak [Prob. 14-7-108].

110 Prove the equivalence of:

(a) X is a GB space.
(b) Every $u \in B(X, c_0)$ is weakly compact.
(c) Every $u \in B(X, Y)$, Y separable, is weakly compact.

[$(a) \Rightarrow (c)$. The unit disc D of Y' is a weak $*$ compact metric space, so $u'[D]$ is. Hence $u'[D]$ is weakly sequentially compact. Apply Eberlein–Smulian and Theorems 11-4-2 and 11-4-4. $(b) \Rightarrow (a)$. Let $f_n \to 0$ weak $*$ in X'. Define $u: X \to c_0$ by $u(x) = \{f_n(x)\}$ so that $f_n = u'[P_n]$. By Theorem 11-4-2 and Eberlein–Smulian $\{f_n\}$ has a weakly convergent subsequence. Apply Prob. 14-7-1.]

111 Deduce Theorem 14-7-2 from Prob. 14-7-110. ⟦Take $u = i$ in (c).⟧

112 Let X be a GB space and Y a closed subspace of finite codimension. Show that Y is a GB space. The next problem improves this. ⟦Say $X = Y + \{x\}$, $u_n \to 0$ weak $*$ on Y. Choose a convergent subsequence of $\{u_n(x)\}$ and apply Prob. 14-7-1.⟧

113 Show that GB is coseparable hereditary, i.e., if X is GB and X/Y separable, then Y is GB. ⟦Let $S \subset X$ be countable with $q[S]$ dense in X/Y. Let $\{u_n\}$ be bounded in X', $u_n \to 0$ on Y. Choose a subsequence which converges on S. Apply Probs. 14-7-1 and 11-3-126.⟧

114 Show that a σ-stonean space is a G space. ⟦See Prob. 14-6-112. In [130] this is extended to zero-dimensional F spaces.⟧

115 Let K be obtained from a G space H by identifying the members of some finite set S. Show that K is a G space. ⟦$C(K) = \{f \in C(H): f(a) = f(b)$ for $a, b \in S\}$. Apply Prob. 14-7-112.⟧

116 Show that the converse of Theorem 14-7-6 (and both parts of Prob. 14-7-114) is false. ⟦Let O, E be the odd and even integers; $a \in \overline{O}\backslash O$, $b \in \overline{E}\backslash E$, $H = \beta N$ with a, b identified. In H: O, E are disjoint open F_σ sets whose closures are not disjoint and H is a G space by Prob. 14-7-115.⟧

117 Accepting the result of Prob. 14-7-303 show that if H is a G space, $C(H)$ has no complemented reflexive (hence no complemented separable) subspace S. This extends Example 14-4-9. ⟦The unit disc D of S is weakly compact in S; hence in H, and so $D = PD$ is totally bounded. See Theorem 6-4-2.⟧

---

201 Is GB coreflexive hereditary? An affirmative answer would extend Prob. 14-7-113 by Theorems 14-7-2 and 14-7-3.

---

301 Let a lcs space X be called TG if weak $*$ convergent sequences in X' are weakly convergent. A separable complete TG space is semireflexive; this extends Theorem 14-7-2. A relatively strong Mazur TG space is sequentially barrelled. If X is a TG space which is separable and metrizable, $X'' = \gamma X$. ⟦See [146], p. 358.⟧

302 Let H be a G space; then $C(H)$ contains no dense proper subspace which can be made into a Banach algebra. ⟦See [148], Theorem 2.5.⟧

303 Let $u \in B(X, Y)$ be weakly compact, where $X = C(H)$ and Y is a Banach space. Then $u(A)$ is totally bounded whenever A is weakly compact. This is related to the Dunford–Pettis property, Prob. 14-2-302. ⟦See [37], p. 494, Theorem 4.⟧

FIFTEEN

BARRELLED SPACES

15-1 BARRELLED SUBSPACES

It is important to recognize when a lc space is barrelled since, for example, the uniform boundedness principle may be applied (Theorem 9-3-4). Significant inclusion theorems also hold for barrelled spaces (Theorem 15-2-1) and various open mapping and closed graph theorems (Theorems 12-4-9 and 12-5-7). Indeed, there is a closed graph theorem which holds only if X is barrelled (Theorem 12-6-3). A second category space is barrelled [Example 9-3-2] and another criterion involving F linked topologies was given in Prob. 9-3-112. A relatively strong space is barrelled if and only if $(X', \text{weak} *)$ is boundedly complete [Theorem 9-3-13] and in many cases it is sufficient that $(X', \text{weak} *)$ be sequentially complete or have the convex compactness property; e.g., Prob. 9-3-302 and Remark 10-4-14.

1. Theorem Let X be a barrelled space and S a dense subspace. Then S is barrelled if and only if every $\sigma(X', S)$ bounded set in X' is $\sigma(X', X)$ bounded.

PROOF $\Leftarrow$: Let $B \subset S'$ be $\sigma(S', S)$ bounded. Since S is dense, $S' = X'$ and so B is $\sigma(X', S)$ bounded. By hypothesis, B is $\sigma(X', X)$ bounded and so, since X is barrelled, B is an equicontinuous set in X' [Theorem 9-3-4]. Thus $B^\circ \in \mathcal{N}(X)$ and so $B^\circ \cap S \in \mathcal{N}(S)$. This makes B an equicontinuous set in S' and so S is barrelled by Theorem 9-3-4. $\Rightarrow$: Let $B \subset X'$ be $\sigma(X', S)$ bounded. Since S is barrelled, B is an equicontinuous set in S', that is, $B^\bullet \in \mathcal{N}(S)$, where $B^\bullet$ is the polar of B in S. It follows that the closure of $B^\bullet$ in X is a neighborhood of 0 in X since S is dense. Thus $B^\circ \in \mathcal{N}(X)$ since $B^\circ \supset \bar{B}^\bullet$ and so B is an equicontinuous set in X; hence it is $\sigma(X', X)$ bounded [Theorems 9-1-8 or 9-1-10].

2. Remark It follows easily that any subspace which includes S (Theorem 15-1-1) is barrelled. A direct proof of this fact is also easy. See Prob. 9-3-124.

3. Example Let m_0 be the span in l^∞ of the set of sequences of zeros and ones as in Example 2-3-15. It is clear that m_0 is a dense vector subspace of l^∞; moreover, m_0 *is barrelled* (with the $\|\cdot\|_\infty$ topology of course). To see this, using Theorem 15-1-1, assume that $B \subset (l^\infty)'$ is $\sigma[(l^\infty)', m_0]$ bounded. Let e be a set of positive integers and x the characteristic function of e so that $x \in m_0$. By Example 2-3-14, we may consider $B \subset \mathrm{bfa}(N)$; for each $\mu \in B$ the corresponding $f \in (l^\infty)'$ has $f(x) = \int_N x \, d\mu = \mu(e)$ and so the hypothesis is that $\{\mu(e): \mu \in B\}$ is bounded. By Prob. 14-4-2, B is bounded in $\mathrm{bfa}(N) = (l^\infty)'$.

4. Example *A closed subspace of a barrelled space which is not barrelled.* Since c_0 is a closed subspace of l^∞, $c_0 \cap m_0$ is a closed subspace of m_0. It is easy to see that this is φ; it was pointed out in Prob. 3-3-4 that φ is not barrelled. A generic example was given in Prob. 13-2-120.

The remainder of this section is devoted to the proof of Theorem 15-1-10. We begin with a special case.

5. Lemma Let X be a barrelled space and S a subspace of finite codimension. Then S is barrelled.

Proof We may assume that S has codimension 1 and that S is not closed. [If S is closed it is complemented by Theorem 6-3-4; hence is barrelled by Example 13-1-14.] Now S is dense [Prob. 4-2-5] so the criterion of Theorem 15-1-1 may be applied. Let $B \subset X'$ be $\sigma(X', S)$ bounded and let $x \in X \backslash S$. Since $X = S + [x]$ it is sufficient to prove that $\{f(x): f \in B\}$ is bounded in order to conclude that B is $\sigma(X', X)$ bounded. If this set is not bounded a contradiction results as follows. Let $f_n \in B$ with $|f_n(x)| \to \infty$. Let $g_n = f_n / |f_n(x)|$ and define g on X by $g(s + \alpha x) = \alpha$ for $s \in S$ and scalar α. For each $s + \alpha x \in X$, $g_n(s + \alpha x) = g_n(s) + \alpha \to \alpha = g(s + \alpha x)$. By the closure Theorem 9-3-7, $g \in X'$. But $S = g^\perp$ is not closed.

6. Example Let X be a Banach space, $f \in X^\#\backslash X'$ and $S = f^\perp$. This subspace of X is barrelled by Lemma 15-1-5 and is a noncomplete normed space since it is dense. It is possible that S is of second category in X [Prob. 3-1-4], and it is a long-standing open problem whether it must be. See [124] for what is known. There are first-category barrelled normed spaces [Probs. 15-1-101, 9-3-116, and 5-2-1].

7. Example Let X, Y be Banach spaces and $u \in B(X, Y)$. Then u is called a *Fredholm operator* if it has finite-dimensional null space and finite co-dimensional range. *Every Fredholm operator is range closed.* ⟦$u[X]$ is barrelled by Lemma 15-1-5. By Theorem 12-4-9, u is open onto its range and, by Corollary 6-2-14, $u[X]$ is complete.⟧ These operators are studied in [128], Chap. 5.

8. Lemma Let X be a barrelled space and S a closed subspace of countable codimension. Then S is barrelled.

PROOF By Theorem 13-3-19 and Example 13-1-14.

9. Lemma Let X be a lcs space such that $(X', \text{weak } *)$ is sequentially complete. Let F be a subspace of countably infinite codimension such that F includes a barrel B which is closed in X. Then F is closed.

PROOF Let $h \in X \backslash F$. Let G be an algebraic complement of F with Hamel basis $\{h_n\}$, $h_1 = h$. For each $\varepsilon > 0$ we prove that there exists $g \in X'$ such that

$$|g(h_1)| > 1, \qquad g(h_i) < \varepsilon \text{ for } i > 1, \qquad |g| < \varepsilon \text{ on } B \qquad (15\text{-}1\text{-}1)$$

We may assume $\varepsilon < 1$. By the Hahn–Banach Theorem 7-3-4 there exists $f_1 \in X'$ with $f_1(h_1) = 2$, $|f_1| < \varepsilon/2$ on B. ⟦$h_1 \notin (4/\varepsilon)B$.⟧ Let $B_2 = B + \{\alpha h_1 : |\alpha| \leq 1\}$, an absolutely convex closed set ⟦Lemma 6-5-11⟧, and there exists $f_2 \in X'$ with $f_2(h_2) = -f_1(h_2)$, $|f_2| < \varepsilon/4$ on B_2. ⟦No multiple of B_2 contains h_2.⟧ Continuing in this way, $B_{n+1} = B_n + \{\alpha h_n : |\alpha| \leq 1\}$, $B_1 = B$, $|f_n| < \varepsilon/2^n$ on B_n, $f_n(h_n) = -\sum_{i=1}^{n-1} f_i(h_n)$. For $x \in X$ set $g(x) = \sum f_n(x)$. ⟦Every $x = \alpha b + \sum_{i=1}^{m} \alpha_i h_i$ with $b \in B$; $|f_n(b)| < \varepsilon/2^n$ and, for $n > i$, $|f_n(h_i)| < \varepsilon/2^n$ since $h_i \in B_n$. Thus the series converges.⟧ The completeness hypothesis implies that $g \in X'$. Also $|g(h_1)| = |\sum f_n(h_1)| \geq 2 - \sum_{n=2}^{\infty} |f_n(h_1)| > 2 - \sum \varepsilon/2^n \geq 1$; for $i > 1$, $|g(h_i)| = |\sum_{n=i+1}^{\infty} f_n(h_i)| < \sum \varepsilon/2^n < \varepsilon$; and for $x \in B$, $|g(x)| < \sum \varepsilon/2^n = \varepsilon$ since $B \subset B_n$ for all n. This proves (15-1-1).

Now in (15-1-1) set $\varepsilon = 1/n$, $n = 1, 2, \ldots$, and write g_n for the corresponding g. Let $t_n = 1/g_n(h_1)$. Then $|t_n| < 1$ and $t_n g_n \to f$ pointwise where $f \in X^\#$, $f = 0$ on B, $f(h_i) = 0$ for $i > 1$, $f(h_1) = 1$. In particular, $f = 0$ on F, $f(h) = 1$. By the completeness hypothesis $f \in X'$ and so $h \notin \bar{F}$.

10. Theorem : S. A. Saxon–M. Levin ; M. Valdivia Let X be barrelled and S a subspace of countable codimension. Then S is barrelled.

PROOF Let B be a barrel in S, $\bar{B}$ the closure of B in X, and F the span of $\bar{B}$. Then F is barrelled. ⟦If F has finite codimension this is by Lemma 15-1-5. Otherwise it follows from Lemmas 15-1-9 and 15-1-8.⟧ Thus $\bar{B} \in \mathcal{N}(F)$ and so $B = \bar{B} \cap S \in \mathcal{N}(S)$.

The proof given is that of Saxon and Levin. An account of Valdivia's proof may be found in [144], p. 54.

PROBLEMS

In this list $X = (X, T)$ is a lcs space.

1 Let X be barrelled and S a dense subspace. Show that S is barrelled if and only if every $\sigma(X', S)$ bounded set in X' is equicontinuous.

2 Let X be fully complete, Y barrelled, and $u: X \to Y$ continuous, linear, and such that $u[X]$ has countable codimension in Y. Show that u is range closed. [Use Theorems 15-1-10 and 12-4-5, as in Example 15-1-7.]

---

101 Show that m_0 is a first-category subspace of l^∞.

102 If X is a lc Fréchet space show that F, in the proof of Lemma 15-1-9, has finite codimension [Theorem 1-6-1].

103 The unit disc in l^∞ is closed, indeed compact [Prob. 6-5-1], in ω. Thus countable codimension cannot be omitted in Lemma 15-1-9. Show also that l^∞ is not a barrelled subspace of ω [Theorem 12-4-9]. Compare Prob. 6-1-101.

104 Let X, Y be Banach spaces. Show that $u \in B(X, Y)$ is a Fredholm operator if and only if it is range closed and both u and u' have finite-dimensional null-space [Lemma 11-1-7].

105 Call a lcs space an FB space if all of its closed subspaces are barrelled. Show that any lc Fréchet space and any vector space with its largest lc topology are FB spaces. In the latter case all subspaces are barrelled [Probs. 13-2-110 and 9-3-106]. Examples of barrelled non-FB spaces are m_0 and certain products of Banach spaces [Prob. 13-2-120].

106 Let S be a subspace of a Banach space of countable codimension. Show that S has no larger complete norm [Theorems 15-1-10 and 12-4-9]. It may have a smaller complete norm—indeed, must if of finite codimension. [Imitate Prob. 3-1-105.]

107 Let $f \in X^\# \backslash X'$, $T_1 = T \vee \sigma f$, $S = (f^\perp, T)$. Show that $(X, T_1) = S \oplus \mathcal{K}$. (See Prob. 13-2-1.)

108 In Prob. 15-1-107, (X, T_1) is barrelled if (X, T) is [Example 15-1-6 and Theorem 13-1-13].

109 Let X be an infinite-dimensional Banach space. Define two strictly larger norms n_1, n_2 such that (X, n_1) is barrelled and (X, n_2) is not. [$n_1(x) = \|x\| + |f(x)|$ as in the preceding two problems; let H be a Hamel basis and for $x = \sum \alpha h$ set $n_2(x) = \sum |\alpha| \|h\|$. Use Prob. 9-4-111.]

110 In Prob. 15-1-109 show that (X, n_1) has codimension 1 in its completion. This gives another proof that it is barrelled.

111 What result is obtained when Theorem 15-1-1 is applied to Nikodym's Theorem 14-4-11? [See Prob. 14-5-101.]

112 Let H be σ Stonean (e.g., extremally disconnected). Let m_0 be the set of simple functions in $C(H)$. Show that m_0 is barrelled. [See Theorem 15-1-1 and the proof of Theorem 14-4-11, taking account of Prob. 14-6-112. As indicated in Prob. 14-7-114, this holds also for zero-dimensional F spaces.]

113 A sequence a in a TVS is called *subseries summable* (*bounded multiplier summable*) if $\sum x_n a_n$ converges for all $x \in m_0$ (for all $x \in l^\infty$). Show that for sequences in a lcs space Y the first condition implies the second. [Define $u_n: l^\infty \to Y$ by $u_n(x) = \sum_{k=1}^n x_n a_n$. Apply Example 15-1-3 and Prob. 9-3-138 and 9-3-104.]

114 (N. J. Kalton; N. Batt, P. Dierolf, and J. Voigt). Accepting the result of Prob. 15-1-303, show that m_0 is not ultrabarrelled. [Apply Prob. 9-3-139 as in the preceding problem.]

115 Let $S \supset c_0$ be a dense nonbarrelled subspace of l^∞ [Probs. 14-4-301 and 13-4-103]. Show that every $\sigma(l, S)$ bounded set is $\sigma(l, l^\infty)$ bounded. Deduce that a result dual to that of Theorem 15-1-1 fails [Theorem 9-3-4].

116 Show that Theorem 15-1-1 fails if applied to an admissible topology for X [Prob. 15-1-115].

---

201 (J. K. Hampson). Let X, Y be FK spaces with X a dense subspace of Y and Y locally convex. If $X' = Y'$, X, with the relative topology of Y, is barrelled, but the converse is false. This is applied to Hardy classes in [9], Theorem 11.

202 Must a second-category lcs space be an FB space (Prob. 15-1-105)?

---

301 Let X be either ω barrelled or such that $(X', \text{weak} *)$ is sequentially complete. Then each subspace of countable codimension has the same property. For "boundedly complete" this follows from Theorem 15-1-10. Compare also Prob. 11-1-115. [See [96], Theorem 4.]

302 A subspace of finite codimension of a bornological space X is bornological. A subspace of countable codimension need not even be quasibarrelled; it need not be bornological even if X is also barrelled. [See [31].]

303 There exists a Fréchet space, necessarily not lc and not locally bounded, which contains a sequence which is subseries but not bounded multiplier summable. [See [118].]

304 Let a be subseries summable in the weak topology of a lcs space X. Then $\sum a_n$ converges in T_k, where k is the collection of weak $*$ compact sets in X'; hence also in $\tau(X, X')$. [See [11].] The latter result is the Orlicz–Pettis theorem; see [74].

15-2 INCLUSION THEOREMS

There are applications of barrelled spaces to inclusion theorems covering various types of spaces arising in analysis, such as the Hardy classes. We refer to [12], p. 514. The basic result is that a map whose range is barrelled is already onto—a fact which follows immediately from earlier results. The main contribution of this section is that the corresponding converse theorem holds, i.e., that barrelled is the "right" assumption. This conception, and, in particular, Theorem 15-2-7, is due to G. Bennett and N. J. Kalton.

Recall that if the range of a map is of second category the map must be onto: the exact statement is Theorem 5-2-5. The solution of the converse problem for this theorem is obtained by replacing "second category" by the weaker property "dense and barrelled." Recall also that an FY space is a Fréchet space which is a vector subspace of Y with a larger topology and a BY space is a normed FY space.

Theorem 15-2-1 is a special case of Theorem 15-2-7 (see Remark 15-2-8), but is given because it has an easy proof.

1. Theorem Let Y be a Banach space and S a dense subspace. The following are equivalent:
(a) S is barrelled.
(b) For every Banach space X and $u \in B(X, Y)$, if $u[X] \supset S$, then $u[X] = Y$.
(c) Same as (b) with X a lc Fréchet space.
(d) If X is a BY space which includes S, then $X = Y$.
(e) Same as (d) with X a lc FY space.

PROOF $(a) \Rightarrow (c)$: Since $u[X]$ is barrelled [Remark 15-1-2], $u: X \to u[X]$ is an open map [Theorem 12-4-9] and so $u[X]$ is complete [Corollary 6-2-15]. $(c) \Rightarrow (e)$: Apply (c) to the inclusion map from X to Y. $(e) \Rightarrow (d)$: Trivial. $(d) \Rightarrow (b)$: $u[X]$ with quotient topology is a BY space [Corollary 6-2-15]; hence is Y. $(b) \Rightarrow (a)$: Let B_1 be a barrel in S, $B = B_1 \cap D$, D the unit disc in Y. Let X be the span of $\overline{B}$ and p the gauge of $\overline{B}$. Then (X, p) is a BY space [Prob. 6-1-3]. By hypothesis, $X = Y$ and so p is equivalent with the norm of Y [Corollary 5-2-7]. Thus $\overline{B} \in \mathcal{N}(X)$ and so $B_1 \supset B = \overline{B} \cap S \in \mathcal{N}(S)$; S is barrelled.

2. Example: G. L. Seever *If a lc FK space X includes m_0 it must include l^∞.* Since m_0 is barrelled [Example 15-1-3], this follows from Theorem 15-2-1 as modified in Prob. 15-2-1.

3. Remark *The condition $\|A\| < \infty$ is redundant in Schur's Theorem 1-3-2.* To see this, let $X = \{x : Ax \in c_0\}$. Then X is a lc FK space [Example 5-5-15; X is a closed maximal subspace of c_A] and $X \supset m_0$; hence by Example 15-2-2, $X \supset l^\infty$. By Example 3-3-8, $\|A\| < \infty$.

4. Example *Suppose that $\{\delta^n\}$ is a bounded set in a BK space X; then $X \supset l$.* This was given in Prob. 9-1-110; a direct proof is easy. Let $t \in l$. Then $\left\| \sum_{k=m}^n t_k \delta^k \right\| \leq M \sum_{k=m}^n |t_n|$ and so the series $\sum t_k \delta^k$ converges in X [Theorem 5-1-2]. Since X is a BK space it must converge to t, so $t \in X$.

5. Example: J. D. Weston *$l^{1/2}$ is a barrelled subspace of l.* (This was also given in Probs. 9-3-115 and 9-3-116.) If X is a BK space including $l^{1/2}$, then $X \supset l$ by Example 15-2-4. [$\{\delta^n\}$ is bounded in $l^{1/2}$; hence in X by Corollary 5-5-8.] The result follows from Prob. 15-2-1.

6. Remark In the next proof we shall need the fact that the completion of a lc metric space X is a lc Fréchet space. This was given in Prob. 6-1-105 and is also immediate from Corollary 10-5-5. [$X \subset X''$ since X is quasi-barrelled.]

7. Theorem: G. Bennett and N. J. Kalton Let (Y, T) be a lc Fréchet space and S a dense subspace. The following are equivalent:
(a) S is barrelled.
(b) For every lc Fréchet space X and $u \in B(X, Y)$, if $u[X] \supset S$, then $u(X) = Y$.
(c) If X is a lc FY space which includes S, then $X = Y$.

PROOF $(a) \Rightarrow (b)$: This is proved in the same way as the corresponding part of Theorem 15-2-1. $(b) \Rightarrow (c)$: Apply (b) to the inclusion map from X to Y. $(c) \Rightarrow (b)$: $u[X]$ with the quotient topology is a lc FY space [Corollary 6-2-15]; hence is Y. $(b) \Rightarrow (a)$: We use the criterion of Theorem 12-6-3. Let Z be a Banach space and $v : S \to Z$ a linear map with closed graph. Let S have the topology $T_1 = T \vee \sigma v$, a lc metrizable topology, and let W be the completion of (S, T_1), a lc Fréchet space [Remark 15-2-6]. Let $u : W \to Y$ be the continuous linear map which is the extension of the inclusion map from (S, T_1) to (Y, T) [Prob. 2-1-11]. By hypothesis $u[W] = Y$ and we next show that u is one to one. [If $u(w) = 0$, let s be a sequence in S with

$s \to w \in W$. Then $u(s) \to u(w) = 0$ in (Y, T). Also $v(s)$ is Cauchy in Z by Theorems 1-6-8 and 1-6-10, and so $v(s) \to z \in Z$. Since v has closed graph, $z = 0$. Since $s \to 0$ in (S, T) and $v(s) \to 0$ it follows that $s \to 0$ in (S, T_1), that is, $w = 0$.⟧ By the open mapping Theorem 5-2-4, u is a linear homeomorphism. Considering, in particular, $u \mid S$ we have $T = T_1$ on S. Since $v: (S, T_1) \to Z$ is continuous, the criterion of Theorem 12-6-3 is fulfilled and S is barrelled.

8. Remark Theorem 15-2-1 is a special case of Theorem 15-2-7 and the fact that W, in the latter proof, is a Banach space if Y is.

PROBLEMS

1 Let Y be a BK space and S a dense subspace. Show that S is barrelled if and only if whenever X is a BK space and $X \supset S$, then $X \supset Y$. ⟦$X \cap Y$ is BK by Theorem 5-5-9; hence an FY space by Corollary 5-5-8. Use Theorem 15-2-1.⟧

2 Prove the result of Prob. 15-2-1 with X, Y lc FK spaces. ⟦Now use Theorem 15-2-7.⟧

3 If X, Y are lc FK spaces with X a dense proper subspace of Y, show that X is a nonbarrelled subspace of Y ⟦Theorem 15-2-7, $(a) \Rightarrow (c)$⟧.

4 Let X, Y be lc Fréchet spaces, $u \in B(X, Y)$, and $u[X]$ a dense proper subspace of Y. Show that $u[X]$ is not barrelled. ⟦Take $S = u[X]$ in Theorem 15-2-7.⟧

---

101 Add to Theorem 15-2-7: the same as (b) with X any fully complete space. ⟦Now use Theorems 12-4-9 and 12-4-5.⟧

102 Show that "lc" cannot be omitted in Theorem 15-2-7 (b) ⟦Example 15-2-5⟧.

103 Let Y be a Fréchet space and S a dense subspace which is ultrabarrelled. Let X be a Fréchet space and $u \in B(X, Y)$ with $u[X] \supset S$. Show that $u[X] = Y$ ⟦Prob. 12-4-107⟧.

104 Let S be a subset of ω. Show that there is a BK space X which includes S if and only if S has a *growth sequence*, i.e., a sequence $\{\lambda_n\}$ such that $s_n = 0(\lambda_n)$ for all $s \in S$. ⟦$\Leftarrow$: $X = \{x: x_n = 0(\lambda_n)\}$. $\Rightarrow$: $|x_n| = P_n(x)| \le \|P_n\| \|x\|$.⟧ Incidentally, this argument also proves that ω has no larger norm.

105 Let $S = \{s \in \omega: s_{2n} + s_{2n+1} \to 0\}$. Show that S is a dense nonbarrelled subspace of ω, a lc FK space, and is not included in any BK space. Thus in Theorem 15-2-7 (b) it is not sufficient to restrict X to be a Banach space ⟦Probs. 15-2-3 and 15-2-104 and Example 5-5-12⟧.

15-3 THE SEPARABLE QUOTIENT PROBLEM

A famous unsolved problem is to decide whether every Banach space has a separable (infinite-dimensional) quotient, i.e., admits a continuous linear map onto a separable (infinite-dimensional) Banach space. (Of course, a map onto a finite-dimensional space always exists.) An affirmative answer to this question would simplify the study of Banach spaces. Compare, for example, the simple solution of Prob. 15-3-111 with that referred to in Prob. 9-5-301.

Some efforts have been expended in searching for barrelled spaces, e.g., in the preceding section. By a twist of fate, in the shape of Theorem 15-3-1, it is now the existence of a nonbarrelled space which is sought.

We shall use again the notion of an FY and a BY space as described in the preceding section and that of a S_σ subspace as given in Definition 13-3-14. Parts of the next result are given in [69] and [126].

1. Theorem The following are equivalent for a Banach space (Y, T):
(a) Y has a separable quotient Z.
(b) Y has a dense nonbarrelled subspace D.
(c) Y has a dense S_σ subspace.
(d) There is a lc FY space which is a dense proper subspace of Y (equivalently, a map u from a lc Fréchet space X onto a dense proper subspace of Y).
(e) Same as (d) with BY and Banach spaces.

PROOF $(a) \Rightarrow (c)$: First, Z has a dense S_σ [Prob. 13-3-10], say $S = \cup S_n$. Then, with v the quotient map, $v^{-1}[S] = \cup v^{-1}[S_n]$ is an S_σ subspace and is dense in Y since v is an open map. $(c) \Rightarrow (b)$: By Corollary 13-3-16. $(b) \Rightarrow (e)$: By Theorem 15-2-1 $[(b) \Rightarrow (a)]$. $(e) \Rightarrow (d)$: Trivial. $(d) \Rightarrow (b)$: By Prob. 15-2-4. The proof is concluded with $(b) \Rightarrow (a)$: Let B be a barrel in D, $B \notin \mathcal{N}(D)$. We may assume that B is closed in Y. [$\bar{B}$ is a barrel in its span D_1, and $B = \bar{B} \cap D$ so $\bar{B} \notin \mathcal{N}(D_1)$. Further, D_1 is a proper subspace of Y since, as just noted, it is not barrelled.] Let $y_1 \in Y \backslash D$, $\| y_1 \| = 1$. Let $f_1 \in Y'$ with $f_1(y_1) = 1$, $|f_1| \leq \frac{1}{2}$ on B [Theorem 7-3-5.] Now assume that $y_1, y_2, \ldots, y_n$ and $f_1, f_2, \ldots, f_n$ have been chosen such that with $B_1 = B$, $B_{k+1} = B + \{\sum_{i=1}^k \alpha_i y_i : |\alpha_i| \leq 1\}$ we have $y_k \notin \operatorname{span} B_k$, $\| y_k \| = 1$, $f_k(y_k) = 1$, $f_i(y_k) = 0$ for $i < k$, and $|f_k| \leq 2^{-k}$ on B_k for $k = 1, 2, \ldots, n$.

The span A of B_{n+1} does not include $\cap \{f_i^\perp : i = 1, 2, \ldots, n\}$. [Suppose it does. Then A is barrelled by Lemma 15-1-5 and so $B_{n+1} \in \mathcal{N}(A)$, taking account of the fact that B_{n+1} is closed by Lemma 6-5-11. Thus $B = B_{n+1} \cap D \in \mathcal{N}(D)$.] Moreover, B_{n+1} is not absorbing in X by a similar argument, so there exist y_{n+1} not in the span of B_{n+1} with $\| y_{n+1} \| = 1$, $f_i(y_{n+1}) = 0$ for $i = 1, 2, \ldots, n$ and f_{n+1} with $f_{n+1}(y_{n+1}) = 1$, $|f_{n+1}| \leq 2^{-n-1}$ on B_{n+1}. Now three sequences $\{y_n\}, \{f_n\}, \{B_n\}$ have been defined by induction. Let $S_n = \cap \{f_k^\perp : k \geq n + 1\}$, $S = \cup S_n$, an S_σ subspace of Y. *We shall prove that S is dense in Y.* It is sufficient to prove that for each $b \in B$ and each n, $d(b, S_n) < 2^{-n+1}$. [This implies that $d(b, S) = 0$; thus S is dense in B. Since S is a vector subspace and D, the span of B, is dense, it follows that S is dense in Y.] Let $\alpha_1 = -f_{n+1}(b)$ and, for $k \geq 2$, $\alpha_k = -f_{n+k}$ $(b + \sum_{i=1}^{k-1} \alpha_i y_{n+i})$. Then $|\alpha_k| < 2^{-n-k+1}$. [For $k = 1$ this is clear. By induction, since each $y_{n+i} \in B_{n+k}$, $|\alpha_k| < 2^{-n-k} + \sum 2^{-n-i+1} 2^{-n-k} < 2.2^{-n-k}$.] Hence $y = \sum_{k=1}^\infty \alpha_k y_{n+k} \in Y$ since $\| y_k \| = 1$ for all k, and $\| y \| \leq \sum |\alpha_k| < 2^{-n+1}$. But also $b + y \in S_n$. [For $k \geq 1$, $f_{n+k}(b + y) = f_{n+k}(b + \sum_{i=1}^{k-1} \alpha_i y_{n+i}) + \alpha_k f_{n+k}(y_{n+k}) + \sum_{i=k+1}^\infty \alpha_i f_{n+k}(y_{n+i}) = -\alpha_k + \alpha_k + 0.$] The last two statements yield that $d(b, S_n) \leq 2^{-n+1}$ and complete the proof that S is dense in Y. Notice also that the dimension of S_n/S_{n-1} is 1 for each n; indeed, $S_n = S_{n-1} \oplus y_n$. Let $Z = Y/S_1$ and all is concluded by showing that Z is separable; precisely $\{v(y_n)\}$ is fundamental, where $v : Y \to Z$ is the quotient map. To this end, let $f \in Z'$ with $f(vy_n) = 0$ for all n. Then $v'f = 0$ on S. [For $s \in S$, say

$s \in S_n$, $s = s_1 + \sum_{i=2}^{n} t_i y_i$ with $s_1 \in S_1$. Then $(v'f)(s) = f(vs) = f(vs_1) + \sum t_i f(vy_i) = 0$ since $vs_1 = 0$.] Since S is dense, $v'f = 0$. Hence $f = 0$. [v' is one to one by Corollary 11-1-8.] The result follows by the Hahn–Banach theorem, Corollary 7-2-13.

2. Example *Every reflexive space* Y *has a separable quotient.* Let X be a separable closed subspace of Y' and $i: X \to Y'$ be the inclusion map. Then $i': Y \to X'$ is a quotient map [Theorem 11-3-4]. Now X is reflexive [Theorem 3-2-11] and so X' is separable [Prob. 9-5-4].

We introduce now an interesting class of Banach space which simultaneously generalizes separability and reflexivity and is a strong enough hypothesis for important theorems about these two types of space. On several earlier occasions we gave a separate proof for each of the two properties. See, for example, Prob. 15-3-110. A Banach space is said to be WCG (weakly compactly generated) if it has in it a weakly compact fundamental set.

3. Remark If X is WCG it has such an absolutely convex set K [Theorem 14-2-4], and in the sequel K will have this meaning.

4. Example *Every reflexive Banach space is WCG* since the unit disc will do for K, by Example 10-2-6.

5. Example *Every separable Banach space is WCG.* Let $\{x_n\}$ be a dense subset the unit disc D and K the absolutely convex closure of $\{x_n/n\}$. Then K is compact by Example 9-2-10.

Further examples and discussion are given in the problems. See also the middle third of [33].

6. Lemma If a Banach space Y is WCG, it has a dense proper subspace which is a BY space.

PROOF $\Rightarrow$: Let X be the span of K (Remark 15-3-3). Then X is a BY space [Theorem 6-1-17 and Prob. 6-1-3] and is dense. This concludes the proof if $X \neq Y$. If $X = Y$, Y is reflexive [by Example 10-2-6, since K is the unit disc of Y] and the result follows by Example 15-3-2 and Theorem 15-3-1.

7. Corollary Every WCG Banach space has a separable quotient.

PROOF By Lemma 15-3-6 and Theorem 15-3-1.

8. Remark Corollary 15-3-7 and Probs. 15-3-104 and 15-3-303 show that Y has a separable quotient (is WCG) if and only if there is a dense proper (reflexive) BY subspace.

PROBLEMS

In this list X, Y, Z are infinite-dimensional Banach spaces.

1 Show that every separable quotient of a GB space (such as l^∞) is reflexive [Theorems 14-7-3 and 14-7-2].

2 If Y has a quasicomplemented subspace, show that it has a separable quotient. The converse is Prob. 15-3-201 [Example 13-4-5 and Theorem 15-3-1].

3 Given any separable Z, show that there exists Y such that Z is not a quotient of Y. [Use Probs. 15-3-1 or 14-1-104 according as Z is reflexive or not.]

4 Show that l^∞ has a separable quotient. Compare Prob. 15-3-102 [Probs. 14-4-301 and 13-4-103 and Theorem 15-3-1]. This also follows from Probs. 15-3-107 or 15-3-302.

---

101 Consider the sequence $\{X^n\}$ where $X^\circ = X$, $X^{n+1} = (X^n)'$. Show that if this sequence contains a space with no separable quotient, there must be one among the first three members. Thus there is no use searching among higher dual spaces for a space without a separable quotient [Prob. 3-2-5].

102 Show that l^∞ is not WCG. More generally this is true of any nonseparable dual of a separable space. [See Prob. 9-5-113. It also follows from Prob. 15-3-105.]

103 Give an example of a WCG space which is neither separable nor reflexive.

104 Suppose that for some reflexive space X there is a continuous linear map from X onto a dense subspace of Y. Show that Y is WCG. See Prob. 15-3-303.

105 Accepting the result of Prob. 15-3-303, show that if X is WCG, the unit disc in X' is weak $*$ sequentially compact [Prob. 14-1-110].

106 If X is WCG, show that $(X'$ weak $*)$ is a Mazur space [Example 14-1-11 and [33], p. 148, Cor. 3.]

107 Accepting the result of Prob. 15-3-303 show that the dual of a WCG space has a separable quotient [Example 15-3-2].

108 Show the equivalence of
   (a) Every Banach space has a separable quotient.
   (b) Every BK space has a separable quotient.
   (c) For every Banach space Y and closed subspace S, Y has a dense proper BY subspace X which includes S.
   (d) Same as (c) with $Y = l(I)$, the summable functions on a set I.
[$(b) \Rightarrow (a)$: Consider the map $y \to \{f_n(y)\}$ for any $\{f_n\} \subset Y'$. $(a) \Rightarrow (c)$: Let Z be a dense proper $B(Y/S)$ subspace and consider $q^{-1}[Z]$ $(d) \Rightarrow (a)$: Let $q: l(I) \to Y$ by Prob. 6-2-202. Choose X for $S = Nq$. Then qX is dense and proper.]

109 If there is a Banach space without a separable quotient, there is one of cardinality $\mathfrak{c}$ [Prob. 15-3-108].

110 Unify the closed graph theorems of Kalton and McIntosh (Theorem 12-5-13 and Prob. 14-1-105) by showing that it is sufficient to assume about Y that the unit disc in Y' is weak $*$ sequentially compact. Compare Prob. 15-3-105 [the proof of Prob. 14-1-105].

111 Let a *normal sequence for* Y be a sequence $\{f_n\} \subset Y'$ such that $f_n \to 0$ weak $*$, $\|f_n\| \not\to 0$. Problems 9-5-301, 9-5-104, and 14-1-102 show the existence of a normal sequence in general, and easy proofs for separable and reflexive spaces. Show an easy proof that any Banach space with a separable quotient has a normal sequence. [See Prob. 9-5-104; consider q'.]

112 Show an easy proof that the existence of a dense S_σ in Y yields a normal sequence. [Let $\|f_n\| = 1$, $f_n = 0$ on S_n.]

113 Show an easy proof that the existence of a dense proper BY subspace X yields a normal sequence. [$i': Y' \to X'$ is one to one but not an isomorphism, by Theorem 11-3-4. By Lemma 11-3-6, there exists f_n with $\|f_n\| = 1$, $i'f_n \to 0$.]

114 Show that Y has a separable quotient if and only if it has a normal sequence $\{f_n\}$ such that

$S = \cup \{\cap \{f_k^{\perp} : k > n\} : n = 1, 2, \ldots\}$ is dense. [S is an S_σ; the converse is by the proof of Theorem 15-3-1.]

115 Show that if $\{f_n\} \subset Y'$ is bounded, then $f_n(y) \to 0$ for $y \in \bar{S}$ (Prob. 15-3-114); hence, if S is dense, $f_n \to 0$ weak $*$. See Prob. 15-3-202.

116 Let $\{f_n\}$ be a normal sequence and suppose that there exists an unbounded sequence t of scalars such that $A = \{y : t_n f_n(y) \to 0\}$ is dense. Show that Y has a separable quotient. See Prob. 15-3-203. [A is not barrelled by Theorem 9-3-4.]

117 Let $u : l \to Z$ have u'' one to one. Show that u is range closed. [If not, apply Lemma 11-3-6 to find $\{y_n\}$ with $\|y_n\| = 1$, $u y_n \to 0$. For $f \in u'[Z']$, $f(y_n) \to 0$ so $y_n \to 0$ weakly, by Corollary 11-1-8.]

118 Suppose that $u : Y \to Z$ is one to one and has dense proper range. [So far $Z = \gamma(Y, n)$ would do, where n is the smaller norm of Prob. 5-2-302.] If, also, u'' is one to one, show that Y' has a separable quotient [Theorem 15-3-1 and Corollary 11-1-8]. This technique fails in the most interesting case by Prob. 15-3-117.

119 Show that if $u : Z' \to l^\infty$ has range a dense proper subspace, u cannot be $\sigma(Z', Z) - \sigma(l^\infty, l)$ continuous [Prob. 15-3-117, Theorem 11-1-6, and Corollary 11-1-8].

120 Let $A : l^\infty \to l^\infty$ be a matrix. Show that its range cannot be a dense proper subspace [Probs. 15-3-119 and 11-1-102].

121 Show that X, Lemma 15-3-6, is a dual space. Compare Prob. 15-3-303. [K° is $\tau(Y', X)$ bounded by Prob. 8-4-2, so $\tau(Y', X)$ is normed by Theorem 4-5-2. Its dual is X.]

---

201 Show that every separable space has a quasicomplemented subspace. It follows that Y has a separable quotient if and only if it has such a subspace [Prob. 15-3-2 and [103]].

202 Let $Y = \{y \in c_0 : \|y\| = \sup |\sum_{k=1}^n y_k| < \infty\}$, $f_n(x) = x_n$. Show that $\{f_n\}$ is a normal sequence, and that S (Prob. 15-3-114) is φ and is not dense; indeed, Y is not separable. This shows that the converse of Prob. 15-3-115 is false.

203 (A. K. Snyder). Show that no sequence t (Prob. 15-3-116) exists for Y (Prob. 15-3-202). However, Y has a separable quotient by Prob. 15-3-4 and the map

$$y \to z \in l^\infty \quad \text{where} \quad z_n = \sum \{y_k : n^2 \leq k \leq n^2 + 2n\}.$$

204 (A. K. Snyder). Show that l^∞ has no dense proper lc FK subspace of the form E_A, $E = l^\infty$, c, ω; i.e., if A is a matrix mapping a dense subspace of l^∞ into E, then A maps all of l^∞ into E.

205 In the spirit of Remark 15-3-8, what is a characterization of spaces with a dual BY subspace?

---

301 Every nontrivial subspace of l^∞ is quasicomplemented. [See [119], p. 178.]

302 Let H be an infinite compact T_2 space. If H contains a perfect set, $C(H)$ has l^2 as a quotient; if not, H must contain a convergent sequence; hence [Theorem 14-7-4] $C(H)$ has c_0 as a quotient. Since $\beta N \backslash N$ is perfect, l^∞ has l^2 as a quotient. Compare Prob. 15-3-1. [See [94].]

303 If Y is WCG there exists a reflexive space X and a map from X onto a dense subspace of Y, that is, there is a dense reflexive BY subspace. The converse is Prob. 15-3-104. [See [25], Lemma 1.]

304 If Y is WCG it has a complemented separable subspace. [See [4].] This improves Corollary 15-3-7; l^∞ does not have such a subspace [Prob. 14-7-117].

15-4 THE STRONG TOPOLOGY

Some interesting properties of the strong topology involve the comparison of $\beta(Y, X)$ with $\beta(Y, S)$ where (X, Y) is a dual pair and $S \subset X$. The specific concept to be studied is spelled out in Definition 15-4-8. We begin with an important type of subspace.

1. Definition Let (X, p) be a normed space and S a subspace of X'. Let $p_S(x) = \sup \{|f(x)| : f \in S, \|f\| \leq 1\}$; S is called norming if p_S is equivalent with p.

2. Remark It is always true that $p_S \leq p$; also p_S is a norm if and only if S is total. It follows that a norming subspace must be total.

3. Lemma Let (Z, p) be a Banach space, X a dense subspace, and S a subspace of $X' = Z'$ which is norming over X. Then S is norming over Z.

PROOF The hypothesis is that p_S and p are equivalent on X. Let us write $p(z) = \|z\|$. By hypothesis there exists $\varepsilon > 0$ such that $p_S(x) \geq \varepsilon \|x\|$ for all $x \in X$. Let $z \in Z$. For $x \in X$ and $f \in S$ with $\|f\| = 1$ we have $|f(x)| \leq |f(z)| + |f(x - z)| \leq p_S(z) + \|x - z\|$ and so $p_S(z) + \|x - z\| \geq p_S(x) \geq \varepsilon \|x\| \geq \varepsilon \|z\| - \varepsilon \|z - x\|$. Since X is dense and x is arbitrary, it follows that $p_S(z) \geq \varepsilon \|z\|$.

4. Example: D. R. Kerr *Let X be a noncomplete normed space. Then X' contains a total closed subspace which is not norming.* Let Z be the completion of X, $z \in Z \backslash X$, $S = \{f \in X' : f(z) = 0\}$. Then S is closed and total. ⟦If, for some $x \in X$, $f(x) = 0$ for all $f \in S$ it follows that x is a multiple of z, hence is 0.⟧ But S is not total over Z so, by Remark 15-4-2 and Lemma 15-4-3, S is not norming over X.

5. Lemma Let X be a barrelled normed space and S a total subspace of X'. Then $\beta(X, S) \subset p_S$.

PROOF Let $B^\circ \in \mathcal{N}[\beta(X, S)]$, B a $\sigma(S, X)$ bounded, hence $\sigma(X', X)$ bounded set. Then B is equicontinuous; hence norm bounded, say $\|f\| < M$ for $f \in B$. Now $p_S(x) \leq M^{-1}$ implies $x \in B^\circ$ ⟦for $f \in B$, $|f(x)| \leq \|f\| p_S(x) < 1$⟧ and so B° includes a p_S neighborhood of 0.

6. Theorem Let X be a Banach space and S a total nonnorming subspace of X'. Then $\beta(X, S)$ is not barrelled.

PROOF This is immediate from Lemma 15-4-5 and the uniqueness of topology result, Corollary 12-5-10.

It remains to give an example to support Theorem 15-4-6.

7. Example *A total nonnorming subspace of a Banach space.* Let $X = c_0$ and let $I, I_1, I_2, \ldots$ be a partition of the positive integers into disjoint infinite subsets. Let $S = \{y \in l : ky_k = \sum \{y_i : i \in I_k\}$ for each $k \in I\}$. To see that S is total, let $x \in c_0$, $x \neq 0$. For some j, $x_j \neq 0$. If $j \in I_m$, let $y_j = 1$, $y_n = -1$ where $n \in I_m$ satisfies $|x_n| < |x_j|$, $y_i = 0$ for all other i. Then $[x, y] = \sum x_i y_i = x_j - x_n \neq 0$

and $y \in S$. If, on the other hand, $x_j = 0$ whenever $j \notin I$, $x_j \neq 0$ for some $j \in I$, let $y_j = 1$, $y_n = j$ where n is some particular member of I_j, $y_i = 0$ for all other i. Then $[x, y] = \sum x_i y_i = x_j + j x_n = x_j \neq 0$ and $y \in S$. Next, S is nonnorming: fix $k \in I$ and let $x = \delta^k \in c_0$. Suppose that $y \in S$ with $\| y \|_1 = 1$. Then $| k y_k | = \left| \sum \{ y_i : i \in I_k \} \right| \leq 1$ and so $\left| [x, y] \right| = | y_k | \leq 1/k$. Hence $\| x \|_S \leq 1/k$. But $\| x \|_\infty = 1$ and so the two norms are not equivalent.

The study of norming subspaces is continued in the problems and in [90].

8. Definition Let (X, Y) be a dual pair and S a dense $(= \text{total})$ subspace of X. Then S is called large if $\beta(Y, S) = \beta(Y, X)$. If X is a lcs space and $S \subset X$, apply this definition with $Y = X'$.

For example, the point of Prob. 14-6-108 is precisely that the linear closure of the point masses is a large set of measures.

9. Remark Since $\beta(Y, S) \subset \beta(Y, X)$, S is large if and only if $\beta(Y, S) \supset \beta(Y, X)$. Note also that dense and large (for subspaces) are both duality invariant.

10. Theorem A subspace S of a lcs space X is large if and only if every bounded set $B \subset X$ is included in the closure of a bounded set $A \subset S$.

PROOF $\Rightarrow$: $B^\circ \in \mathcal{N}[\beta(Y, X)] = \mathcal{N}[\beta(Y, S)]$ so $B^\circ \supset A^\circ$ with A a bounded set in S. Hence $B \subset A^{\circ\circ}$ which is the (weak) closure of the absolutely convex hull of A. $\Leftarrow$: Let $B^\circ \in \mathcal{N}[\beta(Y, X)]$, B a bounded set in X. Let $B \subset \bar{A}$, with A bounded and absolutely convex in S. [Replace A by its absolutely convex hull.] Thus $B \subset A^{\circ\circ}$; hence $B^\circ \supset A^\circ \in \mathcal{N}[\beta(Y, S)]$. The result follows by Remark 15-4-9.

11. Remark Other equivalent conditions are that each B is included in the weak closure of A, or in $A^{\circ\circ}$, for some bounded $A \subset S$. This is clear from the proof of Theorem 15-4-10.

12. Example Let (Y, T) be a barrelled space such that $Y' \neq Y^*$, for example, an infinite-dimensional Banach space [Example 3-3-14]. Let $X = [Y^*, \sigma(Y^*, Y)]$ (see Prob. 8-5-7) and $S = Y'$. Then S is a small subspace of X (i.e., not large) [for $[Y, \beta(Y, S)]' = (Y, T)' = S \subsetneq X = Y^* = [Y, \beta(Y, X)]'$]. This shows that $\beta(Y, S)$ need not even be admissible for (X, Y), although it is a polar topology. It also shows that even a singleton need not lie in any $A^{\circ\circ}$, A a bounded set in S. [Such A lies in an absolutely convex $\sigma(S, Y)$ compact set [hence $\sigma(Y^*, Y)$ compact] since S is boundedly complete and BTB by Prob. 8-2-6. Thus $A^{\circ\circ} \subset S$.]

13. Example Every dense subspace S of a normed space X is large, for $\beta(X', S) = \beta(X', X)$ is the norm topology on X', by Example 8-5-6.

In Prob. 5-1-303 it was pointed out that a lc Fréchet space may have a small subspace.

14. Example *Every dense subspace S of a separable lc metric space X is large.* Let $B \subset X$ be bounded and D a countable dense set in B. ⟦Every subset of a separable metric space is separable; compare Remark 15-4-22.⟧ We may consider $X \subset S''$. ⟦S'' is a Fréchet space by Corollary 10-5-5 and S is a subspace since it is quasibarrelled; since S is dense in X, the topologies match.⟧ Now D is $\beta(S', S) = \beta(X', S)$ equicontinuous ⟦Lemma 10-5-3⟧ so $D° \in \mathscr{N}[\beta(S', S)]$, that is, $D° \supset A°$, with A a bounded set in S. Hence $B \subset D°° \subset A°°$. The result follows by Remark 15-4-11.

15. Definition A lcs X is called *distinguished* if it is a large subspace of $[X'', \sigma(X'', X')]$.

The topology on X is irrelevant as long as it is compatible.

16. Theorem A lcs X is distinguished if and only if $\beta(X', X)$ is barrelled.

PROOF The following statements are equivalent: X is distinguished; $\beta(X', X) = \beta(X', X'')$; $\beta(X', X)$ is absolutely strong; $\beta(X', X)$ is barrelled ⟦Theorem 9-3-10⟧.

17. Example *Every normed space is distinguished,* by Theorem 15-4-16.

18. Example *A nondistinguished space.* Let S be as in Theorem 15-4-6 (e.g., Example 15-4-7). Then (S, T) is not distinguished, by Theorem 15-4-16, where T is any topology for S compatible with (S, X).

Theorem 15-4-20 exposes an important property of distinguished Fréchet spaces. The significance of this result lies in that of Prob. 13-1-301. For more details see [38], Exercises 8.44 and 8.45.

Let X be a lc metric space, $\{U_n\}$ a base for $\mathscr{N}(X)$, Y_n the span of $U_n°$. Each Y_n is a Banach space with topology larger than $\beta(X', X)|_{Y_n}$. ⟦See Theorem 6-1-17; each $U_n°$ is β complete by Prob. 8-5-5 and β bounded by Theorem 9-1-8.⟧ Form $\mathfrak{T}_i$, the inductive limit topology for X', by the inclusion maps $i_n : Y_n \to X'$.

19. Lemma With these notations, $\mathfrak{T}_i \supset \beta(X', X)$ and $\mathfrak{T}_i$ is admissible for (X', X'').

PROOF The first part is because β is a test topology. For the second part we shall prove that $\mathfrak{T}_i$ has a base of neighborhoods of 0 consisting of $\beta(X', X)$ barrels and apply Theorem 8-5-8. ⟦Note that $\mathfrak{T}_i \supset \beta(X', X) \supset \sigma(X', X'')$.⟧ Let $V \in \mathscr{N}(\mathfrak{T}_i)$ be absolutely convex. For each n, $i_n^{-1}[V] \in \mathscr{N}(Y_n)$ and so $V \supset 2t_n U_n°$ for some $t_n > 0$. Let W_n be the absolutely convex hull of $\cup \{t_i U_i° : 1 \le i \le n\}$, $W = \cup W_n$, and $\bar{W}$ the β closure of W. Then $\bar{W} \in \mathscr{N}(\mathfrak{T}_i)$

$[\![$Theorem 13-1-11$]\!]$ and it remains only to show that $\overline{W} \subset V$. To this end, let $f \in X' \backslash V$. For each n, W_n is β closed $[\![$by Prob. 6-5-3 it is weak $*$ compact$]\!]$ and $f \notin 2W_n$ so there exists an absolutely convex $\sigma(X', X)$ closed $N_n \in \mathcal{N}(\beta)$ such that $f + N_n \not\subset 2W_n$ $[\![$the polar of a bounded set in $X]\!]$. Let $H_n = N_n + W_n$, $H = \cap H_n$. Now $f + H \not\subset W$ $[\![$otherwise $f + h = w$, $w \in W_n$ for some n and $h = n_n + w_n$ so $f + n_n = w - w_n \in 2W_n]\!]$ and all comes to showing that $H \in N(\beta)$. For this we apply the criterion of Prob. 10-5-1. Each H_n is closed $[\![$the sum of a closed and a compact set in $\sigma(X', X)$; see Lemma 6-5-11$]\!]$. The proof is concluded by showing that H is a bornivore. It is sufficient to show that H absorbs each W_n. $[\![$If B is $\beta(X', X)$ bounded it is equicontinuous by Theorem 10-1-11 so, for some n, $B^\circ \supset U_n$, that is, $B \subset U_n^\circ$; hence B is absorbed by some $W_n.]\!]$ Now W_n is β bounded $[\![$each U_n° is equicontinuous; hence bounded$]\!]$, so for each k, $W_n \subset c_k N_k \subset c_k H_k$ for some $c_k > 0$. Let $c = \max\{c_1, c_2, \ldots, c_{n-1}, 1/t_n\}$. Then $cH \supset [\cap\{c_k H_k : 1 \le k \le n-1\}] \cap [\cap\{t_n^{-1} H_k : k \ge n\}] \supset W_n$ $[\![$since $t_n W_n \subset W_k \subset H_k$ if $k \ge n]\!]$.

20. Theorem Let X be a lc metric space. The following are equivalent:
(a) X is distinguished.
(b) X' is barrelled.
(c) X' is bornological.
(d) X' is quasibarreled.
(e) X' is the inductive limit of a sequence of Banach spaces.

PROOF $(a) \equiv (b)$ by Theorem 15-4-16. $(b) \Rightarrow (e)$: By Lemma 15-4-19, $\beta(X', X) \subset \mathfrak{T}_i \subset \beta(X', X'') = \beta(X', X)$ $[\![$Theorem 9-3-10$]\!]$. $(e) \Rightarrow (c)$ by Theorem 13-1-13. $(c) \Rightarrow (d)$ by Example 10-1-10. $(d) \Rightarrow (b)$ by Remark 10-4-13.

Part of Theorem 15-4-20 is also given in Prob. 13-1-7.

21. Remark An example of a nondistinguished lc Fréchet space X is given in [88], 31.7, and one with the additional property that $\tau(X', X'')$ is bornological is given in [86]. The dual of such a space is not an (LF) space since it is not barrelled. The second example also reveals the paradoxical fact that a strong topology need not be relatively strong, for if $\tau(X', X'')$ is bornological and $\beta(X', X)$ is not, they must be different.

22. Remark It is worth mentioning that there exists a separable lcs space, indeed the weak $*$ dual of l^∞ $[\![l$ is dense$]\!]$, which has a nonseparable closed subspace. $[\![$See [102].$]\!]$

23. Remark Even though X need not be a large subspace of $[X'', \sigma(X'', X')]$ $[\![$Example 15-4-18 or Remark 15-4-21$]\!]$, the extreme pathology of Example 15-4-12 cannot happen in this case; namely, $\beta(X', X)$ is surely admissible, even compatible, for (X', X''). Moreover, each singleton $z \in X''$ lies in the closure of a bounded set in X. $[\![z^\circ \in \mathcal{N}[\beta(X', X)]$ so $z^\circ \supset A^\circ$ with A absolutely convex and bounded in $X.]\!]$

PROBLEMS

1 Show that a semireflexive space must be distinguished.

2 Show that any subspace which includes a large subspace is itself large.

3 Show that a barrelled dense subspace need not be large ⟦Example 15-4-12⟧.

4 Show that a large subspace of a barrelled space need not be barrelled ⟦Example 15-4-13⟧.

5 Show that a dense maximal subspace need not be large. ⟦Let S be reflexive, $a \in \gamma S \backslash S$ (Prob. 10-3-301), $X = S + [a]$.⟧

6 Let S be a large subspace of X and B a bounded set in X. Show that the natural choice $A = B^{\circ\circ} \cap S$ will not do in Theorem 15-4-10. ⟦Let $S = \varphi$, $X = \omega$, $B = \{t1 : |t| \le 1\}$. See Example 15-4-14.⟧

7 Give a direct proof that φ is a large subspace of ω. ⟦For bounded $B \subset \omega$ (to apply Theorem 15-4-10) try $A = \{b^n : b \in B, n = 1, 2, \ldots\}$ where $b^n = \sum \{b_k \delta^k : 1 \le k \le n\}$.⟧

8 Let X be an infinite-dimensional vector space. Show that X is a small subspace of X^{**}.

9 Let X be quasibarrelled. Show that if $\beta(X', X)$ is bornological it is barrelled. ⟦See Probs. 13-1-110 and 8-4-105 and Theorem 13-1-13. Note that each U° is $\beta(X', X)$ complete, by Prob. 8-5-5.⟧

10 Show that a lc metric space with separable dual must be distinguished. ⟦Apply Theorem 15-4-20. Let B be a bornivore barrel, $\{f_n\}$ dense in $X' \backslash B$. (Separable is G hereditary! Compare Remark 15-4-22.) By Theorem 4-2-6, $f_n \notin B + U_n$. Let $V_n = B + U_n$, $V =$ interior of $\cap V_n$ (not empty by Prob. 10-5-1). Then $X' \backslash V \supset X' \backslash B$ so $B \supset V$. This also follows from Prob. 9-3-302 and Theorem 10-5-4.⟧

11 Show that $\beta(\varphi, \varphi)$ is barrelled but the dual pair (φ, φ) is not barrelled. ⟦$\tau(\varphi, \varphi)$ is metrizable; $B(\varphi, \varphi) = \beta(\varphi, \omega)$ by Example 15-4-14. See Examples 8-5-20 and 13-3-18.⟧

12 Let (X, Y) be a dual pair and S a dense subspace of X. Call S *restrictive* if $\beta(X, Y)|_S = \beta(S, Y)$. Show that the following are equivalent:

(a) S is restrictive.

(b) $\beta(S, Y) \subset \beta(X, Y)|_S$.

(c) Every $\sigma(Y, S)$ bounded set is included in the $\sigma(Y, S)$ closure of a $\sigma(Y, X)$ bounded set.

13 Let X be barrelled and S a dense subspace. Show that S is restrictive if and only if it is barrelled ⟦Theorem 15-1-1⟧.

14 Show that X is restrictive as a subspace of $[X'', \sigma(X'', X')]$ if and only if X is a Banach–Mackey space ⟦Prob. 10-4-101⟧.

15 A restrictive subspace need not be strongly dense. ⟦Apply Prob. 15-4-14 to a normed space.⟧

16 Let X be a Banach–Mackey space and $X \subset S \subset X''$. Show that S is restrictive. This extends Prob. 8-5-115.

Problems on Norming Subspaces

In this list, X is a normed space, S a total subspace of X', and p_S as in Definition 15-4-1.

17 Let $\bar{S}$ be the norm closure of S. Show that $p_S = p_{\bar{S}}$. Thus there is no loss in discussing only norm closed subspaces of X' in the norming theory.

18 Let X be a Banach space, $u : X \to S'$ be $u(x) = \hat{x}|_S$. Show that S is norming if and only if u is range closed.

19 Let A be a subspace of X, $f \in X'$, $p(f) = \|f|_A\|$. Show that $p(f) = d(f, A^\perp)$. It follows that $p_S(x) = d(x, S^\perp)$ in X''. ⟦For $g \in A^\perp$, $a \in A$, $|f(a)| = |f(a) - g(a)| \le \|f - g\| \|a\|$ so $p \le d$. Conversely, extend $f|_A$ to $g \in X'$ with $\|g\| = p(f)$. Then $d \le \|f - (f - g)\| = p(f)$.⟧

20 The *Dixmier characteristic* is defined to be $k(S) = \sup \{\varepsilon : \varepsilon \|x\| \le p_S(x)$ for all $x\}$. Show that S is norming if and only if $k(S) > 0$.

21 Let $k_1(S) = \sup \{\varepsilon : D \cap S$ is weak $*$ dense in $D_\varepsilon\}$. Here $D_\varepsilon = \{f \in X' : \|f\| \le \varepsilon\}$ or $\{x \in X : \|x\| \le \varepsilon\}$, depending on the context; $D = D_1$. Show that $k_1 = k$ (Prob. 15-4-20). ⟦If $0 < \varepsilon < k$, $(D \cap S)^\circ =$

$\{x: p_S(x) \leq 1\} \subset D_{\varepsilon^{-1}}$, so $(D \cap S)^{\circ\circ} \supset D_{\varepsilon}$. Hence $k_1 \geq \varepsilon$ and so $k_1 \geq k$. If $0 < \varepsilon < k_1$, $D_{\varepsilon} \subset (D \cap S)^{\circ\circ}$ so $D_{\varepsilon^{-1}} \supset (D \cap S)^{\circ} = \{x: p_S(x) \leq 1\}$. Hence $\varepsilon \|x\| \leq p_S(x)$ by Prob. 2-2-4. Thus $k \geq \varepsilon$ and so $k \geq k_1$.]

22 Show that S is a large subspace of $(X', \text{weak} *)$ if and only if S is norming. [Apply Theorem 15-4-10 to the preceding two problems.]

23 Considering X as a subspace of X'' total over X', show that $k(X) = 1$. This is the result of Prob. 10-2-3 [Theorem 3-2-2].

24 Let $z \in l^\infty \backslash c_0$, $S = \{y \in l: \sum y_i z_i = 0\}$, $X = c_0$. Show that $k(S) = \limsup |z_n|/(\|z\|_\infty + \limsup |z_n|)$.

25 Show that $k(S) = d(C, S^\perp)$ in X'' where $C = \{x \in X: \|x\| = 1\}$ [Probs. 15-4-19 and 15-4-20].

26 Let $f \in X'' \backslash X$, $c(f) = k(f^\perp)$. Show that $c(f) = \inf \{\|f - x\|/\|x\|: x \neq 0\}$. [Let $x \in X$, $x_1 = x/\|x\|$, $g = (f - x)/\|f - x\|$, $\lambda = \|f - x\|/\|x\|$, $[f] = $ span of f. Then $c(f) = c(x_1 + \lambda g) = d(C, [x_1 + \lambda g])$ (Prob. 15-4-25) $\leq \|x_1 - (x_1 + \lambda g)\| = \lambda$ so $c \leq \inf$. Conversely, if $r > c(f)$ there exists x with $r\|x\| > p_S(x) = \|x - \lambda f\|$ for some λ, by Prob. 15-4-19. So $r\|x\| > \|x - \lambda f\|$ for some $\lambda \neq 0$. Let $x_1 = x/\lambda$; then $r > \|x_1 - f\|/\|x_1\|$ so $r \geq \inf$; hence $c \geq \inf$.]

27 Let A be a closed subspace of X, $b \in X \backslash A$, $B = $ span of b. Show that there exists $\varepsilon > 0$ such that $d(x, B) \geq \varepsilon \|x\|$ for all $x \in A$. [We may assume $X = A \oplus B$ with A complete. The map $x \rightarrow x + B$ from A to X/B has continuous inverse by the open mapping theorem. A direct proof is also easy.]

28 Let $f \in X'' \backslash \gamma X$. Show that $f^\perp$ is a norming subspace of X'. [See Probs. 15-4-26 and 15-4-27. It also follows from Prob. 15-4-31.]

29 Let $f \in X'^*$. Show that $S = f^\perp$ is a norming subspace of X' if and only if f is not $aw *$ continuous. [If $f \in X$, S is not total; if $f \in \gamma X$, X'', X'^*, apply Example 15-4-4 and Probs. 15-4-28 and 15-4-17 respectively.]

30 Show that X is complete if and only if every norm-closed total maximal subspace of X' is norming [Prob. 15-4-29 and Corollary 12-2-16].

31 Let X be a Banach space. Show that S is norming if and only if $S^\perp + X$ is closed in X''. [$\Rightarrow$: Define a complete norm on $S^\perp + X$ by $p(f + x) = \|f\| + \|x\|$. For $x \in X$, $f \in S^\perp$, $\|f + x\| \geq d(X, S^\perp) = p_S(x) \geq \varepsilon \|x\|$ by Prob. 15-4-19. Thus $\|f + x\| \leq p(f + x) \leq 2\|x\| + \|f + x\| \leq (2\varepsilon^{-1} - 1)\|x + f\|$. Thus the original norm is equivalent with p; hence complete. $\Leftarrow$: The two norms are equivalent by the open mapping theorem, say $\varepsilon p(f + x) \leq \|f + x\|$. Then $\|f + x\| \geq \varepsilon \|x\|$ so $p_S(x) = d(x, S^\perp) \geq \varepsilon \|x\|$.]

32 Let S have countable codimension in X'. Show that S is norming over X [Probs. 15-4-17, 15-4-31, and 15-1-102 and Theorem 6-3-3].

33 With S as in Example 15-4-7 show that $S^\perp + c_0$ is not dense in l^∞ (see Prob. 15-4-31). This is another failure in the search for an easy solution of Prob. 15-3-4. (Others were Probs. 15-3-116, 15-3-118, and 15-3-203.)

34 With S as in Example 15-4-4, show that $(X, p_S)' = S$. Hence, if X is barrelled (e.g., Probs. 3-1-5 and 9-3-116 and Example 15-1-3), $p_S = \beta(Z, S)$ [Lemma 15-4-5] and so $\beta(Z, S)$ is barrelled.

35 Let X be a separable Banach space. Then S is norming if and only if S is weak * sequentially dense in X'. [$\Leftarrow$: Problem 15-4-21. $\Rightarrow$: See [35], theorem 5.]

36 Every total subspace of X' is norming if and only if X is quasireflexive. [$\Leftarrow$: Problem 15-4-31. $\Rightarrow$: See [113].]

15-5 MISCELLANEOUS

In the text thus far we have presented the basic information needed for the study of topological vector spaces. Many special directions for continuation have been indicated in the problems. A few more of these will now be given.

Let $\{X_\alpha: \alpha \in A\}$ be a set of bornological spaces and $|A|$ the cardinal of A. Theorem 15-5-1 shows that the question of whether this product must be bornological can be answered purely by knowledge of $|A|$ and that for all practical purposes the product is always bornological. An *Ulam measure* on A is a countably additive set function whose values are 0 or 1 only, 0 on each singleton, and not identically 0. No such measure exists unless $|A|$ is inconceivably large. Possibly no such measure exists at all.

1. Theorem: Mackey–Ulam theorem Let A be given. The following are equivalent:

(*a*) The product $\pi\{X_\alpha : \alpha \in A\}$ is bornological for all sets $\{X_\alpha : \alpha \in A\}$ of bornological spaces.

(*b*) $\mathscr{K}^A$ is bornological.

(*c*) The discrete space with $|A|$ members is real-compact.

(*d*) No Ulam measure exists on A.

PROOF See Prob. 8-4-303; [82], Theorem 19.9; [46], Chap. 12.

2. Example We have seen barrelled, nonbornological spaces in Prob. 10-3-301. The first published examples were done by means of Probs. 8-4-303 and 9-3-301. A rather frivolous example can be given by means of Theorem 15-5-1, namely, take A to have a very large cardinal [Prob. 13-2-205].

3. Remark Lebesgue's bounded convergence theorem suggests the study of sequences which are convergent in one topology and bounded in another. This leads to the study of two-norm spaces and mixed topologies. A special case is the so-called strict topology which is also a T^+ topology in the sense of Probs. 8-4-124 and 8-4-201. We refer to [154], Chap. 5, p. 83; [121], p. 409; and [131]. The last-mentioned reference also discusses conditions under which $\beta(X, Y)$ is relatively strong; see Remark 15-4-21.

4. Remark In Prob. 6-5-111, semi-Montel spaces are defined. This concept arose from classical complex analysis in which *normal families* of functions, i.e., conditionally sequentially compact sets of functions, played a useful role, as in Montel's theorem, Prob. 6-5-301. A *Montel space* is a quasibarrelled semi-Montel space. *A semi-Montel space must be semireflexive.* [Bounded closed sets are compact; hence weakly compact, so Theorem 10-2-4 applies.] Thus a *Montel space is reflexive; hence barrelled.* If (X, T) is semi-Montel, *T and $\sigma(X, X')$ have the same compact sets.* [Every weakly compact set is bounded and closed in (X, T); hence compact.] Thus *they have the same convergent sequences.* For further developments, see [88], 27.2, 27.3, 27.4, 31.5; [58], 3.9.

5. Remark Our final topic is a theory which was developed by G. Choquet. Let K be a compact convex set in a lcs space X and A be the set of real affine continuous functions on K. An *affine* function f is one satisfying $f(ak_1 + bk_2) = af(k_1) + bf(k_2)$ whenever $0 \le a \le 1$, $a + b = 1$. Then A, with

$\| f \|_\infty$, is a Banach space. It is easy to see that such f need not have an extension to an affine continuous function on X; those which have are dense in A. For $k \in K$, let $u_k(f) = f(k)$ define $u_k \in A'$. Such members of A' are characterized by $\| u \| = 1$, $u(f) \geq 0$ if $f \geq 0$. [[The fallacious "proof" of Prob. 9-6-108 works because K is now a convex set in A'.]] In particular, if μ is a probability measure on K, that is, $\mu \geq 0$, $\mu(K) = 1$, then $\int f \, d\mu = f(k_\mu)$ for all $f \in A$, and k_μ is called the *barycenter* of μ. For $k \in K$, let R_k be the set of probability measures μ such that $k = k_\mu$. Then k is on the boundary of K if and only if R_k contains only the point measure at k. The *Choquet boundary* with respect to certain subsets S of A is defined analogously in terms of $R_k(S)$, the set of probability measures μ such that $\int f \, d\mu = f(k)$ for all $f \in S$. We refer to [3] for further study.

TABLES

We use the following abbreviations:

ac	Absolutely convex	codim.	Codimension	seq.	Sequential (ly)
assoc.	Associated	Cor.	Corollary	Th.	Theorem
ba.	Barrelled	Def.	Definition	tot.	Totally
bdd	Bounded	dual	Strong dual	w	Weak
B–M	Banach–Mackey	equic.	Equicontinuous	w*	weak *
bo.	Bornological	Ex.	Example	WCG	Weakly compactly generated
BS	Banach space	L.	Lemma		
cc	Convex compactness	M	Lc metric		
		QB	Quasibarrelled	#	Problem
CG	Closed graph	Rem.	Remark		
		RS	Relatively strong		

Example 1 Must a lcs space be separable if all its bounded sets are separable? See Table 27, col. 2, item 4 or 5; no.

Example 2 Must the completion of a bornological space be bornological? See Table 12, second list; no.

Example 3 Is $\tau(l, \varphi)$ barrelled? See Table 2, Spaces, item 7; no.

Example 4 Must $\beta(X, Y)$ be boundedly complete if $\sigma(X, Y)$ is? See Table 4, col. 1, item 4; yes.

Example 5 Must the dual of a lc metric space be C barrelled? See Table 5, col. 1, item 3; Table 17, col. 1, item 4; yes.

Example 6 Suppose that $\sigma(X, Y)$, $\tau(X, Y)$ have the same convergent sequences and that a set is $\sigma(X, Y)$ compact. Must it be $\tau(X, Y)$ compact? See Table 22, col. 1, item 3, col. 2, item 3; yes if τ is metrizable, but not in general.

Table 1 Banach–Mackey (see Table 12)

Implied by	Not implied by
See # 10-4-101, 15-4-14	Bo. Th. 10-4-12
Ba. Ex. 10-4-4	Mazur Th. 10-4-12
Boundedly complete Th. 10-4-8	
cc Th. 10-4-11	
Dual of Mazur # 10-4-5	Spaces
Every barrel is a bornivore Th. 10-4-7	
M' Cor. 10-4-15	(l, φ) no # 10-4-102
Semireflexive Ex. 10-4-6	
Seq. ba. # 10-4-106	
Seq. complete Th. 10-4-8	
(Y, X) if (X, Y) is B–M Th. 10-4-5	

Table 2 Barrelled (see Table 12)

Implied by	Not implied by
See # 9-3-105, 15-1-109, 15-1-201, 15-4-13	See # 15-2-3, 15-2-4, 15-2-105, Th. 15-4-6
Absolutely strong Th. 9-3-10	Assoc. bo. of ba. [55]
Assoc. lc of ultrabarrelled # 9-3-114	$\beta(X, Y)$ Th. 15-4-16, Ex. 15-4-18
Bo. and dual of QB # 15-4-9	B–M, RS # 10-1-3, Th. 9-2-14
Boundedness criterion Th. 15-1-1	C-ba. RS # 11-1-118
CG criterion Th. 12-6-3	CG Th. to separable BS [75], p. 407
Completion of QB Table 12	Dual of lc Fréchet Rem. 15-4-21
Countable codim. in ba. Th. 15-1-10	Fully complete RS Ex. 12-3-8
Dual of distinguished Th. 15-4-16	$\cap$ 2 ba. subspaces Ex. 15-1-4
Dual of semireflexive Th. 10-2-4	Large $\subset$ ba. # 15-4-4
F linked criterion # 9-3-112	ω-ba. RS # 9-4-301
Has dense ba. subspace # 9-3-124	Open mapping Th. from fully complete
Inclusion criterion Th. 15-2-7	# 13-3-201
(LF) Th. 13-1-13	QB Mazur $[\![M]\!]$
Montel Sec. 15-5	RS (X', w^*) cc Table 18
QB and B–M Th. 10-4-12	RS (X', w^*) seq. complete Table 18
and cc Rem. 10-4-13	Semireflexive complete RS Th. 9-2-14
and dual of Mazur # 10-4-5	Strictly hypercomplete RS Ex. 12-3-8
and semireflexive Rem. 10-4-13	w^* bdd ac M $\Rightarrow$ equic. [75], p. 407
and seq. ba. # 10-4-107	
and seq. complete Rem. 10-4-13	
and has (X', w^*) seq. complete Rem. 10-4-14	Spaces
and has (X', w^*) cc Rem. 10-4-14	
Reflexive Ex. 10-3-3	Certain $C(H)$ # 9-3-301
RS strongly separable (X', w^*) seq. complete	$(l^{1/2}, \|\cdot\|_1)$ yes Ex. 15-2-5
# 9-3-302	m_0 yes Ex. 15-1-3
RS (X', w^*) boundedly complete Th. 9-3-13	M, S_σ no Cor. 13-3-16
RS w^* bdd $\Rightarrow$ equic. Th. 9-3-4	$\beta(\varphi, S), \varphi \subset S \subset \omega$ yes # 15-4-10, 15-4-11
Second-category Ex. 9-3-2	$\sigma(X^*, X)$ yes # 9-3-119
Separable dual of M # 15-4-10	$\tau(l, \varphi)$ no # 9-3-120
	$\tau(\varphi, l)$ no # 3-3-4, Ex. 13-3-18
	$\tau(\varphi, \varphi)$ no # 15-4-11
	$\tau(X, X^*) = $ largest lc yes # 9-3-118
	$X \otimes X, X = $ Hilbert space yes [14], Chap. 5, # 10C

Table 3 Bornological (see Table 12)

Implied by	Not implied by
See #8-4-105, 8-4-106, 8-6-117	See #12-2-114
Finite codim. in bo. #15-1-302	Ba #10-3-301
M Ex. 4-4-7	Bidual of bo. [88], 28.4
QB and dual of M Th. 15-4-20	Complete RS Table 18
Semibo. RS #8-6-116	Countable codim. in ba. bo. #15-1-302
(X', T_N) complete #8-6-301	C-seq. RS #8-6-118
	Dual of Fréchet, τB_0 Rem. 15-4-21
	Dual of reflexive #10-3-301
Spaces	Mazur RS Table 18
	QB (DF) [140]
C (real-compact) yes #8-4-303	Reflexive #10-3-301
(LF) yes Ex. 13-3-20	RS has complete dual #9-3-101
$\sigma(X^*, X)$. See #8-1-106, Th. 15-5-1	$X' = X^*$ [85]
$\tau(\varphi, l)$, $\tau(\varphi, l^\infty)$ yes Table 15	
$\tau(\varphi, c_0)$ no $[\{x: \|x\|_1 \le 1\}]$	
$\tau(X, X^*) = $ largest lc yes #8-4-4	
$\tau(X', X)$, X nonreflexive BS no Table 18	
Test functions yes #13-3-301	

Table 4 Boundedly complete (see Table 12)

Implied by	Not implied by
Complete	Dual of semireflexive [88], 23.8
cc and w* dual of QB Rem. 10-4-14	cc seq. complete Ex. 9-3-14, Th. 14-2-4
F linked criterion Th. 6-1-16	(LB) [89]
Larger admissible #8-5-5	von Neumann complete [150]
Montel Sec. 15-5	w if τ is complete Th. 10-2-4
Semireflexive, Th. 10-2-4, #8-5-5.	
Seq. complete and w* dual of QB Rem. 10-4-14	
w* dual of ba. Th. 9-3-11	**Spaces**
	Nonreflexive BS w no Th. 10-2-4
	$\sigma(X, X^*)$ yes #10-2-101

Table 5 C barrelled (see Table 12)

Implied by	Not implied by
CG Th. to separable BS [75], Th. 3-1	Seq. ba. RS #14-2-103
Larger compatible #9-4-104	
ω ba. #9-4-103	
RS (X', w^*) seq. complete #11-1-115	**Spaces**
	$\tau(l, c_0)$ no #14-2-103
	$\tau(l^\infty, l)$ yes #11-1-118

Table 6 Compact (see Table 22)

Implied by	Not implied by
See Th. 14-1-9, Ex. 14-1-10, #14-1-302, 14-1-303, 14-1-306, 14-2-301	ac closure in lc Table 8
ac hull finite ∪ ac compact sets #6-5-3	Disc in equiv. norm on X' #12-3-105
Balanced hull #6-5-2	Tot. bdd closed #6-4-105
Disc in reflexive BS w Ex. 10-2-6	
Disc in (X', w^*) X normed Th. 9-1-12	**Spaces**
Equic. closed w Th. 9-1-10	
Tot. bdd complete Th. 6-5-7	Disc in l^∞, $\tau(l^\infty, l)$ yes #12-1-122
(U°, w^*) $U \in \mathcal{N}$ Cor. 9-1-11	Disc in l, $\tau(l, c_0)$ no #14-2-107
H if $C^*(H)$ separable #9-6-106	

Table 7 Complete (see Table 12)

Implied by	Not implied by
See #6-1-110, 8-6-301, 12-1-102, 15-4-30	$\beta(X, Y)$ #9-3-2
$\beta(X', X)$ See Dual (below)	Bidual of ba. #10-3-301
Ba M $\tau(X', X)$QB Th. 9-3-13, Rem. 10-4-13	Boundedly complete Ex. 9-3-12
Bidual of M Cor. 10-5-5	Boundedly complete ba. separable #10-3-301
B_r complete Th. 12-4-3	Cat II normed #3-1-5
cc M Th. 12-3-10, #12-3-301, 12-3-110, 11-1-112	Dual of ba. #10-3-301
Compact L. 6-1-18	Montel separable #10-3-301
Dual of Mazur Cor. 8-6-6, #12-2-112	Normed CG Th. $X \to X$ [62]
Finite-dimensional Th. 6-3-2	Normed every f has max on disc [67]
Finite inductive limit Th. 13-4-4	Reflexive separable #10-3-301
F linked criterion Th. 6-1-13, 6-1-16	$\tau(X', X)$ X ba. #10-3-301
Gauge Th. 6-1-17	$\tau(X', X)$ X cc #10-2-106
Grothendieck's Th. 12-2-15, 12-2-16, 12-2-19	$\tau(X', X)$ X normed #8-6-106
Larger admissible #8-5-5	
Largest lc Ex. 13-2-13	
M $\tau(X', X)$ seq. ba. [146], prop 4.2	**Spaces**
$\mathcal{N}$ has a complete set #6-1-107, see also 6-1-113	
Reflexive γX fully complete #12-4-113	$\beta(\varphi, S)$ $\varphi \subset S \subset \omega$ yes Ex. 15-4-14, Cor. 8-6-6
Reflexive M Cor. 10-5-6	$\beta(X, X^*)$ yes #13-2-106
Seq. ba. (X', w^*) Mazur #12-2-102	$C(k)$ yes #6-1-205
Seq. complete ba. Schauder basis [85]	$\sigma(X, X^*)$, $\sigma(X, X')$ no #13-2-105
Seq. complete M Th. 6-1-3	$\sigma(X^*, X)$ yes #8-1-5
$\tau(X', X)$ X cc Mazur Th. 9-2-14	$\tau(l, \varphi)$, $\tau(\varphi, l)$ no #8-6-106
$\tau(X', X)$ X seq. complete, Mazur #11-1-117	$\tau(X, X^*)$ yes #13-2-106

Table 8 Convex compactness (see Table 12)

Implied by	Not implied by
Admissible ⊃ cc Th. 9-2-11	Ba. Th. 12-3-10
Boundedly complete (compatible) Ex. 9-2-10, Th. 14-2-3, #14-2-301	B–M #12-3-111
	Seq. complete #9-2-301

Table 8 Continued

Implied by	Not implied by
N complete #9-2-111	σ if τ has cc #14-6-110
w* dual of ba. #9-3-1	τ if β complete $[\![(l, \varphi)]\!]$

	Spaces
	$\beta(\varphi, l), \beta(l, \varphi)$ yes Table 7
	$\tau(\varphi, l), \tau(l, \varphi)$ no #10-4-102

Table 9 C sequential

Implied by	Not implied by
Bo. #8-4-127	RS #14-2-108
Seq. #8-4-201	Semibo. #8-6-118
RS. Mazur $[\![T_c^+ \tau = T]\!]$	

	Spaces
	Normed weak no #9-1-106
	T^+ yes #8-4-201
	$\tau(l^\infty, l)$ yes # 9-5-108

Table 10 Finite dimensional

Implied by	Not implied by
See #12-4-302	Normed w* $= \tau [\![X = \varphi, \|\cdot\|_\infty).]\!]$
BS w* $= \tau$ Th. 9-2-14 or #10-2-121	Range (closed) of weakly compact map
Bdd if $X' = X^*$ #4-4-15	#11-4-101
BTB normed #6-4-107, Ex. 6-4-14	Weakly complete #8-1-5
Locally compact #6-4-107	
Range (closed) of compact map #11-4-108	
Reflexive BS norm $=$ w (seq.) #14-1-101	
Reflexive BS projection criterion [26], V.4(12)	
Reflexive $C(H)$ #10-2-120	
Reflexive $\subset$ certain $C(H)'$ [119]	
Reflexive $\subset c_0$ #14-1-301	
Reflexive $\subset l$ #14-1-103	
Reflexive quotient of c_0 #14-1-104	
$\sigma(X, X^*)$ complete #13-2-105	
Subspace of l with separable dual #15-3-1	
Subspace of l closed in l^2 or c_0 #14-1-109	
Subspace of L closed in L^∞ [132]	
TB $U \in \mathcal{N}$ Th. 6-4-2	
Weakly seq. complete $C(H)$ #10-2-120	
w* convergent sequences are norm convergent	
BS #9-5-301	

Table 11 Fully complete (see Table 12)

Implied by	Not implied by
See Ex. 12-3-8	CG Th. to BS [116], p. 124
Almost open image [116], p. 113	Complete #12-4-113, 12-4-301, Ex. 13-2-14
Ba. image of fully complete Rem. 12-4-10	γ (reflexive) #12-4-113
Dual of reflexive Fréchet #12-3-108	Largest lc #12-5-1
Larger compatible #12-4-102	QB image of fully complete #12-4-104
lc Fréchet Th. 12-3-3	Strict inductive limit ($\aleph_0$) Fréchet #13-3-302
Open map criterion Th. 12-6-10	$X \times X$, X strictly hypercomplete [109], p. 29
Strictly hypercomplete Th. 12-4-3	
w complete #12-4-108	

Spaces

Distributions no [109], p. 50
$\mathscr{K}^a$ yes [21], p. 279
(φ largest lc) yes #12-3-108
X^* yes Th. 12-4-19

Table 12 Heredity

Properties inherited by:

Closed subspace

Yes
Boundedly complete
BTB
Complete
cc
Equic. $\Rightarrow$ w* seq. compact [39]
Fully complete #12-4-201
G #14-7-102
N seq.
Reflexive M #10-3-108, Th. 3-2-11
Schwartz [92]
Semireflexive #10-2-111
Seq.
Seq. complete
Weakly seq. complete #11-1-3
No
Ba. Ex. 15-1-4
Bo. #8-4-111, 8-4-302
C seq. #7-2-113
(LF) [59], p. 517
Mazur [[MR, vol. 49, no. 1048]]
QB Table 18
Reflexive [88], 23.5
RS Table 21
Separable Rem. 15-4-22
WCG no [120]

Completion

Yes
Ba. #12-2-104
M #6-1-105
Mazur
QB [88], 27.1, [2]
Ultrabarrelled [117], Prop. 14
No
Bo. [88], 28.4 (3)

Dense subspace

Yes
BTB
lc
M
No
Ba. Th. 15-3-1
B–M Th. 10-4-12
Bo. Ex. 15-4-12
Mazur [[MR, vol. 49, no. 1048]]
QB Ex. 15-4-12
Reflexive Rem. 3-2-8
RS #9-2-113 or Table 21

Table 12 Continued

Properties inherited by:

Direct sum

Yes
See "Inductive limit"
Complete Th. 13-2-12
Montel [88], 27.2
Reflexive [88], 23.5 (9)
Semireflexive [82], 20.2

No
Fully complete # 13-2-109
Separable # 13-2-116

Inductive limit

Yes
Ba. Th. 13-1-13
Bo. Th. 13-1-13
Largest lc
Mazur Th. 13-1-8
QB # 13-1-4
RS Th. 13-1-17

No
See "Quotient," "Direct Sum"
($\aleph_0$) M # 13-1-113
Fully complete # 13-3-302

Product (See Th. 15-5-1)

Yes
Ba. # 13-2-205
Bdd # 4-4-13
BTB Th. 6-4-13
Compact # 6-5-1
Complete Th. 6-1-7
Equic. $\Rightarrow$ w* seq. compact [39]
lc # 7-2-5
QB [82], 20.4
Reflexive [88], 23.5 (9)
RS # 13-2-206
Schwartz [92]
Semireflexive # 10-2-109
Totally bdd Cor. 6-4-9
WCG (BS) [39]

No
Fully complete [142]
Largest lc # 4-1-3
Separable
Strictly hypercomplete [109], p. 29

Quotient (separated)

Yes
See "Inductive limit"
Banach space Cor. 6-2-14
BS with separable dual Th. 11-3-4
Equic. $\Rightarrow$ w* seq. compact [39]
Fully complete Th. 12-4-5
GB Th. 14-7-3
lc # 7-1-2
lc Fréchet Cor. 6-2-15
M Th. 6-2-11
Reflexive BS # 11-3-103
Schwartz [92]
Separable
Strictly hypercomplete like Th. 12-4-5
Ultrabarrelled [117], prop. 13
WCG BS

No
Complete # 12-4-301
(*FM*) [88], 31.5
Montel [88], 27.2
Reflexive # 10-2-302
Reflexive Fréchet # 10-2-302
Semireflexive # 10-2-302

Strict inductive limit (sequence)

Yes
See "Inductive limit"
Complete Th. 13-3-13
Montel [88], 27.2

No
Fully complete # 13-3-302

Subspace

For RS see [141]

Sup

Yes
Bdd Th. 4-4-5
BTB Th. 6-4-11
lc # 7-2-3
Tot. bdd Th. 6-4-7

No
Boundedly complete Ex. 5-2-9
Compact # 6-5-104
Complete Ex. 5-2-9, # 6-1-109
Fully complete Ex. 5-2-9
RS Table 21
Semireflexive Ex. 5-2-9

Table 13 Mazur (see Table 12)

Implied by	Not implied by
See #8-6-201, Ex. 14-1-11	Ba. separable #10-3-301
Ba. Schauder basis #9-3-107	(DF) [146], p. 361
Bo. Ex. 8-6-4	Dual of M [146], p. 361
C seq. #8-4-128	Relatively w X' complete #14-7-109
Semibo. #8-6-115	Smaller admissible #9-3-117
Smaller compatible #8-6-1	$\sigma(X', X)$ X BS #14-7-109
$\sigma(X', X)$, X separable complete #12-2-3	$\sigma(X', X)$ X reflexive separable #12-2-102,
$\sigma(X', X)$, X WCG BS #15-3-106	10-3-301
Schauder basis (X', w^*) seq. complete [147]	$\sigma(X', X)$ X separable ba. normed Cor. 8-6-6

	Spaces
	$\sigma(l, c_0)$ yes #12-2-3
	$\sigma(l^{\infty\prime}, l^{\infty})$ no #14-7-109
	$\tau(l, c_0)$ no # 10-4-205 or # 14-2-107
	$\tau(l^{\infty}, l)$ yes #9-5-108
(See also pp. 221–5 in [84])	$\sigma(L, L)$ yes [39], Cor. 1.4

Table 14 Metrizable (see Table 12)

Implied by	Not implied by
ac closure Cauchy seq. [75], p. 402	
Bidual of M #10-1-5	
Bdd $\subset (X, w)$ X normed X' separable #9-5-111	Ba. and bo. Ex. 13-3-20
C (hemicompact) #4-1-105	Bdd $\subset$ (normed, weak) #9-5-112
Equic. in dual of separable Th. 9-5-3	Every bornivore$\in \mathcal{N}$ #8-4-129
First countable Th. 4-5-5	N seq. RS #8-4-129
Inductive limit (finite) #13-4-101	RS X' separable BS Table 18
$\cap(\aleph_0)$ FH Th. 5-5-9	Strictly hypercomplete RS Ex. 12-3-8
Locally bdd Th. 4-5-8	
$\mathcal{N}$ has cofinal chain #4-5-201	
$\pi(\aleph_0)$ #2-1-12	Spaces
$\sigma P(\aleph_0)$ #4-1-5	
H if $C^*(H)$ separable #9-6-106	$\beta(X', X)$, X M not normed no #10-1-106
	Ba $S\sigma$ no Cor. 13-3-16
	Largest lc no #7-1-8, Ex. 13-3-18
	(LF) no #13-3-3, Cor. 13-3-16
	$\sigma(S, \varphi)$, $\varphi \subset S \subset \omega$ yes #4-1-5
	$\tau(l, \varphi)$, $\tau(\varphi, l)$ yes #8-6-106
	$\tau(X, X^*) = \beta(X, X^*)$ no Ex. 8-5-19
	$\cup l^n$ no #13-1-201

Table 15 Normed (see Table 12)

Implied by	Not implied by
See $\#7\text{-}2\text{-}8$	Ba. bo. X'M $\#10\text{-}1\text{-}106$
Locally bdd lc $\#7\text{-}2\text{-}7$	Between 2M $[\![\sigma(\varphi,\varphi)\subset\sigma(\varphi,c_0)\subset\tau(\varphi,l^\infty)]\!]$
M and X' Cat II $\#10\text{-}1\text{-}106$	M $\exists$ bdd convex absorbing set $[\![\sigma(\varphi,\varphi)]\!]$
and X'M $\#10\text{-}1\text{-}106$	$\tau(X,X')$ and $\tau(X',X)$M $\#10\text{-}1\text{-}107$
and X'N seq. $\#12\text{-}3\text{-}302$	
and $\aleph_0$ cobase for bdd $\#7\text{-}1\text{-}201$	
and $\tau(X',X)$ M ba. $\#10\text{-}2\text{-}116$	
$\mathcal{N}$ has bdd convex set Th. 4-5-2	

Spaces	
Analytic functions no $\#7\text{-}2\text{-}115$	σF no Rem. 4-1-14
$B(X,Y)$ yes Th. 2-3-3	$\tau(\varphi,c_0)$ no Table 3
$C(H)$ no $\#7\text{-}2\text{-}114$	$\tau(\varphi,l)$ yes $[\![\|\cdot\|_\infty]\!]$
$L^{1/2},\mathcal{M}$ no $\#2\text{-}3\text{-}121$	$\tau(\varphi,l^\infty)$ yes $[\![\|\cdot\|_1]\!]$

Table 16 N sequential

Implied by	Not implied by
See L. 14-1-5, 14-1-8	See Table 28
	Dual of Montel [146], Prop. 5-7
	Seq. $\#8\text{-}4\text{-}123$, 12-3-113

Spaces	
$\beta(X',X)$, XM not normed, no $\#12\text{-}3\text{-}302$	
φ largest lc no $\#8\text{-}4\text{-}123$, 12-3-113	

Table 17 ω barrelled

Implied by	Not implied by
Ba. $\#9\text{-}4\text{-}103$	C ba. RS $\#11\text{-}1\text{-}118$
B–M (DF) $\#10\text{-}5\text{-}103$	(X',w^*) seq. complete RS $\#9\text{-}4\text{-}107$
$\aleph_0$ codim. in ω ba. $\#15\text{-}1\text{-}301$	
Dual of M $\#10\text{-}5\text{-}104$	
Larger compatible $\#9\text{-}4\text{-}104$	

	Spaces
	$\tau(l^\infty,l)$ no $\#11\text{-}1\text{-}118$

Table 18 Quasibarrelled (see Table 12)

Implied by	Not implied by
See #10-1-109, 10-1-11	See #10-1-3, Table 21
β bdd $\Rightarrow$ equic. #10-1-110	$\beta(X, Y)$ Table 21
Ba. or bo. Ex. 10-1-10	Complete RS Th. 9-2-14, #8-4-6
(DF) separable [88], 29.3	$\aleph_0$ codim. in bo. #15-1-302
Finite codim. in QB [59], p. 518	Closed $\subset$ ba. #7-2-113, 13-2-205
Natural embedding continuous Th. 10-1-8	Dual of lc Fréchet Rem. 15-4-21
Separable dual of M #15-4-10	Mazur RS #9-5-108
$\tau(X, X')$ if X' semireflexive #10-2-113	Natural embedding seq. continuous #10-1-115
	ω ba. Table 21
	RS X' separable BS #8-4-6
Spaces	RS (X', w^*) cc #14-2-106
	RS (X', w^*) seq. complete #9-3-108
$\tau(X', X)$ X nonreflexive BS no #8-4-6	Strictly hypercomplete RS Ex. 12-3-8

Table 19 Reflexive (see Tables 12 and 20)

Implied by	Not implied by
See Th. 10-3-2, #12-4-114	Ba. X' reflexive #10-3-103
$\beta(X, X')$ if $\tau(X, X)$ reflexive #10-3-105	Semireflexive RS $[\![\tau(X', X)]\!]$
Ba. B–M $\tau(X', X)$ QB Th. 10-4-5, 10-4-12	(X, Y) if $(X, Y), (Y, X)$ both M #10-3-101
Boundedly complete RS X' reflexive #10-3-104	
Boundedly complete RS $\tau(X', X'')$ reflexive #10-3-106	
Closed $\subset$ reflexive lc Fréchet [88], 23.5	Spaces
Dual of reflexive #10-3-102	
Montel Sec. 15-5	(X, X^*) yes
QB semireflexive	$\mathcal{K}^a$, $\mathcal{K}^{(a)}$ yes Th. 13-2-15, 13-2-17

Table 20 Reflexive Banach space

Implied by	Not implied by
Banach–Saks, D_1 w* seq. compact [39], Th. 2.7	Closure range of compact map #11-4-103, 11-4-109
bw complete [26], Th. III.4.2	GB Ex. 14-7-7
$C[H]$ weakly seq. complete #10-2-120	Normed every f has max on disc Table 7
Closed range of weakly compact map #11-4-102	Reflexive + reflexive #15-3-201
D weakly complete or compact #10-2-1, Ex. 10-2-6	Span of compact set [139]
Differentiable norm [45]	WCG Ex. 15-3-5

Table 20 Continued

Implied by	Not implied by
Every f has max on disc (BS) [68]	w seq. complete #8-1-11
Every subspace with basis is reflexive [105], p. 77	Separable D_2 w* seq. compact $[\![c_0]\!]$
Fatou's lemma [129]	$\tau(X', X)$M, X normed #10-3-101
Finite dimensional #3-2-107	X and X' separable $[\![c_0]\!]$
GB, D_1 w* seq. compact #14-7-103	$X^{(n)}$ separable $\forall n$ #10-2-301
GB separable Th. 14-7-2	$X \subset Y$, w* seq. closed in Y'' #8-1-11
GB WCG #15-3-105	
Isomorph #3-2-4	
Quasireflexive w seq. complete [19]	**Spaces**
Semireflexive Ex. 10-2-2	
Separable quotient of GB #15-3-1	See #3-2-104, 10-2-120
Summability criterion #3-2-301	$B(l^p, l^q)$, $1 < q < p < \infty$ yes [122]
$\tau(X', X)$ QB (BS) #8-4-6	
Uniformly convex #3-2-302	
w bdd complete #10-2-1	
w seq. complete, D_2 w* seq. compact #14-1-111	
w seq. complete, X' separable Ex. 12-5-15, #10-2-104	
S and X/S reflexive Cor. 11-4-7	
$X \subset Y$, w* closed in Y'' #11-3-118	
X' reflexive Th. 3-2-9	
X' if X reflexive Th. 3-2-9	
X' has reflexive total subspace #11-3-117	
$X \vee Y$, $X \wedge Y$, reflexive BH spaces [127], Th. 2.3	

Table 21 Relatively strong (see Table 12 and [141])

Implied by	Not implied by
M ba. bo. or QB Th. 10-1-9	$\beta(X, Y)$ Rem. 15-4-21
	Closed $\subset$ ba. #13-2-120
	Complete #12-1-102
Spaces	C seq. #8-4-201 [146], p. 344
	Dense $\subset$ ba. bo. Ex. 15-4-12
T_{c_0}, T_K, X reflexive BS no #10-2-118	Mazur #8-6-1
	ω ba. #9-4-301
	Strictly hypercomplete Ex. 12-3-8
	Sup of normed #8-4-112
	T^+ #8-4-126

Table 22 Same compact sets

Implied by	Not implied by
$\beta(X, Y)_a$, $\sigma(X, Y)$ #12-1-201	$\sigma(X, Y)$, $\tau(X, Y)$ Ex. 10-2-6, #6-4-107
Dunford–Pettis #14-2-302	$\sigma(X, Y)$, $\tau(Y, X)_K$ #12-1-301
Same convergent sequences σ, τ if τ is M #14-1-106	Same convergent sequences #14-7-106
Same tot. bdd #12-1-4	$\sigma(c_0, l)$, $\|\cdot\|_\infty$ #8-1-105
σ, τ if X semi-Montel Sec. 15-5	
$\sigma(l, l^\infty)$, $\tau(l, c_0)$, $\|\cdot\|_1$ Ex. 9-5-5, #14-2-107	
$\tau(l^\infty, l)$, $\sigma(l^\infty, l)$ #12-1-122	
$\tau(X', X)$, norm BS X' separable [76]	
$T^{\circ\circ}$, T Cor. 12-1-16	
Weak operator and dual weak operator topology on compact maps between BS [78], Th. 1	

Table 23 Same absolutely convex compact sets

Implied by	Not implied by
$\sigma(X, Y)$, $\tau(X, Y)$ if $\sigma(Y, X)$, $\tau(Y, X)$ #12-1-121	$\sigma(Y, X)_K$, $\tau(X, Y)$ [112], Ex. 3.2
$\sigma(X, Y)$, $\tau(Y, X)_K$ #12-1-119	

Table 24 Same convergent sequences

Implied by	Not implied by
See Table 22, #8-4-126, 9-5-301	Same bounded sets Th. 8-4-1
G space, discrete Th. 14-7-4	$\sigma(X', X)$, norm #9-5-301
Same compact sets	$\sigma(X', X)$, T° #12-1-115
$\sigma(l, c_0)$, T_K #8-6-204	
$\sigma(X, X^\#)$, $\tau(X, X^\#)$ #7-1-112	
T°, $\sigma(X', X)$ if X seq. ba. #12-1-114	

Table 25 Same totally bounded sets

Implied by	Not implied by
$T^{\circ\circ}$, T Cor. 12-1-116	Same compact sets #12-1-4

Table 26 Semireflexive (see Table 12)

Implied by	Not implied by
See [149]	See Table 1
Boundedly complete X' semireflexive #10-2-115	Weakly seq. complete Rem. 10-2-7
Bdd $\Rightarrow$ weakly relatively compact #10-2-102	(X, Y) if (Y, X) semireflexive #10-2-2
B–M $\tau(X', X)$ QB Th. 10-4-5, 10-4-12, 10-2-4	
cbw, boundedly complete or semireflexive [149]	
Duality invariant	**Spaces**
Quotient with conditions #10-2-110	
Semi-Montel Sec. 15-5	(l, φ) no Table 1
Separable complete TG #14-7-301	Largest lc yes #10-2-101
Weakly boundedly complete Th. 10-2-4	$(X, X^*), (X^*, X)$ yes #10-2-101
$X \subset Y$, w* closed in Y'' #11-1-121	
(X, Y) if $\tau(Y, X)$ ba. Th. 10-2-4	

Table 27 Separable (see Table 12)

Implied by	Not implied by
Bidual (w*) of separable QB $[\![X \text{ is dense}]\!]$	Admissible for separable dual pair $[\![\beta(l^\infty, l)]\!]$
Continuous image	Bidual of separable $[\![\tau(l^\infty, l)]\!]$
C (compact metric) #9-6-201	Equic. $\Rightarrow$ w*M #9-5-110
$\aleph_0$ inductive limit #13-1-106	Every bdd set separable #9-6-114
Duality invariant #8-3-103	Every bdd set separable lc Fréchet (CH) #9-6-302
Equic. $\Rightarrow$ w* M, normed #9-6-201	Inner product space with $\aleph_0$ max$l$ orthonormal set [50]
(FM) [88], 27.2 (5)	Montel [88] 27.2
M X' separable #9-5-4	X' separable #13-2-203
Range of compact map #11-4-108	(X', w^*) separable, M #9-5-101
Reflexive $\subset l^\infty$ #11-1-109	(X', w^*) if X separable #9-5-102
Reflexive BX, $X = $ dual of separable normed #11-1-109	X' if $X = $ BS $\neq l$ [99]
Separable dual M #9-5-4	
Smaller topology	
S and X/S separable	**Spaces**
Weakly compact $\subset$ dual of separable normed #9-5-113	
WCG and dual of separable BS #9-5-113	$\beta(\varphi, \omega)$ yes #13-1-106
w* dual of FK $[\![\{P_n\}]\!]$	l^∞ no #2-3-7, 9-5-106, 12-5-101
w* dual of separable M Ex. 9-5-4	$\mathcal{K}^{(\mathcal{X})}, R^{(R)}$ no #13-2-116
	$\mathcal{K}^{\mathcal{X}}, R^R$ yes #13-2-116
	Largest lc, dim $\geq c$ no #13-2-117

Table 28 Sequential

Implied by	Not implied by
See #12-3-302	Bo. #13-3-301
Dual of Montel [146], Prop 5.7	C seq. #13-3-301
M	Mazur $[\![(BS, w)]\!]$
N seq. #8-4-118, 8-4-201	Smaller than compatible seq. #9-1-106
$(X', T°)$ X M ba separable #12-3-112	(X', w^*) X separable BS #15-4-35, 15-4-36, or 9-1-106

Spaces
$\beta(X', X)$, X M Ba separable BTB yes #12-1-5, 12-3-112
$\beta(X', X)$ X (FM) yes #12-1-5, 12-3-112
(BS, w) no #9-1-106
(c_0, w) no #8-4-120, 9-1-106
$C(0, 1)$ pointwise no #8-4-121
Distributions no [36]
φ largest lc yes #12-3-113
$\sigma(X^{\#}, X)$ X BS no #8-4-119
Test functions no #13-3-301
(X', w^*) X nonquasireflexive separable BS no #15-4-35, 15-4-36

Table 29 Sequentially barrelled

Implied by	Not implied by
B–M Mazur RS #10-4-204	Boundedly complete RS [146], p. 360
C ba. #9-4-103	B–M RS $[\![$Boundedly complete$]\!]$
$\aleph_0$ codim. [96]	
Dual of M Table 17	
Larger compatible #9-4-104	Spaces
Mazur TG RS #14-7-301	
RS (X', w^*) has cc (or scc) #9-4-108, 11-1-112	$\tau(l, c_0)$ yes #14-2-103
RS $(w^*)c_0$ compatible #12-1-109	$\tau(l^\infty, l)$ yes Table 5
Seq. complete (DF) [146], p. 361	
Seq. complete Mazur RS [146], Prop. 4-3	
$\tau(X', X)$ X seq. complete #11-1-114	
$T \supset (w^*)_{c_0}$ #9-4-106	

Table 30 Sequentially complete (for M, see Table 7)

Implied by	Not implied by
Banach–Steinhaus criterion [17]	B–M # 12-3-111
Boundedly complete	cc # 14-2-102
Bidual (w^*) of M Th. 10-5-4	Dual of boundedly complete $⟦Y$ ba.,
(BS, w) if X'' w seq. complete # 8-3-119	$X = (Y', w^*)⟧$
CG criterion Th 12-6-4	$\sigma(X', X)$ X M Rem. 10-4-14
$\aleph_0$ codim. criterion # 15-1-301	$\sigma(X', X)$ X seq. ba. # 14-2-102, 14-2-103
F linked criterion Th. 6-1-16	X if $\tau(X', X)$ seq. ba. [146], p. 354
Larger admissible # 8-5-5	
RS inheritance criterion [141]	
Semireflexive Table 4	**Spaces**
$\sigma(X', X)$ if X ba. Ex. 9-3-8	
if X B–M Mazur RS # 10-4-205	$C(H)$ w no # 10-2-120
if X C ba. # 9-4-107	$\sigma(c_0, l)$ no # 8-1-108, 12-5-102
if X seq. ba. Mazur [146], Prop. 4.4	$\sigma(l, l^\infty)$ yes # 8-1-11
$\sigma(X', X'')$ X GB # 14-7-104	$\sigma(l^\infty, l^\infty{}')$ no # 12-5-102
	$\sigma(l^\infty{}', l^\infty{}'')$ yes # 14-7-104
	$\sigma(X, X^{\#})$ yes # 9-3-119, Ex. 9-3-8
	$\sigma(X', X)$ X M, $S\sigma$ no # 13-3-110

Table 31 Totally bounded (see Tables 12 and 25)

Implied by	Not implied by
See L. 12-1-9	See # 12-1-107
$A^{\circ\circ}$ if A tot. bdd in admissible # 12-1-110	$A^{\circ\circ}$ if A tot. bdd # 12-1-111
ac hull in lc Th. 7-1-5	ac hull # 6-4-102
Cauchy sequence # 6-4-105	Not duality invariant # 8-4-101
Closure # 6-4-4	
Compact L. 6-4-1	
Countably compact # 6-4-106	
Every $\aleph_0$ subset tot. bdd # 6-4-104	
Finite $\cup$ small sets Th. 6-4-5	
Grothendieck interchange L. 12-1-14	
Tot. bdd in larger topology	

Table 32 Ultrabarrelled

Implied by	Not implied by
Banach–Steinhaus Th. to every TVS [143], p. 22 F linked criterion #9-3-126 CG criterion #12-6-302 Second category #9-3-113	Assoc. lc of Fréchet #9-3-116 Ba. lc #9-3-116, 15-1-114
	Spaces
	$(l^{1/2}, \|\cdot\|_1)$ no #9-3-116 m_0 no #15-1-114

Table 33 Weakly compactly generated (Banach space) (see Table 12)

Implied by	Not implied by
Separable Ex. 15-3-5 Reflexive Ex. 15-3-4	Dual of WCG $[\![l^\infty]\!]$ X' WCG [99]
	Spaces
(See also MR, vol. 54, no. 5809, 1977)	l^∞ no #15-3-102

BIBLIOGRAPHY

Abbreviations

AM	*Annals of Math*	*DMJ*	*Duke Math. J.*	*PAMS*	*Proc. AMS*
AMM	*Amer. Math.*	*IJM*	*Israel J. Math.*	*PCPS*	*Proc. Cambr.*
	Monthly	*JFA*	*Jour. Funct.*		*Phil. Soc.*
AMS	*Amer. Math. Soc.*		*Analysis*	*PJM*	*Pac. J. Math.*
BAMS	*Bull AMS*	*JLMS*	*Jour. LMS*	*PLMS*	*Proc. LMS*
CJM	*Canad. J. of*	*LMS*	*London Math.*	*PNAS*	*Proc. Nat. Acad.*
	Math.		*Soc.*		*Sci.* (USA)
CM	*Colloq. Math.*	*MA*	*Math. Annalen*	*SM*	*Studia Math.*
CR	*Contes Rendues*	*MR*	*Math. Reviews*	*TAMS*	*Trans. AMS*
	(Paris)	*MZ*	*Math. Zeits.*		

1. Ackermans, S. T. M.: Problem 5094, *AMM*, vol. 71, pp. 445–446, 1964.
2. Agnew, R. P.: Convergence Fields of Methods of Summability, *AM*, vol. 46, pp. 93–101, 1945.
3. Alfsen, E. M.: "Compact Convex Sets and Boundary Integrals," Springer-Verlag, 1971.
4. Amir, D., and J. Lindenstrauss: The Structure of Weakly Compact Sets in Banach Spaces, *AM*, vol. 88, pp. 35–46, 1968.
5. Anderson, R. D., and R. H. Bing: A Complete Elementary Proof that Hilbert Space is Homeomorphic to the Product of Lines, *BAMS*, vol. 74, pp. 771–792, 1968.
6. Bade, W. G., and P. C. Curtis: Embedding Theorems for Commutative Banach Algebras, *PJM*, vol. 18, pp. 391–409, 1966.
7. Baker, J. A.: Isometries in Normed Spaces, *AMM*, vol. 78, pp. 655–658, 1971.
8. Banach, S.: "Théorie des Operations Linéaires," Hafner, 1952.
9. Bennett, G.: Some Inclusion Theorems for Sequence Spaces, *PJM*, vol. 46, pp. 17–30, 1973.
10. Bennett, G., and J. B. Cooper: Weak Bases in *F* and *LF* Spaces, *JLMS*, vol. 44, pp. 505–508, 1969.

11. Bennett, G., and N. J. Kalton: *FK* Spaces Containing c_0, *DMJ*, vol. 39, pp. 561–582, 819–821, 1972.

12. Bennett, G., and N. J. Kalton: Inclusion Theorems for *K* Spaces, *CJM*, vol. 25, pp. 511–524, 1973.

13. Bessaga, C., and A. Pelczynski: On Extreme Points in Separable Conjugate Spaces, *IJM*, vol. 4, pp. 262–264, 1966.

14. Bourbaki, N.: "Espaces Vectoriels Topologiques," Hermann et Cie, 1955.

15. Cambern, M.: On Mappings Between Sequence Spaces, *SM*, vol. 30, pp. 73–77, 1968.

16. Casazza, P. G., C. A. Kottman, and B. Lin: On Primary Banach Spaces, *BAMS*, vol. 82, pp. 71–73, 1976.

17. Ceĭtlin, Ja. M.: The Banach–Steinhaus Theorem, *Vestnik Moskov*, vol. 26, pp. 11–13, 1971. (*MR*, vol. 35, No. 2433, 1973.)

18. Chernoff, P. R.: Problem 5688, *AMM*, vol. 77, pp. 892–893, 1970.

19. Civin, P., and B. Yood: Quasireflexive Spaces, *PAMS*, vol. 8, pp. 906–911, 1957.

20. Cohen, H. B.: A Bound-two Isomorphism Between $C(X)$ Banach Spaces, *PAMS*, vol. 50, pp. 215–217, 1975.

21. Collins, H. S.: Completeness and Compactness in TVS, *TAMS*, vol. 79, pp. 256–280, 1955.

22. Comfort, W. W.: Book Review, *BAMS*, vol. 82, pp. 857–863, 1976.

23. Corson, H. H., and J. Lindenstrauss: On Function Spaces Which Are Lindelöf Spaces, *TAMS*, vol. 121, pp. 476–491, 1966.

24. Darst, R. B.: On a Theorem of Nikodym, *PJM*, vol. 23, pp. 473–477, 1967.

25. Davis, W. J., T. Figiel, W. B. Johnson, and A. Pelczynski: Factoring Weakly Compact Operators, *JFA*, vol. 17, pp. 311–327, 1974.

26. Day, M. M.: "Normed Spaces," Academic Press, 1962.

27. Daykin, D. A., and A. Wilansky: Sets of Complex Numbers, *Math. Mag.*, vol. 47, pp. 228–229, 1974.

28. DeWilde, M.: Quelques Propriétés de Permanence des Espaces à Réseau, *Bull. Soc. Roy. Sci. Liège*, vol. 39, pp. 240–248, 1970.

29. DeWilde, M.: Sur le Relèvement des Parties Bornées d'un Quotient, *Bull. Soc. Roy. Sci. Liège*, vol. 43, pp. 299–301, 1974.

30. Dierolf, P.: Espaces Complets au Sens de Mackey, *CR*, vol. 283, pp. 245–248, 1976.

31. Dierolf, S., and P. Lurje: Deux Exemples Concernant des Espaces (Ultra) Bornologiques, *CR*, vol. 282, pp. 1347–1350, 1976.

32. Diestel, J., and J. J. Uhl: Theory of Vector Measures. *AMS Mathematical Surveys* 1977.

33. Diestel, J.: "Geometry of Banach Spaces—Selected Topics," Springer-Verlag, 1975.

34. Dieudonné, J.: Bounded Sets in *F* Spaces, *PAMS*, vol. 6, pp. 729–731, 1955.

35. Dixmier, J.: Sur un Théorème de Banach, *DMJ*, vol. 15, pp. 1057–1071, 1948.

36. Dudley, R. M.: Convergence of Sequences of Distributions, *PAMS*, vol. 27, pp. 531–534, 1971.

37. Dunford, N., and J. T. Schwartz: "Linear Operators I," Interscience, 1958.

38. Edwards, R. E.: "Functional Analysis," Holt, Rinehart, Winston, 1965.

39. Faires, B.: Varieties and Vector Measures, *Math. Nachrichten*, 1978.

40. Figiel, T.: On Non-linear Isometric Embeddings of Normed Spaces, *Bull. Acad. Polon. Sci.*, vol. 16, pp. 185–188, 1968. (*MR*, vol. 37, no. 6734, 1969.)

41. Garling, D. J. H.: On Topological Sequence Spaces, *PCPS*, vol. 63, pp. 997–1019, 1967.

42. Garling, D. J. H.: The Filter Condition, the Closed Neighborhood Condition and Consistent Seminorms, *DMJ*, vol. 38, pp. 299–304, 1971.

43. Garling, D. J. H.: A "Short" Proof of the Riesz Representation Theorem, *PCPS*, vol. 73, pp. 459–460, 1973.

44. Garling, D. J. H., and A. Wilansky: The Berg–Crawford–Whitley Summability Theorem, *PCPS*, vol. 71, pp. 495–497, 1972.

45. Giles, J. R.: On a Characterization of Differentiability of the Norm. *J. Austral. Math. Soc.*, vol. 12, pp. 106–114, 1971.

46. Gillman, L., and M. Jerison: "Rings of Continuous Functions," Van Nostrand, 1960.

47. Goldberg, S.: "Unbounded Linear Operators," McGraw-Hill, 1966.
48. Goldstein, J. A.: Problem 6009, *AMM*, vol. 83, pp. 666–667, 1976.
49. Grothendieck, A.: Sur les Applications Faiblement Compactes d'Espaces du Type $C(K)$, *CJM*, vol. 5, pp. 129–173, 1953.
50. Gudder, S.: Inner Product Spaces, *AMM*, vol. 81, pp. 29–36, 1974; vol. 82, pp. 251 and 818, 1975.
51. Halmos, P. R.: "Measure Theory," Van Nostrand, 1950.
52. Halmos, P. R.: "Finite Dimensional Vector Spaces," Van Nostrand, 1958.
53. Hammerle, W. G.: Solution, *AMM*, vol. 76, p. 913, 1969.
54. Hampson, J. K., and A. Wilansky: Sequences in l.c. Spaces, *SM*, vol. 45, pp. 221–223, 1973.
55. Haydon, R.: Three Examples in the Theory of Continuous Functions, *Publ. Dep. Math (Lyon)*, vol. 9, pp. 99–103, 1972.
56. Herman, R.: On the Uniqueness of the Ideals of Compact and Strictly Singular Operators, *SM*, vol. 29, pp. 161–165, 1967.
57. Hewitt, E., and K. Stromberg: "Real and Abstract Analysis," Springer Verlag, 1965.
58. Horvath, J.: "Topological Vector Spaces and Distributions I," Addison-Wesley, 1966.
59. Horvath, J.: Book Review, *BAMS*, vol. 82, pp. 515–521, 1976.
60. Husain, T.: "The Open Mapping and Closed Graph Theorems in TVS," Oxford, 1965.
61. Husain, T.: "Introduction to Topological Groups," Saunders, 1966.
62. Iyahen, S. O.: On the Closed Graph Theorem, *IJM*, vol. 10, pp. 96–105, 1971.
63. Iyahen, S. O.: Finite Dimensional Locally Bounded Spaces, *JLMS*, vol. 6, pp. 488–490, 1973.
64. Iyahen, S. O.: Barrelled Spaces and the Open Mapping Theorem, *JLMS*, vol. 11, pp. 421–422, 1975.
65. James, R. C.: A Non-reflexive Banach Space Isometric with its Second Conjugate Space, *PNAS*, vol. 37, pp. 174–177, 1951.
66. James, R. C.: Weakly Compact Sets, *TAMS*, vol. 113, pp. 129–140, 1964.
67. James, R. C.: A Counterexample for a Sup Theorem in Normed Spaces, *IJM*, vol. 9, pp. 511–512, 1971.
68. James, R. C.: Reflexivity and the Sup of Linear Functionals, *IJM*, vol. 13, pp. 289–300, 1972.
69. Johnson, W. B., and H. P. Rosenthal: On ω^*-basic Sequences and Their Applications to Banach Spaces, *SM*, vol. 43, pp. 77–92, 1972.
70. Johnson, W. B., and M. Zippin: Subspaces and Quotient Spaces of $(\sum G_n)l_p$ and $(\sum G_n)c_0$, *IJM*, vol. 17, pp. 50–55, 1974.
71. Josefson, B.: Weak * Sequential Convergence in the Dual of a Banach Space Does Not Imply Norm Convergence, *Ark. Math.*, vol. 13, pp. 78–89, 1975.
72. Kadison, R.: A Representation Theory for Commutative Topological Algebra, *Memoirs of AMS*, vol. 7, 1951.
73. Kalton, N. J.: A Barrelled Space Without a Basis, *PAMS*, vol. 26, pp. 465–466, 1970.
74. Kalton, N. J.: Subseries Convergence in Topological Groups and Vector Spaces, *IJM*, vol. 10, pp. 402–412, 1971.
75. Kalton, N. J.: Some Forms of the Closed Graph Theorem, *PCPS*, vol. 70, pp. 401–408, 1971.
76. Kalton, N. J.: Mackey Duals and Almost Shrinking Bases, *PCPS*, vol. 74, pp. 73–81, 1973.
77. Kalton, N. J.: Basic Sequences in F Spaces and Their Applications, *Proc. Edinburgh Math. Soc.*, vol. 19, pp. 151–157, 1974.
78. Kalton, N. J.: Spaces of Compact Operators, *MA*, vol. 208, pp. 267–268, 1974.
79. Kalton, N. J., and A. Wilansky: Tauberian Operators on Banach Spaces, *PAMS*, vol. 57, pp. 251–255, 1976.
80. Kelley, J. L.: "General Topology," Van Nostrand, 1955.
81. Kelley, J. L.: Banach Spaces with the Extension Property, *TAMS*, vol. 72, pp. 323–326, 1952.
82. Kelley, J. L., I. Namioka, et al.: "Linear Topological Spaces," Van Nostrand, 1963.
83. Klee, V. L.: Convex Bodies and Periodic Homeomorphisms in Hilbert Space, *TAMS*, vol. 74, pp. 10–43, 1953.
84. Klee, V. L.: Exotic Topologies for Linear Spaces, in J. Novak (Ed.), "General Topology and its Relation to Modern Analysis," Prague Symposium, 1961.

85. Knowles, R. J., and T. A. Cook: Non-complete Reflexive Spaces Without Schauder Bases, *PCPS*, vol. 74, pp. 83–86, 1973.
86. Komura, Y.: Some Examples in Linear Topological Spaces, *MA*, vol. 153, pp. 150–162, 1964.
87. Korner, T. W.: Some Covering Theorems for Infinite Dimensional Vector Spaces, *JLMS*, vol. 2, pp. 643–646, 1970.
88. Köthe, G.: "Topological Vector Spaces I," Springer-Verlag, 1969.
89. Köthe, G.: Abbildungen von F. Räumen in (*LF*) Räume, *MA*, vol. 178, pp. 1–3, 1968.
90. Krishnamurthy, V.: Conjugate l.c. Spaces, *TAMS*, vol. 130, pp. 525–531, 1968.
91. Krishnamurthy, V., and J. O. Loustaunau: On the State Diagram of an Operator in l.c. Spaces, *MA*, vol. 164, pp. 176–206, 1966.
92. Kruse, A. H.: Badly Incomplete Normed Spaces, *MZ*, vol. 83, pp. 314–320, 1964.
93. Kuratowski, C.: "Topologie," Monografie Mat. vol. 20, Hafner, 1952.
94. Lacey, H. E.: Separable Quotients of Banach Spaces, *An. Acad. Brasil Cienc.*, vol. 44, pp. 185–189, 1972.
95. Lebow, A., and M. Schechter: Semigroups of Operators and Measures of Non-compactness, *JFA*, vol. 7, pp. 1–26, 1971.
96. Levin, M., and S. A. Saxon: The Inheritance of Properties of l.c. Spaces by Subspaces of Countable Codimension, *PAMS*, vol. 29, pp. 97–102, 1971.
97. Lindenstrauss, J.: On Some Subspaces of *l* and c_0, *Bull. Res. Council of Israel*, vol. 10, pp. 74–80, 1961.
98. Lindenstrauss, J.: On Complemented Subspaces of *m*, *IJM*, vol. 5, pp. 153–156, 1967.
99. Lindenstrauss, J., and C. Stegall: Examples of Separable Spaces Which Do Not Contain *l* and Whose Duals are Separable, *SM*, vol. 54, pp. 81–105, 1975.
100. Lindenstrauss, J., and L. Tzafriri: On the Complemented Subspaces Problem, *IJM*, vol. 9, pp. 263–269, 1971.
101. Lindenstrauss, J., and L. Tzafriri: "Classical Banach Spaces," Springer-Verlag, 1973.
102. Lohman, R. H., and W. J. Stiles: On Separability in TVS, *PAMS*, vol. 42, pp. 236–237, 1974.
103. Mackey, G. W.: Note on a Theorem of Murray, *BAMS*, vol. 52, pp. 322–325, 1946.
104. Maddox, I. J.: Paranormed Sequence Spaces Generated by Infinite Matrices, *PCPS*, vol. 64, pp. 335–340, 1968.
105. Marti, J. T.: "Introduction to the Theory of Bases," Springer-Verlag, 1969.
106. Michael, E.: Selected Selection Theorems, *AMM*, vol. 63, pp. 233–238, 1956.
107. Murray, F. J.: Quasicomplements and Closed Projections in Reflexive Banach Spaces, *TAMS*, vol. 58, pp. 77–95, 1945.
108. Nachbin, L.: TVS's of Continuous Functions, *PNAS*, vol. 40, pp. 471–474, 1954.
109. Nachbin, L. (Ed.): "Functional Analysis and Applications," Springer-Verlag, 1974.
110. Nissenzweig, A.: w^* Sequential Convergence, *IJM*, vol. 22, pp. 266–272, 1975.
111. Noureddine, K., and W. Habre: Sous-algebres l.c. de *C(T)*, *CR*, vol. 286, pp. 1739–1741, 1975.
112. Ostling, E. G., and A. Wilansky: Locally Convex Topologies and the cc Property, *PCPS*, vol. 74, pp. 45–50, 1974.
113. Pličko, A. N.: Reflexivity and Quasireflexivity Conditions for TVS, *Ukrain. Mat. Z.*, vol. 27, pp. 24–32 and 141, 1975. (*MR*, vol. 51, no. 1329, 1976.)
114. Rätz, J.: On Isometries of Generalized Inner Product Spaces, *SIAM J. Appl. Math.*, vol. 18, pp. 6–9, 1970.
115. Roberts, G. T.: Topologies Defined by Bounded Sets, *PCPS*, vol. 51, pp. 379–381, 1955.
116. Robertson, A. P., and W. J. Robertson: "Topological Vector Spaces," Cambridge, 1973.
117. Robertson, W. J.: Completions of TVS, *PLMS*, vol. 8, pp. 242–257, 1958.
118. Rolewicz, S., and C. Ryll-Nardzewski: On Unconditional Convergence in Linear Metric Spaces, *CM*, vol. 17, pp. 327–331, 1967.
119. Rosenthal, H. P.: Quasicomplemented Subspaces of Banach Spaces, *JFA*, vol. 4, pp. 176–214, 1969.
120. Rosenthal, H. P.: The Heredity Problem for WCG Spaces, *Comp. Math.*, vol. 28, pp. 83–111, 1974.
121. Rubel, L. A., and A. L. Shields: The Failure of Interior–Exterior Factorization in the Polydisc and the Ball, *Tohoku Math. J.*, vol. 24, pp. 409–413, 1972.

122. Ruckle, W. H.: Reflexivity of $L(E, F)$, *PAMS*, vol. 34, pp. 171–174, 1972.
123. Rudin, W.: "Real and Complex Analysis," McGraw-Hill, 1974.
124. Saxon, S. A.: Two Characterizations of Linear Baire Spaces, *PAMS*, vol. 45, pp. 204–208, 1974.
125. Saxon, S. A., and M. Levin: Every Countable Codimensional Subspace of a Barrelled Space is Barrelled, *PAMS*, vol. 29, pp. 91–96, 1971.
126. Saxon, S. A., and A. Wilansky: The Equivalence of Some Banach Space Problems, *CM*, vol. 37, pp. 217–226, 1977.
127. Schaffer, J. J.: Function Spaces with Translations, *MA*, vol. 137, pp. 209–262, 1959.
128. Schechter, M.: "Principles of Functional Analysis," Academic Press, 1971.
129. Scheinberg, S.: Fatou's Lemma in Normed Spaces, *PJM*, vol. 38, pp. 233–238, 1971.
130. Seever, G.: Measures on F-spaces, *TAMS*, vol. 133, pp. 267–280, 1968.
131. Sentilles, F. D.: Bounded Continuous Functions on a Completely Regular Space, *TAMS*, vol. 168, pp. 311–336, 1972.
132. Shapiro, H. S.: Problem 5199, *AMM*, vol. 72, pp. 435–436, 1965.
133. Shirota, T.: On l.c. Spaces of Continuous Functions, *Proc. Japan Acad.*, vol. 30, pp. 294–298, 1954.
134. Singer, I.: "Bases in Banach Spaces," Springer-Verlag, 1970.
135. Snipes, R. F.: C-sequential and S-bornological Spaces, *MA*, vol. 209, pp. 273–283, 1973.
136. Snyder, A. K., and A. Wilansky: Inclusion Theorems and Semi-conservative FK Spaces, *Rocky Mtn. J. Math.*, vol. 2, pp. 595–603, 1972.
137. Titchmarsh, E. C.: "Theory of Functions," Oxford, 1939.
138. Tréves, F.: "Topological Vector Spaces, Distributions and Kernels," Academic Press, 1967.
139. Valdivia, M. On Bounded Sets Which Generate Banach Spaces, *Arch. Math (Basel)*, vol. 23, pp. 640–642, 1972.
140. Valdivia, M.: A Class of Quasibarrelled (*DF*) Spaces Which Are Not Bornological, *MZ*, vol 136, pp. 249–251, 1974.
141. Valdivia, M.: On Mackey Spaces, *DMJ*, vol. 41, pp. 835–841, 1974.
142. Van Dulst, D.: A Note on B and B_r Completeness, *MA*, vol. 197, pp. 197–202, 1972.
143. Waelbroeck, L.: "Topological Vector Spaces and Algebras," Springer-Verlag, 1971.
144. Waelbroeck, L. (Ed.): "Summer School on TVS," Springer-Verlag, 1972.
145. Waterman, D.: Reflexivity and Summability, *SM*, vol. 32, pp. 61–63, 1969.
146. Webb, J. H.: Sequential Convergence in l.c. Spaces, *PCPS*, vol. 64, pp. 341–364, 1968.
147. Webb, J. H.: Schauder Bases and Decompositions in l.c. Spaces, *PCPS*, vol. 76, pp. 145–152, 1974.
148. Wells, B. B.: Interpolation in $C(\Omega)$, *PJM*, vol. 18, pp. 391–409, 1966.
149. Wheeler, R. F.: The Equicontinuous w^* Topology and Semi-reflexivity, *SM*, vol. 41, pp. 243–256, 1972.
150. Wheeler, R. F.: The Strict Topology for P-spaces, *PAMS*, vol. 41, pp. 466–472, 1973.
151. Wilansky, A.: Problems 5831, 5937, 6017, 6078, *AMM*, vol. 80, 1973; vol. 82, 1975; vol. 83, 1977; vol. 84, 1978.
152. Wilansky, A.: Summability, the Inset, *DMJ*, vol. 19, pp. 647–660, 1952.
153. Wilansky, A.: "Functional Analysis," Blaisdell, 1964.
154. Wilansky, A.: "Topics in Functional Analysis," Springer-Verlag, 1967.
155. Wilansky, A.: Life Without T_2, *AMM*, vol. 77, pp. 157–161 and 728, 1970.
156. Wilansky, A.: "Topology for Analysis," John Wiley, 1970.
157. Wilansky, A.: Subalgebras of $B(X)$, *PAMS*, vol. 29, pp. 355–360, 1971; vol. 34, p. 632, 1972.
158. Wilansky, A.: Semifredholm Maps of FK Spaces, *MZ*, vol. 144, pp. 9–12, 1975.
159. Wilansky, A., and K. Zeller: Summation of Bounded Divergent Sequences, *TAMS*, vol. 78, pp. 501–509, 1955.
160. Wilbur, J. W.: Reflective and Coreflective Hulls in the Category of l.c. Spaces, *General Topology and Applications*, vol. 4, pp. 235–254, 1974.
161. Wright, J. D. M.: All Operators on a Hilbert Space are Bounded, *BAMS*, vol. 79, pp. 1247–1250, 1973.
162. Zeller, K., and W. Beekmann: "Theorie der Limitierungsverfahren," Springer-Verlag, 1970.
163. Jameson, G. J. O.: Whitley's Technique, *AMM*, vol. 84, pp. 459–461, 1977.

INDEX